MW00388471

THE
PRESIDENTS
AND THE
PLANET

THE
PRESIDENTS
AND THE
PLANET

Climate Change Science and Politics from Eisenhower to Bush

JAY HAKES

Louisiana State University Press

Baton Rouge

Published with the assistance of the V. Ray Cardozier Fund

Published by Louisiana State University Press
lsupress.org

Manufactured in the United States of America
First printing

Designer: Barbara Neely Bourgoyne
Typeface: Adobe Text Pro
Printer and binder: Sheridan Books, Inc.

Jacket images: *Drowning Earth* courtesy Adobe Stock/freie-kreation.
Map courtesy visibleearth.nasa.gov.

Library of Congress Cataloging-in-Publication Data
Names: Hakes, Jay E., author.
Title: The presidents and the planet : climate change science and politics from
    Eisenhower to Bush / Jay Hakes.
Description: Baton Rouge : Louisiana State University Press, 2024. |
    Includes bibliographical references and index.
Identifiers: LCCN 2024010742 (print) | LCCN 2024010743 (ebook) | ISBN
    978-0-8071-8190-4 (cloth) | ISBN 978-0-8071-8313-7 (epub) | ISBN
    978-0-8071-8314-4 (pdf)
Subjects: LCSH: Climatic changes—United States. | Climatic changes—Government
    policy—United States. | Climatic changes—Political aspects—United States.
Classification: LCC QC903.2.U6 H35 2024 (print) | LCC QC903.2.U6 (ebook)
    | DDC 363.7/0561097309045—dc23/eng/20240412
LC record available at https://lccn.loc.gov/2024010742
LC ebook record available at https://lccn.loc.gov/2024010743

Contents

THE
PRESIDENTS
AND THE
PLANET

Prologue

The National Academy of Sciences lobby, at its building on Fifth Street NW in Washington, DC, offers visitors a crash course in science history. Known as the Keck Center, the newer of the two Academy locations in the nation's capital is less known than its sister headquarters on Constitution Avenue. But its site is more easily accessible, just a short walk from Metro stations serving the Red, Yellow, Orange, Silver, and Green Lines.

The curious can linger to gaze at what the Academy calls its "visual encyclopedia." The engraved murals on the tall stone walls depict the long saga of scientific discovery. One etching portrays the development of maize for human consumption, beginning millennia ago. This dazzling celebration of invention—the work of sculptor and multidimensional artist Larry Kirkland—demonstrates that the expansion of human understanding over a very long time has relied on scientists revising how they, and eventually the broader populous, perceived the world.

Some people and events on the walls evoke memories of early days in the classroom. On the left, for instance, Kirkland engraved the first star map drawn with the assistance of a telescope—the work of Galileo Galilei back in 1610. The Italian's observations of outer space through his self-ground lenses advanced a revolution in human thought. The sun doesn't circle our planet. Instead, the earth revolves around the sun—an idea championed earlier by the Polish astronomer Nicolaus Copernicus but one that clerics still considered heretical. A *National Geographic* kids' book about the pioneering astronomer called Galileo "the genius who charted the universe." Still, it noted that government officials at the time were more impressed by his telescope's ability to spot approaching ships than to track distant stars.[1]

1

Further, on the left wall, the image of a motorized glider commemorates the first flight of Wilbur and Orville Wright's airplane near Kitty Hawk, North Carolina. Despite initial doubts about the crude contraption's value, biographer David McCullough portrays their flying machine as part of the progression to more impressive achievements ahead. "On July 12, 1969," the historian wrote, "when Neil Armstrong, another American born and raised in western Ohio, stepped onto the moon, he carried with him, in tribute to the Wright brothers, a small swatch of the muslin from a wing of their 1903 Flyer."[2]

The murals looming straight ahead include epic breakthroughs by two other heralded scientists—each the subject of another talented biographer, Walter Isaacson. Albert Einstein first revealed his elegant theory $E = mc^2$—energy equals mass times the square of the speed of light—in the early 1900s. Isaacson called it "the best-known equation in all of physics."[3] Not far away, visitors can observe a precise rendering of the human spine (1489) by Leonardo da Vinci, described by Isaacson as "history's consummate innovator."[4]

Far down the right wall, observers crane their necks to see a simple graph just below a three-dimensional model of the eight-carbon chain of atoms found in octane—a booster of gasoline performance. The graph—less recognizable than other displays for those long past their days in the classroom—contains a single ascending line. The x-axis on the bottom covers the years 1958 to 2002. The y-axis tracks the accumulation in the earth's atmosphere of carbon dioxide. These measurements were collected by a pioneering American scientist named Charles (Dave) Keeling. Keeling lacks the name recognition of Galileo, Einstein, or Leonardo. He has not (yet) been the subject of a published biography, let alone a *New York Times* bestseller.

During the summer of 1960, Dr. Keeling—a thirty-two-year-old geochemist at the Scripps Institution of Oceanography in La Jolla, California—published a scholarly paper that endures as a paradigm-altering contribution to science. Although still a junior faculty member, Keeling was already the world's leading expert on measuring the carbon dioxide residing in the earth's atmosphere. Now he made his initial findings available to the scientific community.

His data indicated that atmospheric carbon ranged from 306.5 to 317.6 parts per million—a tiny fraction of the earth's atmosphere. Still, these levels intrigued a few scientists because they were higher than those estimated before the Industrial Revolution. Moreover, based on very limited reporting, the data showed an upward trend.[5]

Before Keeling's measurements, there was considerable uncertainty about how much carbon emitted from the combustion of fossil fuels (coal, oil, and natural gas) stayed in the atmosphere. If they were even aware of the issue, most scientists had assumed that the world's oceans absorbed the bulk of the emissions not taken in by plants. But the young professor was carefully documenting that a sizeable amount appeared to be reaching the atmosphere, where it remained for a long time.

Keeling's discovery had significant implications. Scientists had demonstrated that, because of its molecular properties, carbon dioxide traps heat. Thus, emissions into the atmosphere from human activities could, over time, alter the earth's temperature and climate. On a broader scale, his data expanded scientists' understanding of the growing impact of humans on the planet.

The diligent professor continued refining and updating his graph in the following decades. It became known as the Keeling Curve and eventually earned its place at the National Academy among the outstanding achievements of science. On the question of when the modern era of climate change science began, the 1960 publication of Keeling's careful measurements of atmospheric carbon would be a strong candidate.

While researching this book, I was struck by a less remembered Keeling paper published in 1970, a decade after his first (but far from last) article documenting the rising accumulations of atmospheric carbon dioxide. In *The Proceedings of the American Philosophical Society,* he engaged in a rare (for him) reflection on the long-term implications of what he had found. He speculated about critical time lags confronting the prospect of a changing climate.

Keeling observed that most impacts of higher carbon accumulations on the earth's climate would likely not become visible to the human eye until the twenty-first century. He believed that, until then, people would be reluctant to deal with a problem that was foreseen by scientists but entailed effects that were gradual and difficult to observe. The professor predicted, however, that his students would later become part of the first generation "to feel such strong concern for man's future that they will discover means of effective action. This action may be less pleasant and rational than the corrective measures that we promote today, but thirty years from now, if present trends are any sign, mankind's world will be in greater immediate danger than it is today, and immediate corrective actions, if such exist, will be closer at hand."[6] The scholar from La Jolla foresaw a future historical marker when the earth's population worried enough about its inadvertent alterations of the atmosphere to take "corrective actions."

Keeling's comments suggested that climate change requires time perspectives covering decades and even centuries. Two primary time lags of climate change distinguished it from other environmental challenges and impeded efforts to forge responses—one based on science, the other on politics. Carbon dioxide remained in the atmosphere longer than the average human life. Thus, a sudden spurt or, conversely, slashing of emissions would not produce much visible impact on climate in the short term. Later on, however, more perceptible and eventually immense changes would result from trends in emissions. Because the time between emissions and their visible impacts would be long, climate science would require humanity to be futuristic by projecting what was likely but not entirely sure to happen during the lives of later generations. Such time frames would pose formidable challenges for elected officials asked to address risks in the distant future while facing the pressures of short-term politics.

This narrative begins during the presidency of Dwight Eisenhower, when Keeling published the first iteration of his famous curve, and ends with that of George H. W. Bush, when 154 nations signed the first international agreement on climate change. This story introduces us to the curious scholars who helped shape contemporary thought on climate and the prominent political leaders who learned about a new risk to the planet. It addresses five salient questions: What did we know about climate change? When did we know it? Who knew? What did we do about it? And, more importantly, why did we or didn't we?

Trying to find these answers convinced me that a historical narrative about the pioneers of climate change grounded in available but often overlooked records is an indispensable tool for honestly coming to grips with our age's most complex and multidisciplinary challenge. And for discerning where the future may lead.

1

Measuring the Invisible Blanket

The Eisenhower Presidency, 1953 to 1961

In the spring of 1953, a flicker of national news coverage alerted diligent readers to an obscure scientific theory that would bedevil scholars, government officials, and the general public well into the next century.

Customers paying twenty cents for the May 24 Sunday edition of the *New York Times* confronted a whopping 369 pages of news articles, opinion pieces, and advertising. The front page featured two stories on criticism of U.S. management of the Korean War, then almost three years old but heading toward its unofficial end in July. Few people wading through the hefty paper likely noticed a five-paragraph story in the bottom right corner on page eleven in the education section. Nonetheless, its headline appeared dire: "How Industry May Change Climate."

The brief snippet reported on a recent scholarly paper Johns Hopkins University's Gilbert Plass had presented at the annual meeting of the American Geophysical Union in Washington, DC. The thirty-three-year-old physicist from Canada asserted that the burning of coal and oil by humans would lead to a rise in the earth's temperature. The carbon dioxide emitted into the atmosphere created surface warming because, "like gases in a greenhouse," it allowed short-wavelength radiation from the sun to pass through but inhibited the escape of the long waves reflected from the surface (heat). The professor calculated that significant impacts would result from a doubling of carbon in the atmosphere, which he expected to occur sometime around 2080. His time frame gave readers reason to believe the prospect might be ominous but not yet urgent.

The same week, the popular news magazine *Time*—like the *New York Times,* a dominant source of news at a time when most Americans didn't own televisions—ran a story headlined "Invisible Blanket," also based on the Plass presentation. The article in the science section resembled the one in the *Times:* short (four paragraphs) and buried deep in the publication at the bottom of the right column. It also explained the chemistry of carbon dioxide as a "greenhouse gas." The weekly colorfully observed, "In the hungry fires of industry, modern man burns nearly 2 billion tons of coal and oil each year," producing unsightly smoke and soot. Fuel combustion also belched a "spreading envelope" of carbon dioxide around the world. In contrast to the visible effects of energy production, this carbon dioxide remained unseen by the human eye. The magazine noted Plass's prediction that with industrial expansion, the earth would continue to warm.

The easy-to-miss articles in widely read periodicals appeared just over four months after the inauguration of Dwight Eisenhower, Supreme Allied Commander in Europe during World War II, as the thirty-fourth president of the United States. They stand as early mentions in the American mass media of global warming caused by the fuels that powered the Industrial Revolution.[1]

The Plass paper was not wholly original. His work built on the insights of previous scientists laboring in obscurity, going back to the nineteenth century. In the early 1800s, the French mathematician and natural philosopher Joseph Fourier—orphaned at nine and raised by Benedictine monks—tried to explain why the earth's temperature was suitable for human life. Considering the planet's distance from the sun, he theorized, its atmosphere must be allowing some, but not all, of the sun's heat to reflect into space. Otherwise, the earth would be too hot or too cold for human habitation.

During the 1850s, American scientist Eunice Newton Foote—a signer of the Seneca Falls, New York, Women's Rights Convention, along with Lucretia Mott and Elizabeth Cady Stanton—discovered an explanation for Fourier's theory. She used glass cylinders to demonstrate that moist air (water vapor) contains more heat than dry air. Even more striking, a cylinder with carbon dioxide attained much higher temperatures and took much longer to cool.

Foote presented her findings in a paper, "Circumstances Affecting the Heat of the Sun's Rays," at a meeting of the prestigious American Association for the Advancement of Science on August 23, 1856—though a male scientist read it. Foote concluded, "An atmosphere of that gas [carbon dioxide] would give to our earth a high temperature; and if, as some suppose, at one period of its history the air had mixed with it a larger proportion than at present, an increased temperature . . . must have necessarily resulted." Her paper received additional visibility when *The American Journal of Science and Arts* published it in November.[2] Nonetheless, Foote's epic breakthrough in understanding natural variations in climate history was ignored until well into the twenty-first century.

In the 1860s, the Irish-born physicist John Tyndall—a spokesman for the advancement of science and defender of Charles Darwin's controversial theory of evolution—conducted laboratory experiments that shed additional light on the role of the atmosphere. He demonstrated that some components (e.g., oxygen and nitrogen) are transparent and do not block heat reflected from the earth's surface. On the other hand, methane, carbon dioxide, and water vapor are opaque and do trap some of the sun's heat. His findings further advanced the understanding of the planet's human-friendly climate.

Near the end of the nineteenth century, the Swedish scientist Svante Arrhenius—curious about the risk of another ice age—made crude projections of global temperatures if atmospheric carbon dioxide were halved or doubled. He also used the concept of a glass-covered "hotbed" (later called a greenhouse) to explain the atmosphere's effect on surface temperatures. He found it difficult to test his theories. At the time, there was no reliable way to measure the amount of carbon dioxide in the atmosphere.[3]

In the decades preceding the Eisenhower presidency, the study of the atmosphere's impact on the earth's temperature was most closely associated with the rudimentary calculations of Guy Callendar. The quiet and unassuming Londoner worked as a steam technologist in the British Electrical and Allied Industries Research Association and enjoyed spending his free time on the tennis courts. He was also an amateur scientist who concluded that the increasing accumulation in the atmosphere of an opaque gas like carbon dioxide was gradually warming the planet.

A 1938 Callendar paper presented at the Royal Meteorological Society estimated that human combustion of fossil fuels had emitted 150,000 million tons of carbon dioxide over the past half century. He believed that about three-quarters of those emissions remained in the atmosphere. Based on his understanding of the chemistry of carbon and water vapor, he calculated that the earth's mean temperature should have risen very slightly. The patchy temperature data he could find suggested that the pace of warming was a bit faster than his estimates. Callendar conceded that few experts on weather and climate "would be prepared to admit that the activities of man could have any influence upon phenomena of so vast a scale." He hoped his paper would "show that such influence is not only possible, but is actually occurring at the present time."[4]

Callendar's ideas failed to gain much traction with fellow scientists. As an independent researcher, he lacked the academic clout to overturn prevailing assumptions about the atmosphere. Many scholars, including the official discussants of his paper, believed that the bulk of carbon emissions not taken up by forests and other plants ended up in the oceans, making their impacts on the atmosphere slight. Or, they suggested, any link between carbon accumulations and rising temperatures might be coincidental.

Callendar himself offered little reason for concern. On the contrary, he initially thought that the use of fossil fuels would likely produce win-win benefits for humanity. Besides furnishing heat and power, they would help avoid the return of a new ice age.

In the early 1950s, few scientists studying the atmosphere paid attention to carbon dioxide's role due to its relatively low volume, particularly compared to water vapor. Most Americans viewed carbon dioxide as a force for good—an odorless gas with many positive impacts (such as the fizz in their cola drinks). Most importantly, it was an indispensable component in the photosynthesis cycle that made plant and animal life possible on earth.

But the question of how atmospheric carbon might affect temperatures and climate was about to achieve greater prominence. In La Jolla, California, a scientist with the influence Callendar lacked was assembling at the Scripps Institution of Oceanography a team of researchers who would revolutionize how scientists viewed the planet's atmosphere and climate. His name was Roger Revelle.

*   *   *

Revelle was a giant in his chosen field of oceanography. He took to the subject, he would jest, "like a duck to water."[5] As a graduate student at the University of California, Berkeley, he relished extended boat trips in the Pacific Ocean to collect data. Revelle earned his entry into the world of professors in 1936 with a doctoral dissertation that appeared esoteric to most nonscientists—an analysis of sample muds collected from the bottoms of the Pacific. After graduate school, he moved down the coast to Scripps, where he taught and remained an active researcher.

To play his part in World War II, Revelle interrupted his academic career to join the active U.S. Navy.[6] Rising to the rank of commander, the scholar in uniform became, in effect, the navy's chief oceanographer. He arranged for Scripps scientists to work with the nearby naval base in San Diego. In Massachusetts, the Woods Hole Oceanographic Institution—the Atlantic coast equivalent of Scripps—also made its expertise available to the navy.

The wartime oceanography program's contributions to the eventual Allied victory were many. It coordinated research on the quality of radar and sonar equipment on U.S. ships, the effects of winds and currents on life rafts, and the forecasting of wave and surf conditions for amphibious landings. In the months before the war's end, Revelle and his team of scientists helped plan for the ultimately unneeded invasion of Japan.

After the war, Revelle joined a large group of scientists studying the underwater atomic weapons tests at Bikini Atoll. His research at the remote reefs in the central Pacific Ocean helped improve understanding of the environmental effects of radioactive fallout in the ocean and how the seafloor had changed over geologic time. Convinced that American science had provided a competitive edge during the war, the navy chose Revelle to supervise the oceanography, meteorology, and geology programs at its new Office of Naval Research.

Eventually, the navy vet returned to Scripps, where he became director in 1951. While attending to his administrative and classroom duties (teaching courses in oceanography and geophysics), he became, first and foremost, a connector of dots across many scientific disciplines, a person with the big picture. During the war, the value of scientists had derived in part from their collaborations across the traditional silos of academia—a model that Revelle would favor throughout his career.

Revelle's range of interests and affable personality made him a much-sought colleague for a breathtaking variety of academic and governmental projects. In 1955, when the United Nations Educational, Scientific, and Cultural Organization (UNESCO) formed its nine-nation advisory committee on marine sciences, it was Revelle who represented the United States.

During this period, the ever-curious professor became familiar with the work of Callendar and believed that a better understanding of ocean phenomena would help answer the Londoner's question about how much carbon was collecting in the atmosphere. At a December 1955 symposium of the American Association for the Advancement of Science (ninety-nine years after Foote's paper before the same organization), he suggested that, as a buffered solution, oceans would resist changes in their acidity from carbon deposition. Thus, the seas would lose some of their ability to serve as carbon sinks over time and would expel some carbon into the atmosphere. Oceanographers already understood the chemistry of seawater. However, it was Revelle who connected its implications to Callendar's theories.[7]

Revelle's views about the limited ability of the world's oceans to absorb the carbon from fossil fuel combustion might appear to be a natural progression from Callendar's work. Compared to the dominant view of U.S. climate scholars, however, Revelle's stance was highly disruptive.

Just two months before Revelle's AAAS presentation, the prestigious Institute for Advanced Study in Princeton, New Jersey (for years the workplace of Albert Einstein, who had died that spring at the age of seventy-six), along with the U.S. Air Force's Geophysics Research Directorate, convened twenty-six American and two European experts on weather and climate. The institutions with more than one participant included the Massachusetts Institute of Technology (MIT), the U.S. Weather Bureau, the University of California, Los Angeles (UCLA), the Woods Hole Oceanographic Institution, the University of Stockholm, the Institute for Advanced Study, and the University of Chicago. Unlike Revelle, this "who's who" of climate scientists showed little interest in the subject of atmospheric carbon dioxide.

The group's principal goal was to create a new era of climatology based on general circulation models. Such models appeared increasingly feasible with new high-speed (for the time) computers. They hoped to explain weather

problem rather than to try to take a political stand and to try to state the facts as best he can."[13]

Despite his caution, Revelle did, on occasion, challenge government restrictions that hindered scientific progress. During World War II, he battled against the denial of security clearances for U.S. citizens based on ethnic background, arguing that it hampered the use of some of the country's best minds. In the 1950s, he opposed the overclassification of naval research based on fears that the Soviet Union could use the information. Revelle made the counterargument, largely vindicated by history, that restricting the dissemination of scientific findings would unduly limit American access to needed information.[14]

Over his career, Revelle displayed a scholarly temperament. He fought for academic freedom so that scientists could do their work with minimal interference from funders. He also tried to avoid policy advocacy, even when the issue with which he became most associated, human-induced climate change, became the grist of intense political division.

Revelle's reputation soared with his 1956 selection to chair the panel on oceanography for the International Geophysical Year. The position gave him substantial influence over the distribution of research funds. Announced with great fanfare by President Eisenhower, the IGY was a big deal. The highly touted global project, which ran from July 1957 to December 1958, encouraged scientific collaboration across numerous disciplines and nations to better understand the planet. Classrooms around the country received educational materials on the endeavor. The United States minted a special commemorative postage stamp, as did dozens of other countries. In the United Kingdom, Prince Philip gave the enthusiastic support of the royal family. Money flowed into research connected to the IGY.

A *Time* interview with Revelle after the announcement of his international position led to the top article in its May 28 science section, titled "One Big Greenhouse." Revelle said that burning fossil fuels might dramatically affect the earth's climate, including melting the ice caps of Antarctica and Greenland and flooding the earth's coastal lands. According to the magazine, "Dr. Revelle has not reached the stage of warning against this catastrophe, but he and other geophysicists intend to keep watching and recording." When the scientists working on the IGY finished their studies, "they may be able to predict whether man's factory chimneys and auto exhausts will eventually cause salt water to flow in the streets of New York and London."

The hoopla over the International Geophysical Year inspired a ten-program Monday night series on NBC television. The show starred Frank Blair, news chief of the popular morning show *Today,* as host and Revelle as one of the six scientists regularly appearing on camera.[15]

Revelle's comfort with the media was hardly new; he had already become a quotable source for science reporters. His skill in fielding press questions likely traced back to his school days. Before getting hooked on the study of geology at Pomona College, he had intended to pursue a career in journalism. He coedited the school's *Student Life,* which he tried to convert from a weekly to a daily newspaper. He later laughingly remembered that the task sometimes demanded working until three in the morning to "get this god-damned paper out."[16] In any case, Revelle in the 1950s was displaying a knack for the apt sentence or two that helped science writers explain to the public the complexities of nature.

Revelle's flair with words also attracted the attention of congressional committees. In 1956 testimony at one hearing, he described the earth as a "spaceship" that required greater understanding of "its air control system."[17]

As Revelle took on new national and international responsibilities, he recruited scientists for Scripps whose expertise would contribute to a better understanding of the complex relationships among oceans, carbon dioxide, and the atmosphere. In 1955, he hired two pioneers in the emerging field of radioactive carbon dating. Hans Suess, who had migrated to the United States from Austria in 1950, and Harmon Craig, who had participated in the University of Chicago's early work, knew how to use carbon isotopes to estimate the age of objects ranging from trees to Egyptian antiquities. Of greater interest to Revelle, both were refining techniques to determine how long ago various layers of the oceans had taken up carbon. Their work might help the understanding of how much might eventually end up back in the atmosphere, since carbon in the upper levels of the oceans could migrate either back into the atmosphere or (over a very long time) deeper into the sea.[18]

In 1956, Revelle hired Dave Keeling, the first postdoctoral fellow in the new geochemistry program at the California Institute of Technology in nearby Pasadena. Keeling had previously earned his doctorate in polymer chemistry from Northwestern University, with a minor in geology. He had rejected job opportunities in chemical manufacturing after observing his classmates being hired to, in his words, "make breakfast cereals crisper, gasoline more powerful, plastics cheaper, and antibiotics more expensive." He vowed to "take only a job which would allow me to study nature itself, not merely ways of exploiting it."[19] After landing at Caltech, he cultivated his growing interest in geology and began to focus on the equilibrium between water and air by measuring the relative levels of carbon dioxide—a topic that appeared to most people at the time to have no practical value.

Under the leadership of the highly regarded Harrison Brown, Caltech's geochemistry program was a hotbed for developing new understandings of carbon. Its focus drew the attention of Humble Oil (a precursor of Exxon-Mobil), which funded some research there. One Humble scientist communicating with Caltech and Scripps published a scholarly paper in *Transactions of the American Geophysical Union* (1957) on measuring historical fossil fuel emissions using carbon isotopes in tree rings.[20]

Keeling's postdoc research, funded by the Los Angeles Air Pollution Foundation and the Atomic Energy Commission, evolved into an obsession with accurately measuring CO_2 in the atmosphere. He collected air samples from beaches, mountaintops, and even a blimp over Los Angeles. At the Big Sur State Park, Olympic National Park, and numerous other locations, he would get out of his sleeping bag during the night to capture round-the-clock data and fine-tune his understanding of carbon accumulations. He stored his samples in tightly sealed five-liter glass flasks and logged his measurements in a large green notebook. Along the way, he adapted existing equipment to obtain more exact measurements.[21]

Keeling chafed at occasional questions from his funders about the broader purposes of his research (beyond his curiosity about the natural world). In later life, he explained his dogged pursuit of precision, which went well beyond the requirements of his early grants: "I was having fun."[22] Fortunately for Keeling, federal agencies at the time were often willing to support basic research, even when there were limited prospects for short-term payoffs.

Charles (Dave) Keeling with wife, Louise, during his postdoctoral year at Caltech, circa 1955. In Pasadena, Dr. Keeling began his quest to measure the carbon dioxide collecting in the earth's atmosphere. Courtesy University of California, San Diego Library Special Collections.

A Keeling trip to Washington in 1956 (his first plane flight) helped advance his ambitious plans for measuring atmospheric carbon. He met with Harry Wexler at the old Weather Bureau office on Twenty-Fourth and M Streets NW. Wexler, still at the bureau's division of meteorological research, had been in communication with Revelle and was now more open to the idea that atmospheric carbon merited careful attention. Wexler agreed to use a big chunk of his IGY money to fund Keeling's work. They decided that an ideal spot for precise measurements would be a million-year-old Hawaiian volcano rising about 14,000 feet above the Pacific, Mauna Loa—"Long Mountain" in English and visible from the popular tourist destination of Kona. The site had relatively pristine air and a recently built Weather Bureau observatory that Wexler was keen to see utilized. A second site, in Antarctica, and sporadic data collections from airplanes (using Keeling's trusty glass flasks) would help replicate the findings at Mauna Loa.[23] The costs of advanced monitoring equipment and far-flung locations were substantial but not beyond the IGY's resources. The bureau, tight on space, offered Keeling a basement office where he could oversee his project but agreed he could take his grant money to Scripps if Revelle paid his salary.

Revelle's new hires found the temperate climes of La Jolla (part of the city of San Diego) a particularly hospitable locale for pursuing their careers. Keeling's job interview included a lunch in the backyard of the Suess residence, in what he fondly remembered as "brilliant sunshine wafted by a gentle sea breeze."[24] For Keeling, raised in the Chicago suburbs, the allure of a Southern California campus just a two-hour drive from Caltech with views of the Pacific and access to nearby mountain trails trumped the Weather Bureau's DC basement. Another appeal of Scripps was Revelle's ability to use his close ties to federal agencies to garner federal grants for new positions and expensive equipment.

The scientists migrating to Scripps also appreciated the intellectual stimulation of working with their new boss. They viewed themselves as building on Callendar's work, which many in their profession regarded as "way out in left field," and several were corresponding with the Londoner, now in his sixties and still publishing. Revelle's comments on their work often provided new insights and perspectives. After talking science all day in the office, Revelle prowled the labs at night to continue the conversations. If people had gone, he sometimes walked over to their nearby homes. "He always wanted to talk science. He would come to my house, and we'd sit all night, and we'd talk about CO_2 and things like that," Craig later remembered.[25] As science historian Spencer Weart observed, "Revelle habitually collaborated with anybody who came near him. He kept a large number of puzzles simmering on the back burner."[26]

At six foot four, Revelle was a big man on campus in multiple ways. As if he wasn't busy enough upgrading the program at Scripps, he was also leading a massive effort to expand the oceanography school into what would become the University of California, San Diego. He created new departments, hired eminent professors, and navigated the politics of the state's highly regarded system of higher education.

In 1957, with Suess as his coauthor, Revelle published his evolving thinking on the broader implications of seawater chemistry. Their short paper in the Swedish geophysics journal *Tellus,* "Carbon Dioxide Exchange between Atmosphere and Ocean and the Question of an Increase of Atmospheric CO_2 during the Past Decades," built on the work of Callendar, Plass, and Revelle's AAAS paper. It called on scientists to sharpen their still fuzzy estimates of how much carbon ended up in the atmosphere. Less confident than Callendar about the benefits of global warming, it memorably observed, "Human beings are now carrying out a large scale geophysical experiment of a kind that could not have happened in the past nor be reproduced in the future. Within a few centuries we are returning to the atmosphere and oceans the concentrated organic carbon stored in sedimentary rock over hundreds of millions of years."[27]

The article expressed doubts about Callendar's high estimates of carbon accumulating in the atmosphere. But it also challenged the view at the Princeton conference that negligible emissions remained in the atmosphere for very long. Suess and Revelle agreed on the need for an accurate baseline measurement of future changes in atmospheric CO_2 to help assess its potential effects on weather and climate.[28]

Revelle's views on the relationship between oceans and the atmosphere entered the public arena more quickly than would have been the case with a scientist of less stature. His role with the IGY enhanced his ability to disseminate his thoughts on carbon beyond his university-based colleagues and attract support for climate research. In 1958, the congressional subcommittee on military applications of atomic energy selected Revelle for its advisory group of scholars. The following year, he joined the science advisory panel of the House Committee on Science and Astronautics.[29]

The White House also recognized Revelle's role as an eminent American scientist. On Tuesday, February 4, 1958, President Eisenhower and First Lady Mamie Eisenhower hosted a black-tie state dinner not to celebrate a visiting foreign sovereign but to honor American military and scientific leaders. Those enjoying the food, conversation, and music included the secretaries of the Army, Navy, and Air Force, the chairman of the Joint Chiefs of Staff, most members of the President's Science Advisory Committee, and "Dr. and Mrs. Roger Revelle, oceanography, University of California." If there was a gulf between the ivory towers of academia and the halls of political power, it was a narrow one that cold winter evening.

Back on Capitol Hill in 1959, Revelle testified on the results of the International Geophysical Year. He advised the members of a House appropriations subcommittee, "Because we do not have sufficient understanding of the processes that control climate, we are unable to make a forecast. Nevertheless, a prediction of future climate would be of inestimable value to society."[30]

Revelle's efforts to elevate the status of oceanography and the study of climate were bearing fruit. When Eisenhower released his annual budget message in January 1960, he recommended that federal support for oceanography and related marine sciences be "substantially augmented" based on recommendations from the National Academy of Sciences. The Committee on Oceanography that wrote the referenced report included Revelle. Its second sentence asserted that the waters that cover two-thirds of the earth's surface "greatly affect our lives [and] play a major role in governing our climate." It continued three pages later, "We know that the average weather conditions we call climate can change over a few decades, and we suspect that changes in the storage of gases and heat in the oceans will profoundly influence the process." Both comments drew on Revelle's 1957 article, published just months before the committee began working on its report.

The Academy's priority was to expand the fleet of scientific boats collecting data. Oceans, it argued, were home to a growing number of U.S. nuclear submarines and, therefore, more important for national defense than outer space. Nevertheless, the report's linkage of oceans and the atmosphere and the presidential endorsement of its findings meant that the topic was gaining traction in influential places.[31]

Keeling lacked Revelle's celebrity. But after his move to Scripps the meticulous researcher (called by one colleague a "classical pedant" because of his attention to detail) continued to work tirelessly on how best to measure the accumulation of atmospheric carbon.

At first, Keeling and Revelle disagreed on what carbon measurements they wanted. Revelle adopted the conventional view that carbon levels varied around the planet and recommended using boats to collect intermittent data at numerous locations. Keeling now believed that carbon was well mixed in the atmosphere and did not vary much by location. He wanted a few sites with continuous measurements to detect any changes. At first, the plan was to incorporate both approaches.

Due to logistical challenges, it took Keeling longer than expected to get his measuring devices in place and working. But as results started coming in, he convinced Revelle and others that distant places would yield similar results if he monitored air where more transient gases from automobiles or industrial activities were less likely to distort the measurement of longer-lasting carbon. Keeling's research reinforced his position that data from a single site could reflect the atmosphere of the entire planet. Thus, just a few monitoring stations would be adequate for verifying global levels.[32]

Keeling reported his data in the June 1960 issue of *Tellus*. He recalled it as the shortest paper he had ever published—just 1,200 words. Its most eye-catching feature was a graph displaying two years of monthly data from Mauna Loa.[33] A single jagged line unexpectedly portrayed seasonal variations, or what historian Gale Christianson later called "the rhythmic breathing of the planet."[34] Carbon dioxide levels in the atmosphere climbed with the death of northern hemisphere vegetation and then fell due to spring and summer plant growth. The pattern offered a new perspective on the impacts of photosynthesis.

Despite the seasonal variations, annual trends for carbon pointed upward. Keeling thought the pattern would continue and would support Revelle's theory on the limits of ocean absorption. The results so far were, he said, "systematic" and "persistent." At the least it appeared that more carbon was being added than subtracted, potentially contradicting the view of von Neumann and others that carbon levels remained near a state of equilibrium.

Furthermore, Keeling's data made one aspect of the climate puzzle less speculative. Scientists would still need radioactive carbon dating to estimate carbon levels before 1958. Henceforth, Keeling's research would provide the baseline data that Revelle and Suess had sought. These measurements could be considered factual and very difficult for skeptics to dispute.

In August, *Life* magazine, while preparing a review of the International Geophysical Year, wrote Keeling, requesting information on his research. The scientist took seven weeks to craft his reply and placed the letters in a new file labeled "Public Relations." Keeling shared many scientists' belief that it was difficult to explain the abstruse world of science to nonscientists, including the press. When he finally responded to the popular weekly, he enclosed a copy of a speech he had given in Helsinki and warned that it was too early "to reach conclusions about long-range trends" on carbon accumulations. He offered to review any drafts of the magazine's article.[35]

In its November 7, 1960, issue—as the nation prepared to cast its vote for a new president—*Life*'s cover drew attention to its feature story on the IGY. On page 75, it called the project "the single most significant peacetime activity of mankind since the Renaissance." It thanked the Scripps Institution for its help in drawing a series of vivid graphic images of the earth and cited Revelle's work on ocean currents. But, perhaps because of Keeling's sluggish response or the editorial judgment that the specter of massive earthquakes held greater intrigue for readers, the story paid no attention to the buildup of carbon dioxide in the earth's atmosphere.

A related question lingered in the background during the Eisenhower years. If the combustion of coal, oil, and natural gas threatened the earth's climate, were there alternative energy sources without such side effects? The rise of fossil fuels seemed inseparable from the Industrial Revolution and economic growth. Coal had powered new inventions like steamboats, trains, industrial machinery, and electricity. Automobiles, trains, and planes relied on oil products. But fossil fuels had once replaced earlier forms of energy—

such as wood, whale oil, or calories burned to produce animal and human labor. Could substitutes for fossil fuels generate needed energy and reduce pollution?

One alternative to fossil fuels stood out in the Eisenhower years: nuclear power using uranium for fuel.

Public knowledge of the enormous energy generated by splitting tiny atoms came with a jolt on August 6, 1945, the day the *Enola Gay* dropped an atomic bomb on the city of Hiroshima. Albert Einstein had alerted President Franklin Roosevelt in 1939 that scientists might be able to set up a nuclear chain reaction that could lead to the construction of powerful bombs. He warned that Germany might pursue such a weapon and urged support for the American physicists trying to create the reaction.[36] In response, the United States established the later heralded but then ultrasecret Manhattan Project to work on a crash basis to make epic breakthroughs in physics and the weaponry of war.

As it turned out, the two atomic bombs deployed during World War II fell not on Germany, which had lagged in its efforts to develop nuclear weapons, but on Japan, which lacked such capability. This irony led to some remorse among scientists who had helped develop the awe-inspiring technology. Many enthusiastically embraced adapting the atom's power for nonmilitary purposes. They cheered Eisenhower's 1953 backing for an initiative called Atoms for Peace, which declared that "if the fearful trend of atomic military buildup can be reversed, this greatest of destructive forces can be developed into a great boon for the benefit of all mankind."[37] The atom's peaceful uses would demonstrate what one leading scientist called its "redeeming" value and keep the United States on the cutting edge of a strategically important technology.[38]

Even before Eisenhower's announcement, the Atomic Energy Commission and the U.S. Navy were developing nuclear-powered submarines. Besides offering the advantages of quiet operation and much greater distances between refueling, the new subs' reactor technology could be transferred to civilian power plants. Westinghouse Corporation partnered with the government to develop the light-water reactor used for the first nuclear sub, the USS *Nautilus*. The boat's 1952 keel-laying ceremony featured an appearance by President Harry Truman. The meticulous engineer who led the effort to develop nuclear power was the Russian-born captain (later admiral) Hyman Rickover, known for his technical brilliance, demanding style of command, and prickly relations with colleagues. The audience included a twenty-seven-year-old naval officer from Plains, Georgia, named Jimmy Carter, part of the team developing the new subs.[39]

Adapting the nuclear propulsion systems designed for the navy, the nation's first full-scale nuclear electric generating station began operational testing on the banks of the Ohio River at Shippingport, Pennsylvania, in late 1957. The date of December 2 was a noted anniversary in the annals of nuclear energy. Just fifteen years earlier to the day, scientists working in a former squash court under the University of Chicago's football stadium had achieved the world's first (albeit extremely small) self-sustained nuclear chain reaction.[40] Rarely had a breakthrough discovery scaled up so rapidly from a laboratory experiment to commercial application.

The navy's early work helped cut the expense of building the Shipping-port plant. Nonetheless, the cost overruns for the first-of-a-kind project came as a shock to the Joint Congressional Committee on Atomic Energy. In defense of its work, Westinghouse asserted that it could produce future reactors at one-fourth or one-fifth of the cost of Shippingport.[41] The hope persisted in Washington that nuclear power could prove cheaper than electricity produced with coal.

In the Eisenhower era, the federal government supplied substantial resources to advance nuclear electric generation. The distinguished national science laboratories that helped build the atomic bomb continued to work on military applications but also provided their expertise for the advancement of civilian nuclear power. In 1957, Eisenhower signed the Price-Anderson Nuclear Industries Indemnity Act. The legislation foresaw that the extraordinarily high insurance costs of covering worst-case-scenario accidents would discourage private investors in nuclear power. It capped industry liability and made the federal government the insurer of last resort to remove an impediment to rapid expansion.

Eisenhower's annual budget message in January 1960 urged additional federal support for the peaceful use of atomic energy. It requested $185 million for research and development and $65 million to construct additional civilian power reactors. The Atomic Energy Commission, said Eisenhower, would determine "the number, type, and size of reactors built and the nature of the assistance provided . . . after considering the state of technology and the cooperation proposed by industry."[42]

Any linkage between nuclear power and the threat of climate change rarely surfaced in public discussions. But there were a few exceptions. The 1957 article by Revelle and Suess briefly noted that scenarios with a very rapid rise in atmospheric carbon assumed that atomic power would play only a limited role in meeting future energy needs, implying that its rapid growth could slow the buildup of carbon in the atmosphere. In November 1959, Edward Teller—a Hungarian-American physicist who led the development of the hydrogen bomb—made a case for a greater reliance on nuclear power at a conference sponsored by the American Petroleum Institute and Columbia University's School of Business. He told attendees that the world would eventually need to replace fossil fuels because they would run out, and the carbon dioxide they emitted contaminated the atmosphere. Although CO_2 was invisible and nontoxic, the science icon said, its presence in the atmosphere trapped heat and could, in more significant amounts, melt ice caps and submerge New York City.[43]

Nuclear scientists from the atomic energy labs echoed Teller's argument at a March 1960 hearing of the Committee on Atomic Energy. While extolling the advantages of nuclear power, they speculated that the environmental damages from burning fossil fuels might be more severe than from the leakage of radiation or long-lasting waste from nuclear power generation.[44]

Bernard Spinrad, director of reactor engineering at the Argonne lab outside Chicago, testified that rather than considering radiation as a danger, "I personally believe that the degree to which we have been able to contain these products is a basic advantage. In the long run, I think that even increased carbon dioxide from fossil energy sources is perhaps a more disturbing factor in our environment." Responding to a question about whether pollution from automobiles, including carbon dioxide, could be reduced by running them with small nuclear reactors, Spinrad warned against being "overly optimistic" about that approach because of the weight of the shielding needed to avoid the leakage of radiation. "The only way I could conceive of nuclear energy to run automobiles," he elaborated, "is to generate electricity which is used to charge up storage batteries and go in for electric engines."[45]

In a story on the hearing, *Time* magazine once again alerted its readers to the carbon dioxide issue. It reported:

> Some scientists suspect that the ever-increasing amount of fossil fuel that is burned may be increasing the amount of carbon dioxide in the atmosphere. They fear that the added CO_2 will have a "greenhouse effect," trapping solar heat at the earth's surface and raising its temperature. The result may be unpleasant changes of climate, including deserts in many places that are now fertile and a disastrous rise of sea level because of melting icecaps. A cure might be a world agreement to use nuclear reactors wherever possible. They excrete no CO_2.[46]

Another witness at the table with Spinrad was Alvin Weinberg, the head of the Oak Ridge national lab in Tennessee. Several days earlier, Eisenhower had appointed Weinberg to the President's Science Advisory Committee. Though the PSAC dealt almost exclusively with military issues at the time, Weinberg's membership ensured that someone at the White House was aware of ideas connecting fossil fuel combustion and climate change.

The views of nuclear scientists on climate change did not go unchallenged. In 1959, a Shell Oil Company scientist, M. A Matthews, disputed Teller's warnings about the risks of burning fossil fuels. Matthews's article argued that, due to the ocean's ability to absorb carbon dioxide, any temperature change resulting from carbon emissions would not likely "result in any great disturbance either of the ice pattern or the rate of growth of vegetation." Thus, he concluded, "Man's efforts in burning large quantities of fossil fuels are inevitably small compared with the magnitude of Nature's carbon cycles."[47] With the Mauna Loa data not yet published, the Shell analysis was still within the bounds of the prevailing scientific thought.

There was another alternative to fossil fuels—one with more distant prospects for widespread adoption: turning the rays of the sun into electricity.

In 1905, in addition to publishing his theory of relativity, Einstein advanced the view that particles of light (protons) could discharge electrons.

This "photovoltaic effect" created the possibility of solar-generated electricity.[48] German efforts to convert Einstein's theory of quantum physics into practical reality slowed with the approach of World War II. In the 1950s, American scientists had taken the lead in producing solar cells that might someday provide power in significant amounts at competitive prices.

The Bell Telephone Company—just called "the telephone company" at the time, due to its national monopoly status—used its famed Bell Labs to conduct research and development for its own technical needs. But it also gave its scientists wide latitude to pursue other significant work, partly as a form of public service. A team of scientists at the quasi-academic institution had long tried to tweak solar technology to develop photovoltaic cells that might have commercial applications.

The editorial board at the *New York Times* in the spring of 1954 took notice of the progress at Bell. It explained that the lab's breakthrough innovation of replacing vacuum tubes with transistors using silicon was advancing solar technology. The *Times* opined, "We are still far from driving the kitchen refrigerator with sunbeams, but we are getting there rapidly. It may well be [that] we ought to transfer some of our interest in atomic power to sun power."[49]

In 1955, one Bell experiment used a small photovoltaic panel to enable the first solar-powered telephone call.[50] Coincidentally, this pioneering application took place in the south Georgia town of Americus—just ten miles northeast of the even smaller town of Plains, where the same year ex-naval officer and now peanut farmer Jimmy Carter began his first term on the Sumter County school board. As an immature technology with low efficiency in converting sunlight into electricity, these early solar panels were prohibitively expensive, constraining their role in commercial markets.

In the late 1950s, the Vanguard satellite program—a U.S. space exploration response to the Soviet Union's shocking launch of its Sputnik satellite—achieved mixed results. But it did demonstrate the feasibility of using Bell's solar cells to power radio communications in places where the amounts of power needed were minimal and transmission lines unavailable.[51] For the foreseeable future, photovoltaic power appeared to be limited to such tiny niche uses.

In any race to provide nonfossil energy sources, nuclear power was the speedy hare—the federal government's favored energy technology, the recipient of generous funding, and off to a blistering start for a still new technology. Might solar energy, with only minuscule support at the time, play the role of Aesop's fabled tortoise?

2

JFK's Early Mention of Climate Science

The Kennedy Presidency, 1961 to 1963

On a 1958 national speaking tour, the junior U.S. senator from Massachusetts set forth a bold national goal. John F. Kennedy repeatedly called on the United States to lead the world in scientific discovery.

At a Loyola College alumni banquet in Baltimore, a Social Science Foundation lecture in Denver, and other events across the country, the descendant of Irish immigrants cited the importance of a broad array of scholarly disciplines—including biology, meteorology, and even Roger Revelle's field of oceanography. Kennedy linked his references to oceanography and other sciences to the competition with the country's Cold War adversary. The Soviet Union's Sputnik satellites had given it a leg up in space exploration. Suddenly, America's presumed dominance as a science innovator was in doubt. The senator expressed the fear that the Soviets might gain additional advantage with other stunning successes in science.[1]

The forty-one-year-old Democrat's far-flung itinerary added to speculation that he was testing the waters for a presidential bid. Indeed, he went on to win his party's 1960 presidential nomination, running on a party platform that included his pledge to expand scientific research and a specific mention of oceanography's critical role.

After taking his presidential oath in January 1961, Kennedy continued to champion the advancement of American science. This commitment prompted a sound bite that became renowned in the annals of presidential rhetoric. On May 25, 1961, just over four months after taking office, he declared in a special appearance before Congress, "I believe that this nation should commit itself to achieving the goal, before this decade is out, of landing a man on the moon and returning him safely to the earth."[2] He vowed to utilize the nation's vast resources and best minds to achieve what many believed impossible (and a few skeptics still believe never happened).

The race to space overshadowed the president's continued focus on a broad array of scientific endeavors. His first State of the Union address, on January 30, repeated his warnings about the neglect of oceanography. To back up his words, he asked Dr. Jerome Wiesner, Special Assistant to the President for Science and Technology, to develop a plan for better understanding the depths of the world's oceans. Oddly enough, this long-overlooked assignment would lead to the first veiled reference by a U.S. president to the possibility that human pollution might be altering the earth's climate.

Wiesner, who taught electrical engineering at MIT, first met Kennedy, in whose congressional district he resided, in 1952. Several years later, then Senator Kennedy began to rely increasingly on his advice about various scientific and military matters, in phone calls and private meetings. The professor realized only later that he had become the senator's primary source for scientific guidance.[3]

The national eminence of the professor hardly depended on his ties to the new president. His career had checked the significant boxes of an academic held in high esteem. After World War II, he worked with Revelle and other scientists studying the first underwater test of the atomic bomb at Bikini Atoll. Wiesner's work on needed electronic components earned him the President's Certificate of Merit, the second-highest civilian award for service to the country. In 1954, he received a patent for an "apparatus for applying high-intensity pulses to crystal rectifiers" and an award as the outstanding engineering graduate of the University of Michigan.

Besides building highly regarded science programs at MIT, Wiesner performed sensitive work for federal agencies requiring high-level security clearances. He reached the pinnacle of such positions in 1957 with his appointment to the President's Science Advisory Committee. Since the job wasn't considered political, Wiesner could serve in a Republican administration despite twice actively supporting Democrat Adlai Stevenson for president.

Nonetheless, a *Washington Post* report during the 1960 presidential campaign that, if elected, Kennedy intended to name Wiesner as his chief scientist had proved awkward for the professor. Wiesner—not involved in the presidential campaign and previously unaware of Kennedy's plan—sent a message to Eisenhower asking whether he should resign his White House post. Given the scientific nature of the work, Ike saw no conflict and declined the offer.[4]

After joining the Kennedy White House, Wiesner convinced a skeptical JFK to preserve the science committee's nonpartisan role by retaining its previous members and staff. With his pipe and slightly disheveled appearance, the professor looked the part of an academic in the halls of power. More importantly, he enjoyed direct access to the commander in chief. "Not since Merlin," Kennedy speechwriter and close confidant Ted Sorensen later quipped, "has any head of state made greater use of, or relied more upon, his chief science advisor than John F. Kennedy relied on Jerry Wiesner."[5]

The scientist's long relationship with his boss allowed their communications to be more candid than the norm for White House staffers. On one consequential matter, Eisenhower had decided during the 1960 election not to share with Kennedy classified evidence that detected no Soviet deployment of intercontinental ballistic missiles. As a result, the Democratic campaign persisted with its claims of a "missile gap" that threatened U.S. security. In the Kennedy White House, the task of truth-telling fell to Wiesner. Several weeks later, the administration reluctantly accepted the intelligence community's view that the "gap" did not exist.[6]

Kennedy, at times, appeared puzzled when told that scientists disagreed on significant issues. Facts are facts, he would declare. But Wiesner explained that in science, there were many more facts to be discovered than were known.[7] Therefore, scientists had to make recommendations based on the best evidence available.

A science advisor—expected to be an expert on everything but having earned his credentials with specialized studies—needed to know other scientists with the right expertise for the right assignments. To review the oceanography proposal going to the president, Wiesner turned to his old colleague and friend in California, Roger Revelle.

In early March, Wiesner updated Kennedy on the oceanography initiative in anticipation of a message to Congress. His briefing made no mention of any connection between the oceans and climate. But after Revelle's comments and a rewrite from Sorensen's White House wordsmiths, the presidential message three weeks later differed in tone and substance.[8]

The president now described the study of oceans in epic terms. Such knowledge, he said, "is more than a matter of curiosity. Our very survival may hinge upon it." He pointed to the potential for harvesting significant amounts of seafood and tapping vast supplies of raw materials like salt, potassium, and magnesium when resources on land "ultimately reach their limits." He cited, as well, the military advantage of better understanding the marine environment and the ocean floor.

In paragraph 7, the message turned to climate. "To predict, and perhaps someday to control, changes in weather and climate is of the utmost importance to man everywhere," it declared. "These changes are controlled to a large and yet unknown extent by what happens in the ocean. Ocean and atmosphere work together in a still-mysterious way to determine our climate. Additional research is necessary to identify the factors in this interplay."

With the private communications of the period available today, the impact of Revelle's review of the draft message is now clear. The president was implicitly referencing the recent technical work by Revelle and his colleagues on where carbon emissions ended up. The mention was brief and, to most observers, inscrutable. Yet it can be said that on March 29, 1961—less than a year after Dave Keeling published his Mauna Loa data and the year before the publication of Rachel Carson's environmental game-changing *Silent Spring*—a U.S. president for the first time recognized, at least in a fleeting way, the potential impacts of carbon emissions on the earth's climate. A different way of looking at the planet, pushed by Revelle in the late 1950s, had already made a small but discernible impression on presidential rhetoric.

Revelle went on to be an academic superstar in the Kennedy administration. The day after the message on oceanography, the White House announced a national advisory council for the Peace Corps—a new program to put American volunteers to work in impoverished countries around the world. It listed Revelle as one of the thirty-three members. He joined a group of luminaries that included Vice President Lyndon Johnson as chair, Supreme Court Justice William Douglas as honorary chair, and former First Lady Eleanor Roosevelt and singer/actor Harry Belafonte as additional members.

Revelle's national prominence followed a severe rebuff at home. In February, the California Board of Regents had selected Herbert York, a Revelle friend and prominent nuclear scientist, as the University of California, San Diego's first president. Due to Revelle's academic acclaim and leadership in creating the school, he had been the widely expected choice for the post. Revelle's supporters speculated that his occasional skirmishes with the regents when he opposed a loyalty oath for faculty members or fought for sites he thought best for the new campus spurred opposition to his selection. One colleague believed that Revelle's rejection was tied to a regents' meeting at Scripps, when the professor reportedly had kept them waiting outside his office for two hours while he extolled to a young seaman from one of the school's research ships the virtues of going to college. Ever the loyal soldier, Revelle supported the new administration at the school he had played such a large part in building. However, his wife, Ellen, later conceded that the rejection was "a terrible blow."[9]

Revelle alerted Wiesner in the spring of 1961 that it might be a good time to take a break from Scripps if the right opportunity became available. The man from MIT had already convinced Kennedy that cabinet agencies should hire eminent outside scientists for high-level positions. The initiative jibed with Kennedy's broad emphasis on expertise. During a short address in April 1961 at the National Academy of Sciences, for instance, the president expressed his appreciation for its work and acknowledged that nearly every question addressed by public officials had scientific overtones. "For those of us who are not expert," he said, "we must turn, in the last resort, to objective, disinterested scientists who bring a strong sense of public responsibility and public obligation." Once again, Revelle provided input for a Kennedy speech citing the value of oceanography.[10]

Soon after, the White House announced Revelle's selection as the full-time science advisor to Secretary of Interior Stewart Udall. Revelle handled a variety of issues at the Department of the Interior but privately complained that its component agencies were too close to their funding sources in Congress and too set in their ways to take much guidance from a scientist in the secretary's office. Despite Revelle's grumbling about his assignment, his influence was widespread because of his membership on the Federal Council for Science and Technology, whose duties included overseeing Kennedy's oceanography initiative. Some scientists who joined the administration re-membered that time as the "golden years" for many of their disciplines and Revelle's contributions as particularly noteworthy.[11]

In May 1961, Kennedy called Wiesner with an unusual assignment. The president of Pakistan, Muhammad Ayub Khan, was coming to Washington for a state visit and would likely request advanced U.S. weapons. Kennedy wanted to deflect the arms issue with something valuable enough to impress the visitor. Wiesner met with Khan's science advisor to find a suitable alternative and determined that Pakistan was losing massive amounts of agricultural land to

Secretary of the Interior Stewart Udall (*left*) next to Roger Revelle, who is being sworn in as a full-time science advisor in the Kennedy administration, 1961. Courtesy University of California, San Diego Library Special Collections.

waterlogging and salinization. Wiesner talked with Revelle about potential solutions and found him intrigued by the challenge. Wiesner later remembered, "This was Roger's normal reaction to a problem; the tougher it was, the more interesting Roger found it." He told Kennedy that Revelle would head a team of experts to study the problem.[12]

To the delight of the Pakistanis (and the U.S. president), Revelle helped develop programs to improve farmer education, irrigation, pesticides, and fertilizers. These advances led to dramatic improvements in crop yields and served as precursors to the later green revolution that prevented a global food shortage. His success earned his decoration with the order of Sitara-i-Imtiaz (star of distinction) by the Pakistani president for "conspicuously distinguished service in science" and an enhanced sense of what good science could accomplish.[13]

The following year, the Kennedy administration, recognizing Revelle's expanded expertise in South Asia, appointed him as the U.S. member of the Education Commission of the Government of India, which examined and recommended changes in the country's school system.[14] Revelle also continued to receive recognition in his field of specialization. The National Academy of Sciences in 1963 awarded him the Alexander Agassiz Medal—his discipline's top honor. With his increasingly diverse White House assignments, however, the scholar with a dissertation on ocean-bottom muds and influential works on the complex heat flows of oceans was now expanding his horizons and gaining new insights into the nuances of economic development in the world's emerging nations.

Before leaving Interior, Revelle took on another challenging assignment. At the request of the White House science office, he headed an interagency task force studying the viability of employing large-scale nuclear plants to distill saline water for use in coastal areas.[15]

Thanks to Revelle, Keeling, and others, interest in atmospheric carbon dioxide grew during the Kennedy years. In March 1963, the Conservation Foundation convened a small workshop in New York City to consider the state of knowledge and future research needs. The six subject experts included Erik Ericksson with the International Meteorological Institute in Stockholm, Sweden; Gilbert Plass, now working at the aerospace division of Ford Motor

Company in Southern California; and the man with the measurements, Dave Keeling.

The scientists were confident that, despite the atmosphere's longtime stability, carbon emissions from "engines and other sources" over the past hundred years had led to observable but still small consequences. These included slight increases in temperatures and sea levels and minor adjustments in the habitats of animals, including fish and butterflies. It appeared "quite certain" that a continuing increase of atmospheric carbon dioxide would be accompanied by significant warming of the earth's surface, which by melting the polar ice caps would raise sea level and by warming the oceans would considerably change the distributions of marine species. In some scenarios, the melting of polar ice caps would inundate coastal areas, including New York City and London.

They identified factors that might offset the effects of rising carbon accumulations. For example, increased cloudiness and greater productivity of plant life might help restore some equilibrium to the atmosphere, but more research was needed. Hopes for increased absorption of carbon by vegetation, for instance, had to be balanced by recognition of massive global deforestation due to human activities. Moreover, it was hard to predict how fast fossil fuels would be burned. The summary of the conference report concluded:

> The effects of a rise in atmospheric carbon dioxide are world-wide. They are significant not to us but to the generations to follow. The consumption of fossil fuel has increased to such a pitch within the last half century that the total atmospheric consequences are matters of concern for the planet as a whole. Although there is the possibility of capturing CO_2 formed by the burning of fossil fuels and storing it in the form of carbonates, relief is most likely through the development of new sources of power.[16]

After reviewing a draft summary, Keeling confirmed that the use of his data was correct. He suggested, however, a somewhat different tone. He counseled that "the recent increase in atmospheric carbon dioxide, as far as we can tell, is not itself particularly alarming, but is nevertheless symptomatic of a host of potentially disturbing effects which now attend the intensive development of our worldwide natural resources."[17] He remained the cautious scientist.

In September 1963, U.S. senator Edmund Muskie and other government officials lifted off a Wall Street heliport for a forty-five-minute tour of Long Island, Queens, and nearby industrial sites in New Jersey. On this day, strong winds contributed to extended visibility across the New York metropolitan area. That was a problem. The Maine Democrat chaired a recently formed subcommittee on air and water. Sooty air and darkened skies, the norm in the region, would have better illustrated the copter ride's intended message—the nation's need to address the problem of polluted air in urban centers. Making the best of the surprisingly clearer-than-usual conditions, Muskie tried to save the photo op by citing the additional threat of nonvisible pollutants that didn't produce smoke.[18]

Despite the attention given to *Silent Spring*'s warnings about the risks of widely used pesticides, public awareness of the environment as a national political concern was at an early stage of development. Wastes dumped into rivers and chemicals emitted into the ambient air seemed inevitable byproducts of industrialization and economic progress. Muskie wanted to demonstrate that air pollution was a public health problem that crossed state lines, suggesting that states and localities needed the federal government's help to deal with the situation.

Muskie's pioneering focus on protecting air resources found solid support at the White House. In "Special Message to the Congress on Natural Resources" in February 1961, Kennedy had already called urban air pollution "a serious hazard to the health of our people that causes an estimated $7.5 billion annually in damages to vegetation, livestock, metals, and other materials." In a strong statement for the times, he declared, "We need an effective Federal air pollution control program now."[19]

In "Special Message to the Congress on Improving the Nation's Health" in February 1963, the president cited recent scientific reports linking air pollution to "the aggravation of heart conditions and to the increases in susceptibility to chronic respiratory diseases, particularly among older people." To deal with these threats, he recommended legislation authorizing the Public Health Service to "conduct studies on air pollution problems of interstate significance" and "take action to abate interstate pollution."[20] The message reflected a new way of looking at the world. The air pollution to which Americans had become accustomed was no longer acceptable.

Later in the year, Congress moved legislation responding to Kennedy's and Muskie's concerns. On July 24, the House of Representatives passed the Clean Air Act of 1963 by a decisive margin. However, fearful of expanded federal authority, most Republican members voted against the bill. The Senate passed its version with an unrecorded voice vote on November 19, two months after Muskie's aerial inspection of New York's air. A conference committee would work out the differences.[21]

The first of the nation's clean air acts did not address climate change or carbon dioxide. Nor did it have enough teeth to deal with the problems of the dirty, often toxic air that Kennedy and Muskie had identified as significant risks. Still, it is an oft-forgotten milestone in how the government was changing its approach to pollution. Even if timid in its mandates, the new law helped set the stage for more robust controls down the road.

Kennedy's brief reference to climate change in March 1961 garnered scant notice at the time. Given that it would take decades for the issue to earn front-burner status, it is not surprising that historians later overlooked the occasion. But the president's commitment to the study of oceanography (and less directly, climate) persisted. In August 1963, he filmed remarks for a government movie entitled *Oceanography: Science for Survival,* which Revelle helped prepare and in which Revelle briefly appeared. Almost verbatim, JFK repeated his earlier oratory about the critical interplay between oceans, the atmosphere, and climate, and the need for further research.[22]

In October, Kennedy addressed the hundredth-anniversary celebration of the National Academy of Sciences. Clad like his audience in an academic gown, and with Wiesner and Revelle seated in the row of honor behind him, the president again called for a better understanding of the interactions between oceans and the atmosphere and for international cooperation to achieve scientific progress. He urged those packed into Constitution Hall for the grand occasion to think long term on all the big issues of science. He cited the story of French marshal Hubert Lyautey, who told his gardener, "Plant a tree tomorrow." The gardener replied, "It won't bear fruit for a hundred years." "In that case," said Lyautey, "plant it this afternoon." To chuckles and applause from the audience, the president closed: "That is how I feel about your work."[23] Revelle had once again helped prepare a presidential presentation to the National Academy, this time leading to his only face-to-face meeting with Kennedy.[24]

By the time of the Academy event, Wiesner had decided to return to MIT, where he would later become the school's thirteenth president. Wiesner's replacement, the Milwaukee-born Donald Hornig, possessed an impressive resume—part of the Manhattan Project, the last person to see the atomic bomb before its first test in the New Mexico desert, chairman of Princeton's chemistry department, and a science advisor at the White House dating back to the Eisenhower years.[25] In mid-November, Honig called on the country to bridge the cultural gap between scientists and nonscientists. He said that much of the public lacked a good understanding of science but could absorb the needed information if the scientific community spent "more time making itself understood to the entire nation."[26]

The Kennedy years were good ones for scientists, including those at Scripps. The White House sought Revelle's ideas, some of which crept into presidential speeches. Federal agencies prioritized issues championed by Revelle for funding. Moreover, as Keeling's line graph lengthened, the pattern of carbon dioxide accumulating in the atmosphere became increasingly evident to scientists around the world. Before Keeling's dogged research, the scientific consensus was that human activities had little effect on the atmosphere. In the early 1960s, that view began to shift.

On November 22, 1963, Revelle was attending a meeting of the International Council for Science in Vienna. At a party for the delegates hosted by the city's mayor, he received the news of President Kennedy's assassination. Revelle later remembered. "It was a very shocking event, terrible. We all cried."[27]

3

LBJ's War
on Pollution

The Johnson Presidency, 1963 to 1969

On March 18, 1964, Roger Revelle sat in the Fish Room (later renamed the Roosevelt Room), located across from the Oval Office—the inner sanctum of national power. He had recently left his post at the Department of the Interior and returned to the University of California. This day, he was back in Washington, joining other leading scholars for a late-morning meeting with recently elevated President Lyndon B. Johnson.

The event's twenty-seven invitees included prominent intellectual heavyweights ranging from widely read historians Clinton Rossiter (Cornell University) and Richard Hofstadter (Columbia) to liberal economist John Kenneth Galbraith (Harvard) to legal scholar and provost Edward Levi (Chicago) to prolific inventor Edwin Land (founder of Polaroid). Revelle was the only academic representing the physical sciences at the high-level discussion about emerging domestic priorities. The new administration wanted the group to propose grand national initiatives that could follow expected historic legislation on civil rights.

Several in the room thought that the president, who was suffering from a cold and had almost canceled his appearance, looked less than fully engaged. Revelle was especially impressed, however, by twenty-nine-year-old aide from Texas Bill Moyers, who, Revelle observed, appeared to be "to LBJ what Ted Sorensen was to the late President Kennedy." Leaving for the airport later that day, the professor was uncertain about what the group had accomplished or whether it would ever meet again.[1]

Nonetheless, the team of intellectuals did reconvene at a hotel off Scott Circle late on Friday, April 3, and again the following day, in the Old Executive Office Building (later renamed the Eisenhower Building) next to the White House. The Saturday session broke down into small groups to discuss and produce big-picture essays of national significance. Historian and moderator of NBC television's program *The Open Mind,* Eric Goldman, who shuffled between Princeton University and the White House to organize the meetings, hoped the papers would help shape a later presidential address.

Revelle's group—which included Margaret Mead, author of the ethnology classic *Coming of Age in Samoa* and the only woman invited—drafted a short statement on "the moral commitments of the nation." One section, "Natural Resources," described challenges posed by advances in science and technology. It noted that "our power to change nature has grown enormously for both good and ill." In words clearly penned by Revelle, the paper predicted, "Sooner or later, we shall be able to change the very climate itself. We are doing this in spite of ourselves by burning oil, coal and natural gas. But by gaining greater understanding, we will be able to make conscious changes— to bring more water to deserts, to bring cooler summers and warmer winters to the Middle West and the Northeast."[2]

The language implicitly drew on studies at Scripps and elsewhere on the rising amount of carbon dioxide in the atmosphere. It also referenced another branch of science in which Revelle was less directly involved—the effort to eventually understand weather and climate with enough precision to modify them for beneficial purposes.

Revelle was favorably impressed by several ideas from the April 4 sessions. He privately expressed frustration, however, with the tight schedule. He preferred the President's Science Advisory Committee's approach, which provided more time for study and more polished products.[3]

Goldman later expressed more bluntly his reservations about the process. Like Revelle, he found the time frames of politicians and professors out of sync. For the president, the Princeton scholar later observed with some indignation, "An idea was a suggestion, produced on the spot, of something for him to do tomorrow—a point to be made in a speech, an action, ceremonial or one of substance, for him to take promptly, a formula to serve as the basis for legislation to be hurried to Congress." For the first-rate intellectuals in his advisory group, "Over the long pull, instant ideas were not their specialty; indeed, men of this type have little use for them."[4]

Unlike Wiesner in the Kennedy days, Revelle and Goldman failed to grasp the potential impacts of terse communications at the upper levels of government. But even if they did not realize it at the time, the clout of the intellectuals was particularly significant during the early days of an administration trying to shape bold new directions for the country.

Efforts to find the big ideas that would define the Johnson administration culminated in a highly publicized presidential speech on May 22 at the University of Michigan's football stadium, the nation's largest. Clad in a black academic robe, the president addressed a commencement crowd estimated at 80,000. The state's two U.S. senators and Republican governor George Romney joined the school's graduates and their families at the massive convocation.

The president told his audience, "In your time we have the opportunity to move not only toward the rich society and the powerful society, but upward to the Great Society." Though Johnson had used the term "Great Society" previously, he now fleshed out the idea for the first time—with input from Moyers, top speechwriter Richard Goodwin, Goldman, and, indirectly, Goldman's advisory team of intellectuals. Another key figure in crafting the new agenda was Johnson's science advisor, Donald Hornig, carried over from the Kennedy administration. Hornig wrote Moyers that pollution and the long-term problems of the environment were going to be prominent political issues in the coming years and that it was time for the administration to get moving on them. (Hornig later called this "one of the prognostications I've been proudest of.")[5]

Johnson's vision of a Great Society included the protection of natural resources. "We have always prided ourselves," he declared, "on being not only America the strong and America the free, but America the beautiful." He warned that this beauty was in danger. "The water we drink, the food we eat, the very air that we breathe, are threatened with pollution." He added, "Once the battle is lost, once our natural splendor is destroyed, it can never be recaptured. And once man can no longer walk with beauty or wonder at nature, his spirit will wither and his sustenance be wasted." Johnson pledged to draw on the best thinkers from all over the world to find answers for America "on the cities, on natural beauty, on the quality of education, and on other emerging challenges."

Following up on his commitments in Ann Arbor, Johnson created a dozen task forces on the environment alone. The search for new ideas would again lead to Revelle.

In September 1964, Revelle began a new chapter in his life at Harvard University. His return to California had proved disappointing after the intellectual stimulation of Washington. Moreover, when the chancellor's position in San Diego became vacant, the regents again passed him over.

Revelle's decision to leave came as a blow to San Diego. He was the founder of its now thriving university. While building the institution, he had turned down the presidencies of two other major universities. Moreover, he had been a pillar of his community, serving on hospital, bank, and arts boards. His popular wife of more than three decades, Ellen, was the grandniece of the founder of Scripps College, a LaJolla native, and a dynamic force in the community. The editors of *San Diego* magazine lamented, "It is hard on a town to lose a great man."[6] The publication suggested that the University of California, San Diego name what was then known as First College for Revelle, an action ratified unanimously by the regents eleven days later.

In Cambridge, Revelle worked out of an old, yellow, three-story colonial house at 9 Bow Street just outside Harvard Yard—the core area of a campus established in 1636. He served as the school's first Richard Saltonstall Professor of Population Policy in the School of Public Health and director of a new population studies center. The scholar made no pretense of being a demographer. But he saw the center as a vehicle for exploring the relationship among critical issues like population growth, food, and ecology. He looked forward to assembling a team of physicians, economists, social scientists, humanists, and engineers to work on them.

Closer proximity to the national capital eased the travel burden when the professor joined discussions with the National Academy of Sciences and federal officials. The month Revelle took up residence at Harvard, NAS president Fred Seitz and House Committee on Science and Astronautics chair George P. Miller appointed him to a select committee advising Congress on the scientific research needed to address national priorities.[7] As during the Kennedy years, Revelle was a go-to professor for White House assignments. In October, LBJ announced the Harvard prof's appointment as an alternative representative to the United Nations Educational, Scientific, and Cultural Organization (UNESCO).

Revelle also continued to influence presidential statements on the environment. A November presidential policy paper declared, "The oceans and atmosphere are the property of all people."[8] Like similar statements by JFK, this short sentence made a fleeting reference to the emerging body of work on the potential alteration of the climate but didn't alert most readers to the issue of climate change.

A more direct reference came in February 1965. LBJ's highly touted "Special Message to the Congress on Conservation and Restoration of Natural Beauty" added atmospheric carbon dioxide to his list of environmental concerns. The document declared, "Air pollution is no longer confined to isolated places. This generation has altered the composition of the atmosphere on a global scale through radioactive materials and a steady increase in carbon dioxide from the burning of fossil fuels." Though brief, this statement constituted the most explicit linkage by a U.S. president to date of fossil fuels, carbon emissions, and the alteration of the earth's atmosphere. It endures as the first presidential "warning" about potential climate change.[9]

Revelle's primary contribution at the Johnson White House was yet to come—his role in preparing the historic White House report *Restoring the Quality of Our Environment,* released in November 1965 after fifteen months of preparation. The publication—in the view of Johnson's advisors, part of their effort to leave an environmental legacy as part of the Great Society— would become a pivotal landmark in the scientific understanding of the risks of human damage to the natural world.

The president's science advisors oversaw the report and selected fourteen people—including physicians, scientists, and engineers—for an Environmental Pollution Panel that would produce what turned out to be a work of almost three hundred pages. Revelle was one of two members from Harvard. Also represented were Bell Labs, Cornell (also with two members), the University of Chicago, Texas A&M, Johns Hopkins, Caltech, the University of

Texas (Austin), the American Hospital Association, and the Pennsylvania Electric Company.

Given the primitive state of environmental studies at the time, the report had to address the most fundamental issue, defining the subject itself. "Environmental pollution," it began, "is the unfavorable alteration of our surroundings, wholly or largely as a by-product of man's actions, through direct or indirect effects of changes in energy patterns, radiation levels, chemical and physical constitution and abundances of organisms."[10]

The report challenged many practices prevalent at the time. For example, it argued, "Disposal of wastes is a requisite for domestic life, for agriculture, and for industry. Traditionally waste disposal was accomplished in the cheapest possible way, usually by dumping in the nearest stream. This tradition is no longer acceptable." The report suggested a new approach to the problem: "The pressure to pollute in the past has been an economic one; the pressure to abate must in the future also be economic."[11]

Restoring the Quality of Our Environment addressed myriad other issues at the nascent stages of understanding. These included the health effects of lead, asbestos, pesticides, and urban smog. In many areas, the regulation of pollution was particularly challenging due to the lack of baseline measurements.

The report gave carbon pollution of the atmosphere prominent attention. On page 1, it declared that pollutants had "altered on a global scale the carbon dioxide content of the air." An introductory section, "Climate Effects of Pollution," eight pages later, noted that burning coal, oil, and natural gas added about 6 billion tons of carbon dioxide to the earth's atmosphere each year. It warned, "By the year 2000 there will be about 25% more CO_2 in our atmosphere than at present. This will modify the heat balance of the atmosphere to such an extent that marked changes in climate, not controllable through local or even national efforts, could occur."

The bulk of the document consisted of reports by eleven subpanels with additional analysis and documentation. Revelle chaired the one on atmospheric carbon dioxide.[12] His group included two former colleagues from Scripps, Dave Keeling and Harmon Craig.

Another member of the subpanel was Joseph Smagorinsky. "Smag" (as he was known to his associates) was the son of parents who escaped anti-

Semitism in Belarus during the early twentieth century. In the early 1950s, he became a pioneer user of the new ENIAC computers to forecast weather. At that time, he had shared his colleagues' view that discounted atmospheric carbon's impact on climate change. In his current position at the U.S. Weather Bureau, he directed the development of complex weather and climate modeling and accepted the new perspectives based on the Keeling Curve. His models were based on fluid mechanics (the movement of liquids and gases under various forces) and considered factors such as the earth's rotation and cabon dioxide accumulation. He remained an influential member of the group of scientists believing that the power of computers would soon allow humans to understand natural forces with statistical precision.

The fifth member, Wallace Broecker, worked as a geologist at Columbia University's Lamont-Doherty Earth Observatory, perched high on the cliffs of the Palisades overlooking the Hudson River, ten miles north of New York City, where scientists purified elements and measured their isotopic composition. Their revolutionary techniques for carbon dating helped track environmental trends as far back as 40,000 years.

Like Keeling a product of the Chicago suburbs, Broecker spent three years at Wheaton College ("Billy Graham's school") in Illinois. After a stimulating summer job at Lamont, he transferred to Columbia for his senior year. A few years later, his doctoral dissertation there used carbon dating to better understand the Ice Age—the most significant change in the earth's environment in millions of years. He remained at Lamont to pursue his professional career.[13]

The twenty-two-page appendix written by Revelle's group explained the basics of climate science, noting that carbon dioxide was far from the only gas in the atmosphere that trapped heat. Water vapor, for instance, dwarfed CO_2 in volume. More precisely, "Only about one two-thousandth of the atmosphere and one ten-thousandth of the ocean are carbon dioxide. Yet to living creatures, these small fractions are of vital importance." Although CO_2 had many positive impacts on human life, the burning of fossil fuels brought historic change in nature's balance. "Within a few short centuries," the scientists observed (quoting Revelle's 1957 article), "we are returning to the air a significant part of the carbon that was slowly extracted by plants and buried in the sediments during half a billion years."[14]

The subpanel made detailed estimates of carbon dioxide emitted over time by the burning of fossil fuels that "power the worldwide industrial civilization of our time." Comparing these data with Keeling's measurements at Mauna Loa, they calculated that roughly half of carbon emissions ended up in the atmosphere—lower than Callendar's estimates but much higher than presumed by most scientists before Keeling published his findings.

Based on recent studies and its own rough projections, the subpanel anticipated that a 25 percent rise in CO_2 levels by the end of the century would produce, according to the models available, warming near the earth's surface of between 0.6 and 4 degrees Celsius (1.1 to 7 degrees Fahrenheit). However, it expressed its skepticism about the high end of the range.

Global warming would not be spread evenly. The scientists expected temperature rise to be much more pronounced in the Arctic. The report identified other possible effects of increasing carbon dioxide emissions, including the warming of seawater, higher acidity of freshwaters, and increased photosynthesis. It also identified the possibly "catastrophic melting of the Antarctic ice cap, with an accompanying rise in sea level" as the direst threat of atmospheric carbon dioxide. It projected that the melting of the entire ice cap would raise sea levels by 400 feet, but this process would likely take centuries.

The subpanel suggested some innovative human engineering to limit the possible "deleterious" impacts of carbon emissions. It urged consideration of spreading tiny light-reflecting particles in the oceans to enhance, in the jargon of scientists, the earth's albedo (ability to reflect light). "Considering the extraordinary economic and human importance of climate," the estimated cost of about $500 million a year did "not seem excessive."

The subpanel's report drew on the emerging scientific literature at the time. In one area, however, it was incredibly bold. It embraced Smagorinsky's view that sophisticated modeling would soon bring much greater scientific precision. It acknowledged that it was not yet possible to predict the multiple effects of global warming quantitatively. It suggested, however, that recent advances in mathematical modeling using large computers "may allow useful predictions within the next 2 or 3 years."

*　*　*

The overall report gave the White House imprimatur to what scientists had discovered and wanted to know about pollution and established a foundation for new directions in environmental policy. In his brief introduction, Johnson declared, "Pollution now is one of the most pervasive problems of our society." A more detailed White House press release included a summary of the report's findings on the impacts of burning fossil fuels on the earth's atmosphere. The release predicted, "It may be very important to find means of preventing or counteracting the changes, or to change our source of energy to one that produces less carbon dioxide." The president was expected to say more on the subject in early 1966.

Restoring the Quality of Our Environment's path-breaking analysis drew some notice in the Sunday papers. Under the headline "Johnson Panel Urges 'Polluters Tax,'" the *New York Times* focused on the report's declaration "There should be no right to pollute." It noted the suggestion at a news conference by panel chair John Tukey of Bell Labs that cigarette smokers and automobile drivers would be taxed for adding pollutants to the air if the "tax the polluter" philosophy took hold. The story, appearing on page 79 (out of 706), was unlikely to garner much attention. Moreover, it ignored the work of the Revelle subpanel, the first report on climate change released by any White House. The *Washington Post* was more generous in its coverage, including a front-page story with several paragraphs on "the threat of carbon dioxide."[15]

The White House document quickly drew the attention of an industry it would likely affect. At its annual meeting the following week, the American Petroleum Institute took note of the report's alarms about gasoline pollution. Its executive vice president, Frank Ikard—a well-connected former Democratic congressman from Texas and now chief spokesman for the oil industry—summarized for members the Revelle panel's findings on carbon dioxide and did not question the underlying science. He did quote with some concern a suggestion from page 11 of the report. Considering the emissions, including carbon dioxide, from vehicles, it said, "The pollution from the internal combustion engine is so serious, and is growing so fast, that an alternative non-polluting means of powering automobiles, buses, and trucks is likely to become a national necessity." Sensing the potential threat to the oil industry, Ikard warned, "This report unquestionably will fan emotions, raise fears, and bring demands for action. The substance of the report is that there is still time to save the world's people from the catastrophic consequence of pollution, but time is running out."[16]

While White House staff worked on a presidential response to *Restoring the Quality of Our Environment,* another prominent study addressed the topic of climate change. In January 1966, the National Academy of Sciences published *Weather and Climate Modification: Problems and Prospects.* The report updated work going back to the Eisenhower presidency on how the government could better predict the weather, and even alter the weather and the climate for beneficial purposes—the idea Revelle had raised when Goldman's intellectuals were advising the president. If ever achieved, such control over the environment could prove extremely valuable for numerous human endeavors, ranging from farmers needing more rain for their crops to battlefield commanders seeking a tactical advantage.[17]

This topic overlapped with the work of Revelle's subpanel on atmospheric carbon dioxide. Smagorinsky, unsurprisingly, served on both groups. The new study devoted considerable attention to what it called "inadvertent atmospheric modification." It quoted with approval the 1957 statement by Revelle and Seuss: "Human beings are now carrying out a large-scale geophysical experiment." The weather modification experts observed, "We are just now beginning to realize that the atmosphere is not a dump of unlimited capacity, but we do not yet know what the atmosphere's capacity is or how it might be measured. The overriding immediate concern is for greatly improved and expanded methods of detecting man-made alterations in the composition and energy budget of the atmosphere."[18]

Despite the parallels, the two groups examined the impacts of carbon emissions from different perspectives. Revelle and his colleagues emphasized long historical patterns in climate that demonstrated a relationship between carbon emissions, carbon accumulations in the atmosphere, global warming, and observed effects, often using carbon dating. Though better models to understand complex phenomena were desirable, they were not mandatory for understanding the general scientific parameters. For the weather and climate modification specialists, high levels of certainty were indispensable to avoid potentially dangerous side effects from human interventions. These scientists also required high-level security clearances, given the military sensitivities of their work.

In February 1966, President Johnson's transmission to Congress of the National Science Foundation's *Annual Report on Weather Modification* cited the National Academy's work. His brief statement adopted an upbeat outlook on making humans the masters of their environment. "Even now," he stated, "men are dreaming and planning of projects that will someday enable us to mitigate the awesome and terrible forces of hurricanes and tornadoes. Such a time is still far off, but perhaps not so far off as we thought only a few years ago."[19]

In early 1966, the rhetoric of environmental protection moved from the purview of White House science advisors to Johnson aides with a more political bent. The "Special Message to Congress Proposing Measures to Preserve America's Natural Heritage" would constitute the president's response to *Restoring the Quality of Our Environment.*

A draft message circulated to the White House science office for review emphasized clean water. It totally ignored the climate issues raised by the earlier panels of scientists. The science office expressed its concerns about the omission of a salient part of its report. It suggested a small insertion to ensure at least some mention of the atmosphere and returned it to top aide Joseph Califano.[20]

As delivered on February 23, the presidential message asserted, "We see that we can corrupt and destroy our lands, our rivers, our forests and the atmosphere itself—all in the name of progress and necessity. Such a course leads to a barren America, bereft of its beauty, and shorn of its sustenance." The message exhibited a rousing commitment to the protection of the natural environment. It asserted that "the ultimate cost of pollution is incalculable" and "no person or company or government has a right in this day and age to pollute, to abuse resources, or to waste our common heritage." Protecting the atmosphere had earned a brief mention but failed to retain its earlier prominence.

The follow-up to the president's statement produced thousands of pages of analysis by federal agencies. But any mention of carbon dioxide, the atmosphere, or climate was hard to find. The demotion in emphasis was hardly surprising. Johnson was trying to shape an emerging federal role in environmental protection, starting almost from ground zero. The vast array of challenges made it impossible to confront them all at once—a problem in later years called "limited political bandwidth." As the president's team assigned priorities, it emphasized the protection of public health and nature's beauty.

Climate change was, in effect, competing with other environmental issues for attention. Carbon emissions had one major advantage over most other forms of pollution. Because of the brilliant foresight of Revelle and Keeling, there was a solid baseline measurement of CO_2 accumulations. In the realm of politics, however, the focus would be on environmental challenges that people could see and that had viable solutions.

Climate lagged on both counts. Toxic gases that blackened urban skies and damaged eyes and lungs (sometimes with lethal consequences), rusting junk automobiles strewn near the nation's highways, and contaminated rivers that sometimes caught fire—all prominent features of 1960s America—were hard to ignore. By comparison, the changes taking place in the atmosphere remained invisible. Threats to the atmosphere were also resistant to quick fixes. The decades and centuries involved in climate science were hard to synchronize with elected officials' shorter time frames. Besides, the solutions proposed by science experts were either too speculative (such as the novel but untested idea of spreading particles across oceans to reflect heat) or too disruptive (such as the phasing out of the internal combustion engine) to have much political appeal. For the most part, the scientists advising the White House were calling for more studies, an easier ask for politicians but not a path promising quick results.

One agency did emphasize the dangers of carbon emissions. The Atomic Energy Commission continued to promote nuclear power as a viable technology for cutting the various forms of pollution created by the combustion of fossil fuels. In 1966, its chair, Glenn Seaborg—a Noble Prize–winning chemist whose theory of elements had forced a realignment of the periodic table—gave several public addresses emphasizing that fossil fuel plants discharged greenhouse gases and toxic chemicals into the atmosphere but nuclear reactors did not. In a June commencement speech at San Diego State College, for example, Seaborg warned that carbon emissions could produce marked changes in the climate that "we might have no means of controlling."

He jested, "I, for one, would prefer to travel toward the equator for my warmer weather rather than run the risk of melting the polar ice and having some of our coastal areas disappear beneath a rising ocean." He regarded it as fortunate that "the peaceful atom" could someday replace fossil fuels and "release no combustion products into the atmosphere." Converting cars and trains to nuclear-generated electricity, he added, could further reduce pollution.[21]

The AEC's unpublished response to the Office of Science and Technology on *Restoring the Quality of Our Environment* complained that it hadn't been very involved in the implementation process. It wrote, "Increased use of nuclear power as an alternative source of electric power production offers a means of reducing environmental pollution from carbon dioxide and other combustion pollution."[22]

Seaborg's optimism about the future role of nuclear energy was far from idle speculation. In 1967, U.S. electric utilities awarded contracts to construct thirty-one nuclear reactors, the first year that most orders for new electricity plants were nuclear, a notable milestone for a technology still in its infancy. This upward trend rested on the assumption that oil resources would start running out in the coming decades and that nuclear power was economically attractive—that is, cheaper than coal. By the end of Johnson's presidency, the Atomic Energy Commission was estimating a quarter of electricity generation would be nuclear by 1980. The share, it predicted, would rise to half by 2000, and virtually all electric plants built in the twenty-first century would be nuclear.[23] If the momentum toward nuclear power continued and spread to other nations, the worst-case climate scenarios from Revelle's panel could be taken off the table.

One other federal agency also displayed some interest in the atmospheric impacts of carbon emission. The Commerce Department—home of the Weather Bureau—told the White House of its plans to augment CO_2 monitoring and evaluation "through the development of necessary sampling techniques and design of the required global network."[24]

Though interest in climate declined at the White House after 1965, Johnson's signing of the Air Quality Act of 1967 in November demonstrated his steady commitment to environmental protection. This bill was the third of his presidency. It was Johnson who signed the Clean Air Act of 1963, since the final bill did not emerge from the conference committee until eighteen days after the death of President Kennedy.[25] At the signing of a second bill on the subject, the Clean Air Act of 1965, Johnson complained that pollution had continued to grow worse since the Industrial Revolution. "We have reached the point," he said, "where our factories and our automobiles, our furnaces and our municipal dumps are spewing more than 150 million tons of pollutants annually into the air that we breathe." He claimed that new controls on the exhausts of carbon monoxide (unlike carbon dioxide, a poisonous gas) from automobiles would enhance the health of the American people.[26]

Johnson arrived late for the official signing of the 1967 air quality legislation due to a longer-than-expected meeting with General William Westmoreland about the war in Vietnam. At the ceremony, he agreed with the bill's principal sponsor, Senator Edmund Muskie, that the three clean air bills during his administration were "baby steps" but credited the man from Maine for "shoving me as no person has, all these years, to do something in the pollution field."[27]

Despite the limited aspirations of the clean air acts enacted under Johnson, they did break new ground in creating a federal role in combating pollution. Moreover, the research they authorized would prove helpful if Congress considered more stringent legislation down the road.

Government regulation of pollution was moving at a faster pace in California, where in August 1967, Governor Ronald Reagan had signed a bill creating a powerful Air Resources Board to reduce severe smog in major cities. He appointed as the first chair a Caltech chemistry professor known for his tough stances on pollution. A federal waiver granted that year allowed the state to adopt stricter rules than national standards, suggesting the new agency would have wide latitude in addressing environmental challenges.

Revelle continued to be a frequent visitor to Washington during the Johnson years, often on matters unrelated to climate. His testimony in 1966 before the House Science Committee had called for more support for the social sciences to better understand the human impact of technology on the environment. Later in the year, he testified before a Senate subcommittee on the wise use of new technologies. In February 1967, he completed service on the presi-

A frequent visitor to the Johnson White House, Roger Revelle (*left*) shaking hands with the president, 1967. Courtesy Lyndon Baines Johnson Presidential Library.

dential committee on the National Medal of Science, which made recommendations for the prestigious award, and attended the White House ceremony for that year's recipients. Four months later, he met with LBJ and others in the White House Cabinet Room for the presentation of the "Special Report by the President's Science Advisory Committee Panel on World Food Supply," which he had helped write. He also worked on a National Academy of Sciences study about human impacts on plant life, commenting, "Our goal should not be to conquer the natural world but to live in harmony with it." The national press frequently reported on the professor's pronouncements.[28]

On March 31, 1968, Johnson announced that he would not seek reelection. By this time, numerous national challenges had eclipsed climate in the hierarchy of policy concerns. Still, government and industry were paying it some attention, and there was little controversy about the underlying science.

A joint congressional resolution adopted in May 1968 endorsed U.S. participation in an international system to observe and analyze the global atmosphere. This effort, it was hoped, would lead to the capability for "long-range weather prediction and for the theoretical study and evaluation of inadvertent climate modification and the feasibility of intentional climate modification."[29] The growing availability of satellite data was welcome news for scientists trying to better understand both weather and climate.

In June, Johnson's top science advisor addressed the Edison Electric Institute, the American electric utility industry's research arm, on "Future Energy Needs vs. the Environment." Hornig included the threats of sulfur oxides and particulate matter in his list of concerns. He also summarized the findings on carbon dioxide from *Restoring the Quality of Our Environment.* "Carbon dioxide is not toxic," he noted, "but it is the chief heat-absorbing component of the atmosphere." Changing levels of carbon from the burning of fossil fuels "might, therefore, produce major consequences on the climate—possibly even triggering catastrophic effects such as have occurred from time to time in the past" due to natural variations. He said considerable work lay ahead to achieve a scientific basis "for sound prediction."

Hornig enunciated two general principles for dealing with the problems of pollution. First, its cost should be incorporated into the cost of production. Second, good policy should try to keep the price of power as low as possible. Reflecting the thrust of Johnson's environmental policy, he advised, "If we must choose between clean air and an increase in the cost of electricity, I believe it has already been decided that we will pay the price for clean air."

He added that national policies should rely on what was then known, even though scientists were generally at the early stages of understanding. "Scientific knowledge," he said, "is the touchstone for establishing air pollution controls; as our knowledge becomes more complete in the future there will be opportunities for adjustment."[30]

Hornig's message was hardly a surprise for the electric industry. One of its large utilities had helped write *Restoring the Quality of Our Natural Environment.*

The oil industry's thinking had also evolved since 1958, when a Shell Oil scientist had brushed off the idea that carbon emissions could significantly impact the atmosphere. With the data from Mauna Loa now available, the American Petroleum Institute struck a different tone in 1968 and 1969. The API commissioned and distributed a report from the Stanford Research Institute that summarized in an evenhanded manner the existing knowledge about carbon dioxide's impact on the atmosphere. Consistent with the conclusions of *Restoring the Quality of Our Environment,* the authors considered carbon dioxide "the only air pollutant" with a global impact "that could change man's environment." Half of CO_2 emissions, it reported, came from coal; about 30 percent came from oil and 10 percent from natural gas. (The remainder came from other forms of combustion, such as forest fires and waste incineration.) The study utilized Keeling's data as the basis of historical and expected future carbon accumulations. The report concluded, "It is rather obvious that we are unsure as to what our long-lived pollutants are doing to our environment; however, there seems to be no doubt that the potential damage to our environment could be severe."[31] At this point in history, there was considerable overlap between reports from the White House science office and the oil industry on the emerging climate science.

Word spread at Harvard: Professor Revelle was a scholar with plenty of government experience at the highest levels. The transplanted Californian regularly offered Natural Sciences 118, a course open to undergraduate and graduate students and focusing on the implications of the rapid increase in the world's population.

Revelle later realized that during his years in Cambridge he had been training future world leaders. He remembered one student who, because of her hair color, was nicknamed "Pinkie" Bhutto.[32] In 1988, Benazir Bhutto became prime minister of Pakistan, the first woman to head a democratic government in a Muslim-majority country. While earning his doctorate in experimental physics at Harvard, Ashok Khosla served as one of Revelle's star teaching assistants. After unsuccessfully urging Khosla to remain at Harvard, Revelle began writing laudatory letters of recommendation for Khosla's employment in his native India.[33] Khosla later became a senior official at the United Nations Environment Programme and a significant figure in international climate negotiations.

For the fall semester of 1968, the fifty-student roster for Revelle's course included Albert Gore Jr., class of 1969. A graduate of the prestigious St. Albans School in Washington, Gore was the son of the longtime Democratic U.S. senator from Tennessee, Albert Gore Sr.

As was typical on campus, the first meeting was in effect part of a "shopping period" when many students sampled various classes before finalizing their enrollment. In his initial lecture, Revelle said that he would discuss global resources and that one fundamental fact to keep in mind was that per capita resources were total resources divided by population. He wrote a simple equation on the board to summarize the point. After class, one student commented that they had planned to take the course until they found out it would be a "math class."

Leading intellectuals in the United States and around the world were arguing that the planet faced an urgent crisis because its resources could not keep pace with population growth. Revelle assigned readings on several doomsday scenarios, including the 1798 Thomas Malthus classic *An Essay on the Principle of Population.* He also cited current dire warnings about overpopulation, such as ecologist Paul Ehrlich's *The Population Bomb,* published the same year Gore took Natural Sciences 118. Yet it was clear that Revelle came down mainly on the other side of the debate. Well versed on the potential of the emerging "green revolution" to dramatically expand world food supplies, and having implemented some of these practices in Pakistan, he espoused optimism about the ability of a growing number of humans to thrive on earth.

Indeed, Revelle held firm views on population. He privately decried "fanatics" that he regarded as more concerned about environmental damage from overpopulation than about people. Eight years later, he told an interviewer that the world "isn't going to hell because of overpopulation. It may be going to hell, but for other reasons." He felt that the focus should be on economic development to combat global poverty and cited his previous work in South Asia as an example of what should be done.[34]

As the Harvard course reached its closing weeks, lecture twenty (out of twenty-four) dealt with the esoteric subject "Long-Term Environmental Perturbations: Carbon Dioxide, Increasing Turbidity, and Possible Climate Change." (Scientists used the term "perturbation" to refer to human interference with natural cycles.) The assigned reading for the day was the section of *Restoring the Quality of Our Environment* dealing with atmospheric carbon dioxide, the report Revelle helped write.[35]

Years later, Gore recalled being impressed by the consistent upward trend in Keeling's carbon measurements from Mauna Loa. He learned from Revelle that a global average temperature change was quite different from seasonal shifts in temperature. Gore noted that "the massive transformation of the earth's climate system that we call the ice ages took place after average global temperatures dropped by only a few degrees." He said he was raised to believe that "the earth is vast and nature so powerful that nothing we do can have any major or lasting effect on the normal functioning of its natural systems." But the course taught him that humans could "actually change the makeup of the entire earth's atmosphere in a fundamental way."[36]

Zooming in on the rise of the climate issue during the Johnson presidency illuminates its often forgotten roots in U.S. politics. Because of the influence of Revelle and other prominent scientists, the topic moved briskly from academic journals read by university professors to its inclusion in White House deliberations.

But zooming out, climate change gets lost in a profusion of issues considered more urgent at the time. Johnson has been remembered for his historic civil rights legislation, a war on poverty, the creation of Medicare, and a military quagmire in Vietnam. In the broad context of a presidency with grand ambitions, including reining in massive pollution, the risks of altering the earth's climate disappear from view.

It is instructive to observe, for instance, that the later books chronicling the Johnson years, even one published in 1987 with two chapters on environmental protection, failed to mention carbon dioxide in the atmosphere or climate. Not until the 2022 publication of Douglas Brinkley's *Silent Spring Revolution* did a presidential or environmental historian mention the LBJ administration's contributions to climate science.[37]

4

The Global Cooling Detour

The Nixon and Ford Presidencies, 1969 to 1977

Richard Nixon aspired to build a presidential legacy based on bold achievements in foreign affairs. Rising concerns about environmental degradation also made the list of priorities for the fifty-six-year-old former U.S. senator from California and Eisenhower vice president. Two stalwarts of the 1968 campaign—land use lawyer John Ehrlichman and PhD geologist and land surveyor John Whitaker—promoted aggressive environmental action from their perches in the new administration. They argued that Nixon would benefit by pushing a popular cause that had gained strength during the decade. Nixon sometimes balked privately at their environmental advocacy. But his public utterances consistently touted the importance of protecting nature.[1]

Like presidents before him, Nixon recruited prominent academics to join his team, one of them a surprising choice. Daniel Patrick "Pat" Moynihan took a leave from Harvard's sociology department to become Nixon's assistant for urban affairs and, unofficially, the administration's "intellectual in residence." Despite Moynihan's history as a Democrat in previous administrations, the professor and the president had substantial face time to discuss national and world affairs.[2]

In September 1969, the jovial Irishman sent a memo on "the carbon dioxide problem" to Ehrlichman, Nixon's domestic policy coordinator. After explaining the chemistry of CO_2 as a greenhouse gas and the impact of burning fossil fuels on the atmosphere, Moynihan said it was "pretty clearly agreed that the CO_2 content will rise 25% by 2000." He defined the possible impacts in stark terms: "This could increase the average temperature near the earth's surface by 7 degrees Fahrenheit [and] raise the level of the sea by 10 feet. Goodbye New York. Goodbye Washington, for that matter." He added wryly, "We have no data on Seattle," Ehrlichman's hometown. He acknowledged potential countervailing factors. For instance, increasing "dust" in the atmosphere might have a cooling effect. A "mammoth" effort to stop burning fossil fuels would slow rising CO_2.[3]

Moynihan urged the administration to get involved by supporting upgrades in worldwide weather and climate monitoring. He attached a Xerox copy of the climate section of *Restoring the Quality of Our Environment* (allowing careful readers to observe that he was exaggerating its findings). His familiarity with the report that Roger Revelle helped write was hardly coincidental. At Harvard, he had joined Revelle's advisory board at the Population Center. The group met every few months, giving the two plenty of opportunities to exchange ideas.

Following custom, Nixon retained many employees in the traditionally nonpartisan science office. In October, David Freeman—leader of its energy group—circulated a paper assessing the new administration's energy challenges. The United States, he wrote, led every other country in energy use, consuming about a third of the world's total despite having only 6 percent of the population. Moreover, he predicted that the nation would probably consume more energy in the next twenty years than during the previous seventy.[4]

At the time, the views of "peak oil" theorists predicting sharp drops in available reserves in the ground held considerable sway. Freeman, however, adopted a somewhat rosier outlook on the availability of fossil fuels. He foresaw that U.S. oil and natural gas resources were "sufficient to meet growing needs for decades, but certainly not centuries." By contrast, coal reserves could last for hundreds of years. He added that if conventional oil resources ran short, it would be possible to tap vast deposits in U.S. shale formations or Canadian tar sands. Another option would be to produce synthetic liquid coal to replace oil.

Freeman's outlook on future energy supplies contained a significant caveat. "Unfortunately," he declared, "our use of energy is also a measure of the rate at which we are damaging the environment." Such damage included health risks for miners extracting dusty coal beneath the earth's surface, scars on the natural landscape from strip mining, and deterioration of marine environments and shorelines from offshore drilling and oil tankers. In addition, "The burning of fossil fuels—whether in automobiles, industrial plants or otherwise—is the major element in the nation's air pollution."

Freeman pointed to other environmental threats, including the climate problem. "The consumption of large amounts of fossil fuels," he wrote, "tends to increase the concentration of carbon dioxide in the atmosphere. Some scientists believe this may upset the earth's heat balance and produce irreversible ecological changes of major consequence." Freeman suggested developing technologies that could produce electric power more efficiently.

The Moynihan and Freeman memos didn't signal that climate change was a top-tier issue in the Oval Office. Neither of them likely came to the president's attention. Still, the effect of energy use on climate was on the radar for some senior staff.

After being assigned Moynihan's note, the White House science office responded in January 1970, initially by forwarding a copy of a recent *Washington Post* editorial. On January 13, the paper noted that many scientists believed that "excess carbon" in the atmosphere created a "greenhouse effect and might increase the mean annual temperature by 3.6 degrees Centigrade in the next half-century."[5] On the other hand, the *Post* wrote, other scientists suggested that data since 1950 indicated that a new ice age might have already begun. These experts thought "the effect of man-made smog will be cooler instead of hotter weather because it will screen out some of the sun's radiation."

Deputy science director Hubert Heffner (on leave from Stanford University, where he had been a professor of electrical engineering and dean of research) replied to Moynihan more fully on January 26. Heffner acknowledged that humans were altering the atmosphere, but what this meant for climate remained a daunting unknown. "The more I get into this," he joked, "the more I find two classes of doom-sayers with, of course, the silent majority in between. One group says we will turn into snow-tripping mastodons because of the atmospheric dust and the other says we will have to grow gills to survive the increased ocean level due to the temperature rise from CO_2." Heffner stressed that the potential consequences made it important for scientists to determine "the true situation."[6]

The sharp contrast between the possible directions of global temperatures added a new twist to the discussion of climate change and would stymie initiatives to develop coherent policy responses. Nonetheless, the administration followed the advice of Moynihan and Heffner by supporting the launch of additional satellites to monitor global trends in weather and climate.

The year 1970 turned out to be the twelve most momentous months in U.S. environmental history.

On New Year's Day, Nixon—at his home in San Clemente, California—signed his first (but not last) historic environmental legislation of the year, intentionally his first official act of the new decade. Congress had led the effort for the National Environmental Policy Act. The president, however, gave it a warm endorsement. NEPA required extensive environmental reviews for major federal projects and established a Council on Environmental Quality at the White House to oversee national policy. He named Russell Train—former head of the World Wildlife Fund and the Conservation Foundation and, at the time, number two at Nixon's Department of the Interior—to head the council.[7]

Nixon burnished the scientific credentials of the CEQ with his appointment of Gordon MacDonald—vice chancellor for research and graduate affairs at the University of California, Santa Barbara. MacDonald's parents had migrated to Mexico, where he spent his early years. At the age of ten, his diagnosis of polio led to a relocation to Texas, where better medical treatment was available. He became determined that his physical condition would be no barrier to success. MacDonald went on to earn a doctorate in geology at Harvard and, in 1960, coauthor the authoritative book *The Rotation of the Earth* with the Scripps Institution's Walter Munk. He held various positions on the President's Science Advisory Committee during the 1960s, working on the oceanography project that connected oceans and climate and overseeing the *Restoring the Quality of Our Environment* report. MacDonald wrote extensively on weather and climate modification during the Johnson years and chaired a National Academy panel on the subject. It was Revelle who first alerted him to Keeling's data collection on atmospheric carbon and its importance for understanding oceans and climate.[8]

MacDonald first met Nixon in 1969, when they strolled the beaches of Santa Barbara to inspect the impacts of the disastrous oil spill there. The personal connection helped pave the way for his White House appointment. Shortly after joining the administration, MacDonald lectured at the Industrial College of the Armed Forces about the challenges for scientists trying to figure out if the earth was warming or cooling and the importance of dealing with either problem.[9]

In his January State of the Union address, Nixon maintained the momentum from his NEPA signing with another strong push for environmental protection. After a boisterous welcome and twenty minutes on other matters, he announced, "I now turn to a subject which, next to our desire for peace, may well become the major concern of the American people in the decade of the seventies." He said this question was "Shall we make peace with nature and begin to make reparations for the damage we have done to our air, to our land, and to our water?"

In a firm voice, Nixon pledged to devote the resources necessary to address the problem: "The program I shall propose to Congress will be the most comprehensive and costly program in this field in America's history." Then, in language resembling LBJ, he declared, "We can no longer afford to consider air and water common property, free to be abused by anyone without regard to the consequences. Instead, we should begin now to treat them as scarce resources, which we are no more free to contaminate than we are free to throw garbage into our neighbor's yard."

Nixon sounded an upbeat note about balancing economic growth and the quality of life. "The answer is not to abandon growth but to redirect it," he said. "Continued vigorous economic growth provides us with the means to enrich life itself and to enhance our planet as a place hospitable to man." He urged each individual to enlist in the fight to protect nature.[10]

Many Americans joined the president's fight for environmental protection. The first national celebration of Earth Day on April 22, 1970, attracted millions of participants and massive media coverage. The countrywide teach-in shut down New York City's Fifth Avenue. The Philadelphia event organized by local universities featured a diverse list of speakers, including Senator Muskie, consumer activist Ralph Nader, Nobel Prize biochemist George Wald, poet Allen Ginsberg, and Senate Republican Leader Hugh

Scott of Pennsylvania. Elsewhere, Gaylord Nelson, Democratic U.S. senator from Wisconsin and known as "the father of Earth Day," spoke at Berkeley. The author of the 1968 book *The Population Bomb,* Paul Ehrlich, appeared at Iowa State. Secretary of the Interior Walter Hickel took part in events at his home state's University of Alaska.

In July, Nixon expanded his ambitious environmental agenda by sending Congress a proposal for a new Environmental Protection Agency that consolidated federal programs regulating pollution.[11] He also called for the creation of the National Oceanic and Atmospheric Administration (NOAA) within the Department of Commerce. Its "atmospheric" functions included weather and climate. After a mandatory period during which Congress could have blocked the reorganization, the man from California created both high-impact agencies by executive order later in the year.

Moving quickly for a new agency, the Council on Environmental Quality issued its first annual report in August. The document laid out in considerable detail the country's environmental challenges. Its transmittal to Congress included an eleven-page message from Nixon, adding to its stature as a presidential document. Influenced by MacDonald's previous work, the report dealt extensively with climate. The first chapter warned, "On a global scale, air pollution could trigger large-scale climatic changes."[12]

The report confronted head-on doubts about whether the planet was warming or cooling. A chapter titled "Man's Inadvertent Modification of Weather and Climate" expressed the view that delicate balances within the atmosphere and the history of past climate change suggested that through their inadvertent actions, humans "may be driving the atmosphere either to a disastrous ice age—or as bad—to a catastrophic melting of the ice caps." The understanding of climate science had matured since *Restoring the Quality of Our Environment.* Thus, the CEQ could offer new insights on why such contradictory scenarios were possible.

Discontinuities in global temperature records justified the hesitancy of scientists to come down firmly on the matter. In the later decades of the nineteenth century, as industrial uses of fossil fuels accelerated, the average temperature began to climb (albeit irregularly). By 1940, it averaged 1.1 degrees Fahrenheit higher. Then, from 1940 to 1970, the average temperature fell by about half a degree. "Thus," according to the report, "during the past three decades, one-half of the warming that occurred during the preceding six decades has been erased."

CEQ agreed adding carbon dioxide to the atmosphere would, all things being equal, warm the planet. During historical periods of rising temperatures, the earth experienced the receding of polar ice caps and increasing aridity in noncoastal areas, leading to dust bowl conditions. Geologic records created the expectation of similar effects from future warming.

Five years earlier, *Restoring the Quality of Our Environment* had focused almost exclusively on carbon dioxide when it discussed the impacts of humans on climate. CEQ pointed to other results of human activities that might affect temperature. Chief among these was "particle pollution" from various sources. These particles, sometimes referred to at the time as "dust," blocked some sunlight. So, at sufficient levels they could lead to global cooling. Historical records indicated that during periods of global cooling, polar ice caps expanded and interior areas enjoyed higher humidity and rainfall. The chapter on weather and climate change ended with a plea for more funding for research.

The year's environmental capstone came with Nixon's December 31 signing of the Clean Air Act of 1970.[13] The legislation established national standards and enforcement to deal with the clouds of pollution that darkened skies and spread noxious odors to areas with industrial activities and traffic congestion, ranging from Pittsburgh and Gary to Corpus Christi and Los Angeles. Unlike the clean air legislation of the Johnson years, it required extensive changes in transportation, oil refining, and electric generation.

During a September debate on the Senate bill, floor manager Ed Muskie asserted that, like the space program, strict new air quality standards would require technologies not yet fully developed. "This may mean," the Maine Democrat acknowledged, "that people and industries will be asked to do what seems to be impossible at the present time." One critic of the bill, GOP whip Robert Griffin of Michigan, complained that it went too far and that the standards for automobiles could lead to the loss of 15 million jobs. Muskie countered that, despite the legislation's toughness, all the Democrats and Republicans on his subcommittee supported it.[14]

With continued uncertainty over whether the planet was warming or cooling, the bill concentrated on "the air that we breathe" and the direct impacts of pollution on human health. Thus, climate change did not play a prominent role in the discussions. However, the debate preceding final adoption, and the text itself, demonstrated that climate was on the mind of the bill's authors.

On the Senate floor, Muskie warned that emissions from human activities "threaten irreversible atmospheric and climatic changes." Caleb Boggs of Delaware—ranking Republican on Muskie's subcommittee (and the victim of an upset victory by Joe Biden in the 1972 elections)—cited the recent CEQ report and its linking of pollution and alterations in climate.

Reports from the Johnson administration and Nixon's CEQ built the scientific foundation for the Clean Air Act. Both landmark White House studies left no doubt that the top scientists at the White House over two administrations considered growing CO_2 in the atmosphere—despite its lack of visibility, odor, or toxicity—an integral aspect of human-generated pollution.[15]

The bill's text didn't attempt to list every pollutant. Still, it did provide for additions "from time to time" of "each pollutant" (or in another section of the bill, "any pollutant") that, in the "judgment" of the EPA administrator, had "an adverse effect on public health or welfare." As to defining the effects on public welfare, the act was clear. Section 15 defined such impacts broadly to include but not be limited to those on weather, visibility, climate, and the deterioration of property. Scientists studying carbon dioxide already foresaw potential impacts on weather, climate, and coastal properties abutting rising oceans. Thus, the legislation appeared to provide ample authorization to address it, as the list of pollutants underwent periodic reviews.

Despite the Clean Air Act's ambitious scope, its final passage proved remarkably nonpartisan. When the compromise version worked out by a conference committee reached the House and Senate floors, both bodies approved the bill with voice votes.[16]

In June 1971, Nixon released a message on "energy resources"—an early call for the nation to develop new technologies to lessen its dependence on foreign oil. The *New York Times* captured the thrust of the message in its front-page headline: "Nixon Offers Broad Plan for More 'Clean Energy.'"[17]

Though the theme of clean energy was consistent with the president's stance on environmental protection, the message was not the one he had instructed his staff to prepare.

In the months leading up to the plan's release, Nixon had become enamored with huge nuclear reactors called fast breeders. Though not yet demonstrated at scale, breeders created more fuel than they used. They could also power massive water desalinate plants—an idea Nixon thought would have great political appeal. Atomic Energy Commission chair Glenn Seaborg made a case for the breeder that included its lack of CO_2 emissions. He declared in epic terms that Nixon's support "would be the most decisive single step that could be taken now toward assuring an essentially unlimited energy supply, free from problems of fuel resources and atmospheric contamination."[18]

Nixon wanted breeder reactors to be the sole focus of his message. But the energy industries consulted during the preparation of the speech balked at an energy strategy relying solely on unproven technology. Even Westinghouse and General Electric, the dominant U.S. manufacturers of nuclear reactors, were wary of the breeders' massive upfront costs when they were still endeavoring to make existing light-water reactors profitable. As a result, the president's staff wrote the "clean energy" message advocating a wide variety of energy technologies, which almost led to its cancellation when Nixon discovered on the morning of its delivery that his wishes had been ignored.[19]

Nonetheless, Nixon directed the bulk of federal energy research and development funding to the breeder. Without the private matching funds promised in the early pleas for the program, Nixon's priority became a heavy drain on the energy research budget. Civilian nuclear power continued to dominate federal energy spending, reducing the money that might have gone to other energy technologies. Moreover, it diverted money from improving light-water reactors, which still faced competition from coal plants and questions about their cost, reliability, safety, and contaminated waste.

A July 1971 article in *Science* intensified alarms that human pollution could lead to global cooling. Authors Ichtiaque Rasool—an expert on the atmospheres of planets at Columbia University and number two at NASA's Goddard Institute in Manhattan—and his graduate assistant Stephen Schneider used a primitive model to predict the relative impacts of carbon dioxide and aerosols on the earth's climate. Consistent with the prevailing scientific consensus, additional CO_2 in the atmosphere led to warmer temperatures. In contrast, aerosols—particles trapped in gases, like dust or sulfur dioxide—reflected incoming sunlight, leading to cooler temperatures.[20] The scientists concluded that the effects of aerosols were proving more powerful and creating a cooling trend. "If sustained over a period of several years," they projected, "such a temperature decrease over the whole globe is believed to be sufficient to trigger an ice age."[21]

The national press pounced on the scholarly article on the day of its release. A story in the *Washington Post* headlined "U.S. Scientist Sees New Ice Age Coming" declared the disaster could be "as little as 50 or 60 years away."[22]

Though influential, Rasool and Schneider failed to convince a majority of their fellow climate scientists that the earth was racing toward another ice age. In a 1972 article in *Nature,* John Stanley present a more conventional view of the effect of carbon accumulation on the planet, based on the work of Guy Callendar, Bert Bolin, and Syukuro Manabe. The predominate effect would be an increase in average world temperature. This article came to the attention of at least one major official in the United States, who underlined the salient points. It was Jimmy Carter, then in his second year as governor of Georgia.[23]

At the beginning of his second term, in 1973, Nixon continued to extol environmental protection. In a February radio address, he touted the successes of the legislation he had signed. "Day by day, our air is getting cleaner; in virtually every one of our major cities, the levels of air pollution are declining."[24]

Later in the year, however, Nixon became increasingly distracted by the rising power of oil exporters in the Middle East, which led to massive lines to purchase gasoline in many sections of the country and sparked America's "first energy crisis." Raging inflation and investigations of the Watergate scandal added to the president's woes and deflected attention away from the natural environmental.[25]

Roger Revelle continued to be a significant force in American science during the Nixon years. In 1974, he became president of the American Association for the Advancement of Science. The Harvard professor also continued to write and speak about diverse topics, including climate. In many areas of science, he had searched for technically feasible solutions to formidable problems. But the climate challenge appeared more daunting, with a greater need for understanding than corrective action. However, in a research paper released in May, he identified a possible solution for protecting the atmosphere. "Most potential climate effects could be avoided," he said, "if sunlight could be utilized as the principal source of energy for human use. Solar energy research and development should be given high priority in mankind's continuing search for new energy sources."[26]

Five months later, Nixon proposed legislation (later signed by his successor, Gerald Ford) elevating the status of federal research on solar technology. A new Energy Research and Development Administration took over the civilian and military nuclear research and development functions of the old Atomic Energy Commission. ERDA also acquired a recently established solar energy development program from the National Science Foundation. Although solar power remained far from the "high priority" envisioned by Revelle, the reorganization was not a trivial matter. New funding poured into energy research by scientists at the laboratories that had once implemented the Manhattan Project. These labs could now extend their energy programs beyond a sole focus on nuclear power.[27]

In November, Revelle offered additional thoughts on the world's energy options at a symposium called "Beyond Today's Energy Crisis," hosted by the Lyndon Johnson Presidential Library and the University of Texas. Focusing on the energy challenge, Revelle said that Western Europeans used only half as much energy per capita as Americans and "fare very well." "Nothing is more ridiculous, more murderous, more insane than the American habit of driving at 60 to 70 miles per hour in two tons of steel with one person sitting in the car." He said smaller, European-type models would be safer and cut U.S. energy demand by almost 20 percent. Revelle also cautioned, however, against going too far in criticizing fossil fuels. They had, after all, made possible the substitution of mechanical energy for human labor, more productive agricultural practices, and a massive rise from poverty.

Despite the contributions of fossil fuels, the world needed to change how it used energy and recognize "the interdependence of the peoples of the earth." Revelle complained that the international conferences he had attended in the past year had "been most depressing experiences because of the complete inability of participants to agree on any feasible action." He mused that a deep sense of responsibility for the welfare of all might come from some sort of religious revival. This change would likely require people learning "to see their own interests in a somewhat longer time perspective than the very short-range perspective common to most of us."[28]

On August 1, Nixon transmitted to Congress the administration's annual plan for U.S. participation in the World Weather Program, which contained his first and last public reference to climate change. He predicted that the increased use of more powerful computers would "in time, produce long-term gains in both immediate and extended range prediction of global weather conditions and in the assessment of the impact of man's activities upon climate and weather."

At about the same time, the National Academy of Sciences published a hefty report, *Weather and Climate Modification: Problems and Progress*. The multiyear project—chaired by prominent meteorologist and dean of the graduate school at the University of Connecticut, Thomas Malone—focused primarily on intentional efforts to alter weather but also included the inadvertent impacts of human activities on weather and climate. Based on the emerging body of scientific literature, it cautiously concluded, "It is known that the world's climate is sensitive to those features that affect the radiation balance, the extent of cloud cover, the concentration of carbon dioxide in the air, the particulate concentration, and the albedo at the earth's surface. Man is contributing to changes of all these quantities. What is not clear is whether these contributions account for a substantial portion of the observed fluctuations in climate."[29]

Given the erratic data on global temperatures, these views were consistent with the general understanding of the scientists advising the Nixon administration.

Though easy to overlook at the time, there was a new wrinkle in Nixon's message on weather. It talked about better understanding weather but not about controlling it. Weeks earlier, the U.S. president and Soviet leader Leonid Brezhnev had signed a joint communique in Moscow charting a new direction. They called for "the most effective measures possible to overcome the dangers of use of environmental modification techniques for military purposes." The willingness of the world's two superpowers to limit work on modifying weather and climate eventually led to a 1978 United Nations convention banning the use of environmental modification for military advantage, which the U.S. Senate ratified by a vote of 98–0.[30]

The Academy report talked only about the peaceful purposes of weather modification, such as seeding clouds to help farmers or reducing the impacts of hurricanes and tornadoes. It would prove difficult, however, to segregate the military and nonmilitary aspects of intentional weather and climate modification ("geoengineering," in modern parlance). This overlap eventually stymied modification efforts across the board. The U.S. government continued its international efforts to better understand weather and climate, but the goal of manipulating them for human benefit faded.

On the day of Nixon's message on weather and climate, facing impeachment charges in the House of Representatives based on attempts to cover up the Watergate scandal, he began preparations for his possible resignation, which he made official the following week.

When Vice President Gerald Ford of Michigan became the thirty-eighth president of the United States, in August 1974, the National Academy of Sciences was finishing a new study on what scientists knew and wanted to know about climate change. In January 1975, it published a substantial report, *Understanding Climate Change: A Program for Action.*

In the forward, Verner Suomi—chair of the Academy's Global Atmospheric Research committee, a meteorologist at the University of Wisconsin, and inventor of the instruments that generated the time-sequence images of clouds later used by TV weather forecasters—welcomed the "tremendous improvement" in the ability to track weather with data from satellites. By the decade's end, he expected such satellites would be able to monitor the entire earth. Moreover, these same satellites could provide data essential for better understanding "those parameters that we now believe control the climate machine: the sun's output, the earth's albedo, the distribution of clouds, the fields of ice and snow, and the temperatures of the upper layers of the oceans." Suomi was encouraged by "a new generation of atmospheric scientists" who understood the new tools of computers, numerical models, and satellites.[31]

The report—cochaired by W. Lawrence Gates of the Rand Corporation and Yale Mintz of UCLA—made it clear that the "action" referenced in the title was additional study. Its stated goal was "to increase significantly our understanding of climate variation" and reach a point where scientists could predict these variations and subject their "ideas to quantitative test whenever possible." The data from climate research, it said, should lead to global climate simulation models with predictive capabilities. Such ambitious models would have to integrate smaller models dealing with the multilayered atmosphere (the "earth's gaseous envelope"), the hydrosphere (largely oceans, the depths of which had "thermal adjustment times of the order of centuries"), the cryosphere (the world's ice masses and snow deposits), the lithosphere (the earth's landmasses), and the biosphere (plants and animals).[32]

The modeling challenges were enormous. The scientists conceded, "While the local details of weather do not appear to be predictable beyond a few weeks' time, the consequences of this fact for climatic variations are not clear." They acknowledged many barriers to predicting climate change with precision, even with ever more powerful computers and vast amounts of new data.[33]

The report expressed concern about human impacts on the environment but cautiously stated them in the conditional, noting the "increasing realization that man's activities may be changing the climate." Global patterns of food production and population were based on the climate of recent centuries and could be seriously disrupted if the climate changed as much as it had due to natural variations. "This dependence of the nation's welfare, as well as that of the international community as a whole, should serve as a warning signal that we simply cannot afford to be unprepared for either a natural or man-made climatic catastrophe." The latter statement triggered a *New York Times* headline "Climate Changes Called Ominous: Scientists Warn Predictions Must Be Made Precise to Avoid Catastrophe."[34]

A big energy question affecting climate remained whether future electric power plants would be coal fired or nuclear. In the Johnson and early Nixon years, the odds appeared to favor nuclear. From 1969 to 1974, the number of operating nuclear reactors rose from seventeen to fifty-five, pushing its share of electric generation to 6 percent. Amid the dramatic growth of nuclear power, however, lurked several signs of trouble. In 1972, electric utilities permanently shut down a seventh completed reactor due to poor performance and canceled seven previously ordered reactors under construction. In 1973, the government reported that nuclear plants were producing at only 54 percent of capacity. With high capital and low fuel costs, economics suggested that these plants should always be the generators of first choice.

Moreover, the Nixon years spawned sharp criticism of nuclear power. The Union of Concerned Scientists—formed by faculty members at MIT in 1969 to use their expertise to support environmental issues—took on the nuclear industry over the adequacy of emergency cooling systems that protected the core of nuclear reactors from a meltdown. To attract more publicity, it allied itself in 1972 with consumer crusader Ralph Nader. In his view, "the risks of something going wrong with these nuclear power plants are so catastrophic that they are not worth the benefit."[35]

The Ford years brought more bad news for nuclear power. In 1974 and 1975, the tide turned sharply against the industry, despite the completion of fifteen new reactors and an increase to 9 percent of electric generation. In 1975, new orders dropped to four, and cancellations rose to thirteen. Two more operating reactors were permanently shut. With delays due to environmental challenges and poor technical performance, nuclear power's appeal to private investors plummeted.

Nuclear's loss of its early luster elevated the prospects for coal. Plentiful supplies of the fossil fuel in solid form were available at a low cost, and newly developed scrubbers could clean up much of the visible pollution. Both Nixon and Ford proposed moving beyond conventional coal. They touted the chemical conversion of coal to liquid and gaseous synthetic fuels, which could, among other uses, fuel vehicles then using oil products. No existing technology could prevent carbon dioxide emissions from coal, but this limitation was rarely mentioned.

However, Wally Broecker—part of Revelle's team working on the 1965 report for LBJ—was more willing than most scientists to share his concerns about the direction of U.S. energy policy. The Lamont geochemist, considered "curmudgeonly" by his peers, was still using pencil and paper to solve problems after others relied on computers. But his deep understanding of historical climate changes revealed by carbon dating gave him a keen sense of where the increased burning of fossil fuels would lead. His article in the August 1975 issue of *Science* concluded that recent global cooling was due to short-term effects. This recognition meant that increases in atmospheric carbon dioxide would, by early in the next century, drive "the mean planetary temperature beyond the limits experienced during the last 1,000 years."[36]

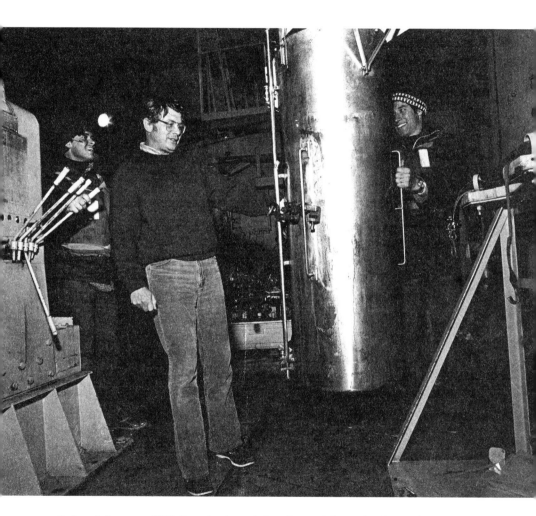

Carbon dating expert Wally Broecker (*center*) aboard a research vessel in the Pacific Ocean, bringing up a cylinder of seawater for analysis of carbon isotopes, circa 1973. Courtesy Lamont-Doherty Earth Observatory.

Broecker dipped his toe into the world of advocacy in a March 1976 letter to Robert Seamans, administrator of the new Energy Research and Development Administration. Broecker argued that the long persistence of carbon dioxide in the atmosphere would make it difficult to rectify later any bad policies of the 1970s, even if their effects would not be felt for a long time. He wrote, "I agree with others that the mean residence time of the CO_2 we release will be hundreds of years. Our present course toward a coal economy will almost guarantee that we will push the atmospheric CO_2 content to at least twice its preindustrial value by the end of the next century." He added that the best estimates suggested that a doubling of atmospheric CO_2 content from preindustrial levels would raise the mean temperature of the planet by about 3.5 degrees Celsius.

He acknowledged that models weren't yet adequate to assess whether the environmental changes would prove to be "bad" or "good." But, he warned, "our energy habits during the next hundred years will dictate the climate of our planet for many tens of generations thereafter." Broecker's private communication (a copy of which he sent to Revelle) may have been the first attempt of the pioneering scientists of climate change to connect their findings so directly to national debates over energy policy.[37]

In the spring of 1976, a new U.S. House subcommittee on the environment and the atmosphere began six days of public hearings on the National Climate Program Act. The schedule focused on a bill introduced by Philip Hayes, Democrat of Indiana, to coordinate climate change activities across federal agencies. However, as subcommittee chair George E. Brown Jr. noted, May 16 was the first congressional hearing to concentrate on climate change. As a result, he wanted to "review the scope and organization of the nation's current activities in climate research, monitoring, prediction, and impact analysis." Later, others would claim to have conducted the first congressional hearing devoted to climate change, but the honor belongs to the gentleman from Southern California.

Brown, a Democrat representing a Republican-leaning district, had majored in industrial physics at UCLA and was a vigorous proponent of science. He often devoted hearings to esoteric (at the time) topics like hydrogen fuels and solar energy. He and his dozens of witnesses in May generally followed the findings of the previous year's Academy report on the status of climate science.

The chairman expressed his concern that "mankind may be inadvertently modifying the climate by polluting the environment with carbon dioxide, particulates, and heat, but there seems to be no agreement as to whether the net result will be to heat us up or cool us off." Therefore, he added, "perhaps we should be satisfied if we can determine what additional research is needed to answer some of the most important and pressing questions."[38]

A recurring theme among the witnesses was the threat of climate change to global food production. Climate fluctuations had occurred over geologic time, but NOAA head Robert White said two things were new. "First, even small natural fluctuations in climate can have disastrous consequences because of the sensitivity of a world under population pressure. Secondly, man now has reached a point in his industrialization and technological development where his own acts can change the climate."[39] Whether the earth was warming or cooling, a shift in climate would alter the optimal locations for growing crops. Humans were no longer nomads who could migrate to previously unoccupied lands.

Although Revelle didn't testify during the six days of the Brown hearings, he remained a frequent link between academia and government. Now sixty-seven years old and still maintaining a schedule that would exhaust a younger man, he reflected on what he had learned about these two worlds during a lengthy oral interview in the summer of 1976.

Politicians and scientists were, he opined, "quite different in many important respects." Politicians, Revelle found, generally made short-term decisions, preferably based on knowledge. But sometimes, when time did not allow for sufficient study, they had to act relying only "on an intuition, or hunch, or a gamble." Long-range decisions were different. Politicians, he said, "try to leave future options open as much as they can—the only thing they can do." A democratic system required that "the poor bastards have to run for election every two years so that some people say that infinity to a politician is the election after the next one." On the other hand, Revelle observed, "The scientist never wants to act, he always wants to find out more before acting, and he always feels that he doesn't know enough to make a positive statement."[40]

To bridge the gap between politicians asking for certainty and scientists reluctant to provide it, Revelle favored a joint commitment to think in terms of statistical probabilities and ranges of possibilities. Meteorologists had already demonstrated, he said, how to incorporate probabilities into their forecasts, as when they predicted a "percent chance" of rain. People learned to make decisions about whether to continue with plans for an outdoor picnic based on such probabilities.

Revelle noted another contrast between politicians and scientists. He had found that politicians, many trained as lawyers, were most comfortable with adversarial debate. Scientists, on the other hand, preferred committee discussions that would lead to a consensus based on compromise. Revelle believed scientists should not "simply present the arguments of both sides and leave it to the politician to decide."

Revelle struggled with the competing demands on scientists. Adhering to scientific norms, he did not claim "certainty" when explaining the findings of climate science. But the frequent advisor to Congress also recognized the need to communicate with politicians in a language they could understand. Thus, his position was that he was "almost certain" that we were modifying climate by adding carbon dioxide to the atmosphere.

The big story for climate change during the Nixon and Ford years was the confusion over whether the earth was warming or cooling. The failure of scientists to agree on even the direction of the shift proved awkward. The problem wasn't just the creation of a few sensational media accounts, as some would prefer to remember. In fact, many major press outlets—including *Time* and *Newsweek*—ran multiple stories on the controversy based on interviews with a cross section of scientists.

By 1976, the tide had turned toward the conclusion that the earth was indeed heating up due to human pollution. One factor was that global temperatures were trending upward again. Another was a better understanding of the various forms of pollution that affect climate. While most witnesses at the Brown hearings in May avoided tilting toward the cooling or warming thesis, Murray Mitchell Jr., the senior research climatologist at NOAA, came down firmly on the side of global warming. In a letter to the subcommittee, he wrote that the one problem that frightened him was "not the threat of an onrushing ice age." It was, instead, "the inexorable accumulation of carbon dioxide in the atmosphere that may end up carrying the world too far away from an ice age before we can stop the problem at its source, with consequences to society no less disruptive than an ice age." He recommended that work should begin immediately on alternatives to fossil fuels.[41]

A belated understanding of massive errors in Rasool and Schneider's warning of an impending ice age also helped resolve the warming versus cooling debate. The model used for their 1971 article had severe limitations. For instance, it went only ten miles (sixteen kilometers) high, so it did not include the stratosphere. This omission sharply cut the model's climate sensitivity to carbon dioxide, leading to a glaring underestimation of its impact. In addition, the scientists had assumed that aerosols resembled CO_2 in having global coverage. After correcting their calculations to reflect that aerosols' effects were regional, it became clear that their model had grossly overestimated their cooling effect.[42]

When the need for these corrections became apparent, the consensus among scientists moved toward Broecker's and Mitchell's view that carbon dioxide would have the dominant impact on climate. The terms "climate change" and "global warming" had, in many respects, become interchangeable.[43]

5

Energy and Climate in the Carter Years

The Carter Presidency, 1977 to March 1979

In late 1976, advice poured into the transition office of former Georgia governor and current president-elect Jimmy Carter. The deluge made it easy to overlook isolated suggestions that he address the risks of changes in the earth's atmosphere.

A few recommendations did trickle in. John (Jack) Gibbons—chair of the Panel on Demand and Conservation at the National Academy of Sciences and director of the University of Tennessee environment center—wanted the new administration to investigate how much carbon dioxide the world's atmosphere could absorb without unacceptable climate modification.[1] George E. Brown—still chair of the renamed House subcommittee on the environment and the atmosphere—warned, "Large-scale change in the world's climate could override all other environmental problems." The congressman urged Carter to consider climate change as "primarily a research frontier" requiring greater understanding.[2]

The dangers of relying on carbon dioxide–emitting fuels had already become a dominant theme in Carter's emerging energy plan. But the risks of coal, oil, and natural gas were many. The chief driver of U.S. energy policy remained the Arab oil embargo of 1973–74. Determined to never again be held hostage by Middle East politics, the president prioritized slashing oil imports. Another spur to action: U.S. oil and natural gas resources appeared insufficient to keep pace with rising energy demand. Most (though not all) experts at the time believed that the depletion of domestic reserves would soon create a need for alternative fuels. These concerns commanded more attention than changes in the atmosphere, the effects of which might become more noticeable in the next century.

Whatever the motivations, there was broad national support for developing new energy sources. Not surprisingly, advocates sparred over which ones the new energy plan should favor. At a private three-hour meeting with White House staff in March 1977, dozens of nuclear industry executives pled the case for their technology. According to an internal White House summary, they failed to argue that nuclear power, like coal, could supply around-the-clock (baseline) power but, unlike coal, didn't emit carbon dioxide.[3]

A few days later, twenty environmental organizations shared their views at a similar (this time five-hour) meeting. They pushed for energy conservation and challenged the need for nuclear power as a substitute for fossil fuels. Several attendees called nuclear a greater danger than coal. One exclaimed, "I'd rather have a coal plant next door to me than see Uganda have nuclear technology." Another group, the Sierra Club, called for added emphasis on solar energy. Again, a detailed meeting summary contained no mention of carbon dioxide or climate.[4] Even among environmental advocates, the specter of rising global temperatures from burning fossil fuels was hardly on the radar.

In mid-April, Carter delivered two prime-time televised addresses unveiling a comprehensive energy plan he declared "the moral equivalent of war." One solution to America's energy challenges, he suggested, was "strict conservation" to reduce the waste of fuel. He called for cutting the growth in energy demand to less than 2 percent a year, well below expectations at the time. He added that reducing gasoline use by a tenth would help rein in growing energy consumption. So would insulating nine out of ten U.S. homes and all new buildings. "Our energy problems have the same cause as our environmental problems—wasteful use of resources," he asserted. "Conservation helps us solve both problems at once."

Another way to deal with uncertain oil and gas supplies was to expand the use of renewable energy. Carter called for increased research on advanced solar technologies, tax breaks for commercially available solar equipment to heat water, and most dramatically, installing solar panels on 2.5 million rooftops. Left unsaid was the notion that conservation and renewable energy could help slow the buildup of atmospheric carbon dioxide.

Carter also proposed making "the most of our abundant resources such as coal." He called for increasing U.S. production by two-thirds—yet another bold goal. White House staff admitted privately that coal created more environmental damage than oil or gas. However, they calculated that coal mining regulations, requirements for "scrubbers" on new coal plants, and reductions in energy demand could reduce sulfur dioxide and particulate pollution. They didn't mention the CO_2 emissions that would accompany greater coal use.

For careful readers, tucked away in the plan's 113 proposals was a $3 million study of the long-term effects of carbon dioxide on the atmosphere.[5] This idea was consistent with the advice from Dr. Gibbons and Chairman Brown.

During the same month as the president's energy addresses, the White House Council on Environmental Quality released a report that linked the need for alternatives to traditional fossil fuels with the threat of climate change. *Solar Energy: Progress and Promise* asserted that the prospects for "solar energy"—defined, as was standard at the time, to include all forms of renewable energy—were "brighter than most imagine." It estimated that with a serious commitment of resources, solar technology could meet a quarter of U.S. energy needs by the year 2000 and more than one-half by 2020. Of the various forms of renewable energy, photovoltaic cells had the most significant long-term potential, followed by wind turbines.[6]

The CEQ's rationale for renewable energy resembled the one found in Carter's energy plan—the expectation that world production of oil and gas would level off. It also made a brief environmental argument for solar energy. It declared, "Unlike coal, solar poses little risk to climate and creates little direct air pollution."

In March 1977, as the Carter team worked on the energy plan, the White House appointed William (Bill) Nordhaus to the Council of Economic Advisers. The Yale professor's soft voice and gentlemanly manner combined with powerful analytic abilities. Ten years after earning his doctorate from MIT, the macroeconomist was already a prolific writer on a broad range of topics.

At the time, Nordhaus was the American Economic Association's representative to the American Association for the Advancement of Science. In the 1970s, the National Science Foundation and Revelle at the AAAS were calling for greater coordination between the physical and social sciences. A skilled statistician and developer of models, Nordhaus was the kind of social scientist they were looking for.

One topic capturing the attention of the scholar from New Haven was the connection between the emerging understanding of climate change and economic growth. He quickly became the first prominent economist to tackle the subject. In June 1975, he delivered a paper, "Can We Control Carbon Dioxide?," at a conference in Austria. He followed up with a second paper, published in the February 1977 issue of the *American Economic Review,* titled "Economic Growth and Climate: The Carbon Dioxide Problem."[7] The professor's economic jargon and equations created barriers for lay audiences. But Nordhaus also had a knack for explaining the complexities of carbon dioxide pollution more clearly than most natural scientists.

In his 1977 article, Nordhaus wrote that some people might regard the claim that human economic activities could reach a scale where they would significantly affect the global climate as crying "wolf." "Unlike many wolf calls," he asserted, "this one should be taken very seriously." Carbon dioxide appeared to be "the first man-made emission to affect climate on a global scale with a significant temperature increase by the end of the century."

Like most work on climate, his analysis built on Keeling's data from Mauna Loa. In Nordhaus's view, upward trends left little doubt about the link between emissions from energy consumption and rising accumulations of carbon. Uncertainties about "proximate effects" such as temperature, rainfall, and ocean levels remained. Those about "ultimate effects" such as agricultural production and the destruction or creation of usable land and capital were even more formidable.

Nordhaus believed doing nothing about carbon emissions or relying on future development of some way to extract carbon from the atmosphere were not credible options. An uncontrolled approach that allowed energy markets to operate freely with carbon emissions priced at zero could eventually lead to a "danger zone" for life on earth. The idea of cleaning up carbon already emitted had "the odor of science fiction." The only realistic approach for dealing with the climate problem was to reduce carbon emissions.

Emission controls did not need to start right away. Nordhaus used his linear programming model to identify "efficiently designed" scenarios that would keep accumulations within acceptable limits while facilitating economic growth. He admitted that he found the results of his cost–benefit analysis "surprising." Energy production emitting carbon did not need substantial change until roughly 2020. The model allowed "a comfortable amount of time to continue research and to consider plans for implementation of carbon dioxide control," if deemed necessary. Adhering to the canons of academic transparency, he acknowledged the limitations of his crude model and its reliance on several critical assumptions.

First, without clear foresight on how higher temperatures would affect life on the planet, Nordhaus needed a proxy value for "acceptable" climate change. He solved this intellectual puzzle by positing that "the climate effects of carbon dioxide should be kept well within the range of long-term climatic variation" before they reached the danger zone. Based on global average temperatures over the past 100,000 years, he argued that global warming of 2 degrees Celsius (3.6 degrees Fahrenheit) fell within such variation. This increase was expected by the time the carbon in the atmosphere doubled from preindustrial times—a likely outcome, if present trends of energy consumption continued, sometime around 2030. Nordhaus regarded his standard as "extremely tentative." He emphasized that his definition of a "reasonable upper limit" for carbon accumulations was *deeply unsatisfactory, both from an empirical point of view and from a theoretical point of view.* Unmentioned, most climate models were showing that less than a doubling of atmospheric carbon would likely bring a 2 degree increase.

Second, Nordhaus adopted a common assumption of economists that the costs of remedial actions should be estimated based on "present value." Thus, to reflect the time value of money, he built a 10 percent discount rate (similar to compounded interest) into his model. At this level (or even lower), the rate heavily favored actions taken later over those begun earlier, because they would have reduced impacts on approaches with less time between costs incurred and expected benefits.[8] Nordhaus's calculations were consistent with the practice of academic economists and the federal government. Nixon's budget office had directed agencies to adopt 10 percent discount rates for most federal projects, a practice that continued into the Carter years.[9]

The third assumption was about the effectiveness of carbon taxes in changing energy consumption. Nordhaus foresaw the implementation of a low tax on carbon in 2020, with a gradual ramp-up to much higher levels. The tax would eventually increase the cost of natural gas by 50 percent, oil by 100 percent, and coal by 250 percent. This assumption, too, came with caveats. Nordhaus warned that it was hard to predict the future of nuclear power, how climate might be better understood, and whether nations with widely divergent interests could agree to an international control strategy.

Like Revelle, Nordhaus had the stature to ensure that his pioneering views would gain notice both within and outside his academic discipline.

After his White House appointment, the economist's views attracted the attention of the *New York Times*. A June article headlined "Climate Peril May Force Limits on Coal and Oil, Carter Aide Says" reported on a Nordhaus presentation to the American Geophysical Union summarizing his earlier papers.[10] Despite the proviso that he was not presenting government policy, the fact that a warning that fossil fuels might have to be curtailed sometime in the next century came from a presidential advisor made the story somewhat newsworthy. Editors placed the piece on page 13 of the D section, one page after the obituaries.

The occasional references to climate change in the first half of 1977 constituted a minuscule portion of the national energy debate. However, the Brown subcommittee held eight hearings about the new energy plan's environmental impacts that included climate among its concerns. Most witnesses treated the issue gingerly.

Ralph Rotty—a mechanical engineer at the Oak Ridge Institute for Energy Analysis and author of a text on thermodynamics—referred, at least obliquely, to the looming conflict between increasing coal use and limiting CO_2 emissions. At a June hearing, he testified that the fit between increased energy consumption and atmospheric carbon accumulations appeared to confirm a close connection. He calculated that in 1976, global carbon dioxide emissions stood at more than 5 billion metric tons. Of this, 27 percent came from the United States. He projected that by 2025 the total would grow more than fivefold, with the developing countries and "communist Asia" producing over half the global total. "The challenge to the United States," said the former air force meteorologist for the Pacific Theater during World War II, "is to develop energy supply systems not based on fossil fuels which can and will be used by the developing nations."[11] Previous strategies dealing with local and regional pollutants would be inadequate for the new challenges of a global pollutant.

During the summer, as Carter's energy plan sped (by usual legislative standards) through the House of Representatives, the National Academy of Sciences released *Energy and Climate: Studies in Geophysics,* a project begun during the Ford administration.[12] Roger Revelle chaired the study, aptly named given the close connection between fossil fuel combustion and climate. Joining his committee of fifteen were two other veterans of the panel that drafted the atmospheric carbon dioxide section in the pioneering 1965 *Restoring the Quality of Our Environment,* Dave Keeling and Joe Smagorinsky. Three U.S. presidents later, the new report would demonstrate how much climate scientists' views had evolved.

For the first time, the panel included economists, who would try to project future energy use, estimates needed to discern future carbon emissions. Hans Landsberg and Harry Perry, both with the think tank Resources for the Future, foresaw that the world's population would reach 10 billion and energy consumption would be five times current levels by the late twenty-first century. They projected that fossil fuels, mainly coal, could provide the needed energy. In his overview, Revelle emphasized the "fragility" of these estimates and suggested that attempting them was probably "foolhardy." However, he acknowledged that it would be difficult to attach timelines to future temperature increases without such projections.[13]

Calculating future energy use wasn't the only hurdle for the report. One theme was a sharp decline in expectations that climate models would soon provide policymakers with precise guidance on future temperatures and impacts (such as melting polar ice). The skepticism struck a different tone than the rosy assertion of the 1965 report for President Johnson that advances in mathematical modeling might allow useful predictions "within the next 2 or 3 years." Contributors to the 1977 report concluded that even recent, more sophisticated models did not adequately account for the complex interactions between the atmosphere and oceans. Nor did they include the impacts of clouds that could amplify or slow global warming. More daunting, the modelers listed dozens of other subjects that cried out for more research and analysis. The chapter by NOAA's Murray Mitchell asserted, "Making reliable projections of future climate of any kind will likely remain an elusive goal."[14]

The report chose its words carefully on whether the current understanding of climate science was sufficient to affect energy policy. Revelle's introduction asserted that the limitations of climate models did not justify postponing action. Despite the need to understand additional factors, carbon dioxide would exert the dominant influence on climate change. Its long persistence in the atmosphere meant that action could not be delayed until its effects became visible. "If the decision [to reverse the upward trend in the consumption of fossil fuels] is postponed until the impact of man-made climate has been felt," he warned, "for all practical purposes, the die will already have been cast."

Revelle opened his summation: "Worldwide industrial civilization may face a major decision over the next few decades—whether to continue reliance on fossil fuels as principal sources of energy or to invest the research and engineering effort and the capital that will make it possible to substitute other energy sources for fossil fuels within the next 50 years." If conversion to nonfossil fuels became necessary by 2025, alternatives not yet "satisfactory for universal use" would need to be developed in the interim. The best option to slow potential climate change, Revelle thought, might be "increasing reliance on renewable resources."[15]

Unknown to the people reading the report, the Academy's reviewing committees had forced Revelle to alter his original plans for the document and suppressed parts of what the report's public face wanted to say about energy and climate. In draft number two (already toned down from an earlier submission), for instance, he had written:

> The probability of such large future increases [in the carbon dioxide content of the atmosphere] has caused justifiable concern among many scientists and citizens that the future climate of the earth may be changed in ways that would be harmful to human beings. Such concerns are made more acute by the apparent narrowing of the margins for human survival resulting from recent and future growth of human populations and their economies, and the accompanying pressures on the earth's resources.

He also talked about the need for additional analysis of the impacts of climate change and for building "more resilient systems" to adapt to these impacts, particularly for agriculture.[16]

His Academy overseers vetoed language like "probability," "harmful to human beings," and "narrowing of the margins for human survival." Ever the team player, Revelle never complained publicly about the organizational shackles on his views. Only those who later plowed through his archives at the University of California, San Diego would discover his intent to be more forceful in his warnings.

The new Academy report was not ignored. White House science advisor Frank Press briefed Carter on it two weeks before the official release. Another heavyweight scientist appointed to the post, Press had served as department head of Earth and Planetary Sciences at MIT. His résumé also included time as a science advisor to Presidents Kennedy and Johnson. Press was already familiar with Revelle's work, having served with him on the Academy's Geophysics Research Board. At the time of his appointment, the *New York Times* wrote, "There are few men who have known the pulse of the earth so intimately as Frank Press." It noted that his expertise on sensitive issues like underground nuclear testing and earthquake prediction, combined with his time in Washington, helped him understand "not only the forces that have shaped the earth but also those that shape political and scientific decisions."[17]

Press's one-page memo to Carter on *Energy and Climate*—"Release of Fossil CO_2 and the Possibility of a Catastrophic Climate Change"—struck a dire tone.[18] The chief scientist emphasized that a doubling of atmospheric carbon dioxide could come in sixty years. He warned, "The urgency of the problem derives from our inability to shift rapidly to nonfossil sources once the climatic effects become evident not long after the year 2000; the situation could grow out of control before alternative energy sources and other remedial actions become effective." He observed that carbon dioxide's chemical properties create a unique long-term risk: "Natural dissipation of CO_2 would not occur for a millennium after fossil fuel combustion was markedly reduced."

Press told the president that the current state of knowledge was insufficient to justify limiting the consumption of fossil fuels "in the near term."

However, the administration did need to "take the potential CO_2 hazard into account" in developing long-term energy strategies. Fuel conservation and nuclear power would help but might not be sufficient. "As insurance against over-reliance on a nuclear economy," he added, "we should emphasize targeted basic research which could lead to breakthroughs for solar electric, biomass conversion, or other renewable energy sources." He asked Carter to approve his working with federal agencies on research and development plans responding to the threat of carbon accumulations in the atmosphere.

Carter—the first president to be briefed so directly on the climate challenge—initialed his agreement. His approval of expanding research on alternatives to fossil fuels, based partly on their atmospheric pollution, constituted another historical marker. It was the first official policy explicitly linked to the desire to slow climate change but that went beyond recommending more data collections and academic studies.

After the publication of the new Academy report, the *New York Times* covered the event with the headline "Scientists Fear Heavy Use of Coal May Bring Adverse Shift in Climate." This story moved up to page 1.[19]

In May 1978, Carter celebrated another milestone that might lead to an energy future less reliant on fossil fuels when he dedicated the new Solar Energy Research Institute in Golden, Colorado. SERI (later the National Renewable Energy Laboratory) would join the ranks of the Department of Energy's famed labs and focus on renewable energy research and development. Four new regional sites would work on the commercialization of emerging technologies.

The lab was still working out of rented space with a skeleton staff of about forty people. According to one senior scientist there, industries and universities largely ignored the development of solar technology at the time because "it was considered to be like a hippie thing . . . and wasn't taken seriously."[20] For that reason, the visit by the president was a big deal, conveying a heightened appreciation for the lab's work.

Having officially declared the date Sun Day, the president joked that he might have to reconsider the selection of the Golden site, given the heavy clouds overhead. When someone in the crowd yelled, "The wind's blowing," Carter grinned and responded, "That's enough; you qualify." Indeed, he noted later in his remarks, "Larger and more efficient windmills are being designed, including one with a rotor as long as a football field."[21]

But there was no mistaking that the biggest bet on the future would be "advanced solar," including photovoltaic cells. Carter said that Native Americans in Colorado, 10,000 years earlier, had used passive solar techniques to heat their dwellings. If the Energy Department's estimate that modern PV technology could be cost-competitive with conventional energy sources "perhaps as early as 1990" panned out, Carter foresaw "the dawning of the second solar age."

Carter briefly cited the environmental benefits of solar power: "It will not pollute the air; it will not poison our waters. It's free from stench and smog." However, the primary motivation for heavy government investments continued to be the danger of depending on fossil fuels, as demonstrated by the Arab oil embargo and its aftermath. He declared, "Nobody can embargo the sun. No cartel controls the sun. Its energy will not run out." The challenge of climate change went unmentioned at the Golden event.

Still, the growth of renewable energy that Carter envisaged would have a sizeable impact on carbon emissions. He touted the CEQ estimate that renewable energy could furnish one-quarter of all energy demand by the end of the century. Carter proclaimed, "American technological genius can bring the same blessings that the rural electrification program brought to me and millions of others when I lived as a small boy in Plains, Georgia."

The president's massive energy plan remained stalled in the Senate well into 1978. Late in the year, however, Congress finally completed legislation that included, hidden among its many provisions, a national climate action program that required federal agencies to produce five-year plans.

After its tortuous passage through Congress, the energy package also contained much that would, as a side effect, cut carbon dioxide emissions. Its conservation provisions, combined with Carter's stringent implementation of auto efficiency standards (passed under President Ford) could help reduce the use of fossil fuels. The State of California had concluded that energy efficiency standards would not increase the cost of new refrigerators while reducing consumers' electric bills and overall energy demand. Based on this rationale, the state began to regulate refrigerators in 1977.[22] With authorization for appliance efficiency standards in the new national legislation, prospects brightened for saving energy with more efficient equipment across the country. Carter's energy legislation also provided incentives for renewable energy sources like solar that could substitute for more polluting fuels.

Some aspects of the energy legislation would have the opposite effect. It encouraged greater use of domestic coal. It even allowed the federal government to order electric power generators to substitute coal for oil generation and severely limited the use of natural gas. Though gas was the least polluting of the fossil fuels, many experts believed that supplies were too limited to help meet the growing demand for power.

Carter reflected in his daily diary that the new energy legislation provided "about 60–65 percent of the energy savings we had projected originally." He blamed the shortfall on "the refusal of Congress to authorize a tax on oil that could be refunded to the American people."[23] During the Ford presidency, a gasoline tax with strong momentum due to memories of the recent oil embargo had been crushed on the House floor. Carter's proposed gasoline and crude oil levies had also proved too tough for Congress to stomach. Nordhaus relied on an enormous tax on carbon sometime in the twenty-first century, in the United States and worldwide, to reverse the tide of rising emissions. Would the political support for new taxes be different then?

Carter also came to believe that the public appetite for conservation might, like the tolerance for energy taxes, be limited if associated with sacrifice. Undeterred, he determined that conservation and sacrifice were essential parts of a comprehensive energy strategy, whatever the political risk.[24]

In the early months of 1979, international institutions and the U.S. president addressed the climate issue in new ways.

In mid-February, the World Meteorological Organization, in collaboration with three United Nations agencies, held the first World Climate Conference in Geneva. Participants from fifty-three countries represented an impressive array of experts—including, amid the Cold War, a senior scientist from the Soviet Union. Robert White, head of the U.S. National Oceanic and Atmospheric Administration, chaired the proceedings.

The various reports generally reflected the views of many U.S. scientists studying the issue. Lawrence Gates, a climatologist at Oregon State University and contributor to several major climate studies, opined that global warming due to CO_2 emissions "may amount to an environmental catastrophe." Edward Munn of the University of Toronto and Lester Machta, head of the Air Resources Laboratory at NOAA, said that "few, if any, scientists believe the carbon dioxide problem in itself justifies a curb, today, in the usage of fossil fuels or deforestation." They warned, however, that governments would face a "crossroad" for enacting policies to deal with the problem within five to ten years. Ralph D'Arge—distinguished professor of economics at the University of Wyoming and editor of the *Journal of Environmental Economics and Management*—called for studies on the costs and benefits of climate change for agriculture, fisheries, human health, and other sectors.[25]

Revelle, who chaired one session, told a press conference that the world faced "a Faustian bargain" on climate. Solar power would help reduce carbon emissions, but the technology was not yet ready (that is, cheap enough) for widespread adoption. He said that another alternative to fossil fuels, nuclear energy, raised the formidable problem of radioactive waste disposal. He lamented, "Whatever you do is bad."[26]

A month later, Carter forwarded to Congress a message on science and technology. His March 27 statement—overshadowed by the signing of a peace treaty between Egypt and Israel on the White House lawn the day before and a core meltdown at the Three Mile Island nuclear plant in Pennsylvania the next day—addressed both energy and climate. He called the federal investment in energy research and development "one of the most important ways to prepare for the future."[27]

In a section titled "Understanding the Forces of Nature and Man-Induced Environmental Changes," the president declared, "Man exists on this planet only with the consent of Nature." He noted that over the past two years, he had more than doubled funding for the National Climate Program Act. He also warned of "the risk that man's own activities—now significant on a global scale—might adversely affect the earth's environment and ecosystem. Destruction of the ozone layer, increase in atmospheric carbon dioxide, and alteration of oceanic flow patterns are examples of the problems we must understand before changes are irreversible or the consequence inevitable."

Though he continued to support the need for more American coal, he also declared that, for the longer term, solar energy and nuclear fusion could provide "virtually inexhaustible supplies of electricity." But for this to happen, "we must invest now." He added that such investments had already begun. The solar research and development budget had grown from almost nothing two years earlier to an annual appropriation of $850 million.

Carter's push to study climate more carefully and invest in lowering the cost of solar power suggested pathways responsive to the comments of Revelle and other scientists in Geneva.

6

A Jolt from JASON

Carter, April to July 1979

In Carter's third year in office, scientists issued in rapid succession a barrage of studies that raised the visibility of climate change at the White House. In April, a group known as JASON—a Department of Defense self-perpetuating scientific advisory committee named after the mythic Greek hero—weighed in on the issue.

Most JASON studies were highly classified and rarely garnered much public attention, though the group had drawn fire from activists criticizing the conduct of the Vietnam War. At the highest levels of government, the views of its eminent scientists carried considerable weight. Gordon MacDonald—a member of Nixon's Council on Environmental Quality and a longtime JASON—chaired an effort, begun two years earlier, that resulted in the 1979 report *The Long Term Impact of Atmospheric Carbon Dioxide on the Climate*. It was funded by the Department of Energy, whose predecessor agency had received Wallace Broecker's letter requesting such a study.[1]

The group thanked the pioneers of climate science like Roger Revelle, Dave Keeling, Harmon Craig, and Broecker for their advice, and its findings built on the established scientific literature. The JASON scholars, however, existed to advise policymakers. For them, understanding climate science wasn't just an intellectual puzzle or an end in itself. This report would address issues affecting government policy more bluntly than its predecessors.

JASON's general assessment of climate science didn't differ much from earlier studies. Its report projected that if the global use of fossil fuels continued to rise at the then current rate of 4.3 percent a year, the amount of carbon in the atmosphere would double from preindustrial levels by 2035. If the increase in fossil fuel use dropped to one percent a year, the doubling would be delayed until 2065. It foresaw that a doubling of carbon would lead to a temperature increase of 2 to 3 degrees Celsius.[2] Temperatures would rise much more quickly near the poles and at higher altitudes, with increases three to four times greater than average.

JASON considered the potential melting of the West Antarctica ice sheet "perhaps the most ominous feature" of global warming. Like previous reports, MacDonald and his colleagues pulled back from predicting the timeline for such a process. They noted, however, that a complete melting of this ice would raise worldwide sea levels by five meters, or more than sixteen feet, "with consequent major disruptions in the world's coastal regions."[3]

Compared to previous studies, the scientists—known in government circles for their self-assurance (some called it cockiness)—put less emphasis on uncertainties than the National Academy of Sciences. They admitted to weaknesses in their relatively simple model and the difficulty of making long-term predictions. However, various models produced similar results, increasing confidence in their reliability. MacDonald's summary acknowledged what was not known about the effects of carbon on climate. However, it did regard it as "highly probable" that increases in the worldwide use of fossil fuels of 1 percent to 3 percent per year would lead to climate change in the second half of the twenty-first century.[4] The magnitude of this temperature increase would be twice that of the decrease during the Little Ice Age during the sixteenth and seventeenth centuries.

The JASON report didn't hesitate to connect its findings to the ongoing debate over U.S. energy policy. It observed, "The Arab Oil Embargo, President Nixon's Project Independence, and increased emphasis on the use of coal as a major energy resource have focused renewed attention on the possible consequences of increased levels of carbon dioxide in the atmosphere." The report emphasized that not all fossil fuels were equal in their environmental impacts; natural gas emits the least carbon and coal the most. It also warned against the expanded use of synthetic fuels. Because of the needed processing, oil or gas produced from coal would emit even more carbon than traditional coal. JASON calculated that synthetic oil would emit 50 percent to 100 percent more carbon than conventional oil.[5]

Solutions were available to delay climate change. JASON concluded, "The increased use of natural gas, if available, would lengthen the time available for a shift to non-carbon-based fuels. Clearly, a significant shift to a nuclear or solar energy economy would postpone carbon-induced shifts." There might be other ways to address the problem. But moving to nuclear or solar power appeared more economical.[6]

JASON distributed more than two hundred copies of its report to scientists, industry and environmental organizations, and government officials, including at the White House.

The JASON report could hardly have arrived at a more awkward time for the Carter administration. Renewed turmoil in world oil markets was intensifying pressure to utilize more domestic coal.

On April 5, 1979, the president delivered another prime-time televised speech amid severe energy shortages following a revolution in oil-exporting Iran. With new gasoline lines looming on the horizon, Carter called on "all citizens to honor, and all states to enforce, the 55 mile-per-hour speed limit" to save fuel. He announced he would phase out federal price controls on domestic crude oil to encourage higher production and discourage demand. He also urged Congress to pass a windfall profits tax on oil production and promised a careful investigation of the nation's most serious nuclear accident just the week before.

Carter's proposals included creating an Energy Security Fund that would use American technology to solve America's challenges, patterned after the Apollo program that sent men to the moon. Massive revenues from a windfall oil profits tax would boost energy research and development funding. Carter called for investments in a diverse array of alternative fuels, including synthetic oil and gas. During his first two years in office, Carter had emphasized coal in its natural state. Now he was calling for chemically altered coal—the fuel JASON just said would have the most adverse impacts on climate—to cut U.S. dependence on foreign oil.

Carter's stress on coal was consistent with the advice of his science advisor in a March 24 memo. Frank Press told the president that during a "transition period" lasting until 2020, "coal will be our most important fuel, both in direct use and as a source of synthetic fuels." The reign of coal would not last forever. Press foresaw that after 2020 "coal utilization will fall under competition from advanced solar and other technologies."[7]

Press didn't mention the environmental implications of increased coal use, but the JASON report had elevated concerns about the push for synthetic fuels. MacDonald visited Press in his office several weeks after the report's release, accompanied by Rafe Pomerance, a lobbyist for Friends of the Earth and perhaps the first prominent environmental activist to prioritize climate change. The White House visitors were an odd couple. At six foot four, the lanky Pomerance, with an undergraduate degree in history, towered over the scientist who had written the book on the earth's rotation.

Gus Speth (*left*), staff attorney for the Natural Resources Defense Council and later chair of the Council on Environmental Quality, with President Carter, 1977. The CEQ became the administration's strongest advocate for solar energy and other responses to the threat of climate change. Courtesy Jimmy Carter Presidential Library.

MacDonald did most of the talking at the meeting, delivering to his colleague on numerous scientific studies over the years what Pomerance remembered as a lengthy "sermon" about climate science. Neither visitor left the meeting knowing how Press would react.[8]

As it turned out, Press did follow up on the visit. On May 22, he wrote the National Academy, asking that the Climate Research Board conduct a quick assessment of what was known about the impacts of carbon dioxide on climate in light of the JASON and other recent reports.[9]

Press wasn't the only White House official the MacDonald/Pomerance duo lobbied. Another target of their missionary work was J. Gustave (Gus) Speth—acting (soon to be confirmed) chair of the Council on Environmental Quality. The thirty-seven-year-old lawyer was a former law clerk of Supreme Court Justice Hugo Black and a founding member of the Natural Resources Defense Council. Speth, admittedly new to the subject, told MacDonald that if he was to brief the president he needed a scientifically reliable memorandum on the problem, which MacDonald promised to provide quickly.[10] With the additional reports requested by Press and CEQ, the White House would now have access to three new studies on climate change within four months.

The JASON report had brought a new burst of attention to the climate issue and inspired an influential environmental organization to advocate for cutting carbon dioxide emissions.

Days after the release of the JASON report, eighty-five experts from the physical, biological, and social sciences descended on the hotels of Annapolis, Maryland, after braving the chaos resulting from a work stoppage at United Airlines and a shortage of rooms aggravated by an unexpected extension of the state legislative session. During the first week of April, the scientists, including some from other countries, spent five days on the conference topic: "Environmental and Societal Consequences of a Possible CO_2-Induced Climate Change." These interdisciplinary workshops have been less remembered than other studies produced during the year. Still, the wide-ranging discussions would substantially impact the framing of the climate issue in later years, in ways many historians would overlook.[11]

The leader of this effort, once again, was Roger Revelle, who had for several years argued for more analysis of climate change impacts. The American

Association for the Advancement of Science had appointed its ex-president to head a new twelve-member committee on climate. The panel set high, idealistic goals for itself. One member said that "the forcing function of knowledge" could help the international political community face the difficult decisions ahead on climate. The scientific community would need to demonstrate "an unusually high degree of unanimity and clarity . . . to obtain from political leaders the decisions and actions required."[12]

Revelle's committee included climate luminaries like U.S. Representative George E. Brown Jr., Bill Nordhaus (back at Yale), NOAA's Murray Mitchell, and Robert White (now at the University Corporation for Atmospheric Research). The group organized the Annapolis panels consistent with Revelle's belief that the implications of climate change extended far beyond the field of climate science. The Department of Energy provided financial support for the conference, including honoraria for participants. Invitations went out to a more diverse array of participants than a National Academy or JASON study. The invitation list included representatives of universities, businesses, and environmental organizations.

Not every invitee believed the conference was worth their time. Freeman Dyson—a brilliant statistician and geophysicist at the Institute for Advanced Study in Princeton, where his career overlapped with those of Einstein and Oppenheimer—sent Revelle a friendly note of regret. Dyson suggested that the world might already be "hooked on" rising levels of CO_2 for agricultural yields. The contrarian believed that stopping carbon emissions from industrial activities, therefore, would not be a good thing.[13] Conference organizers were also disappointed with the absence of many environmental organizations. As historian Joshua Howe noted, most of these groups still "ranked the study of climate change as a low priority and remained aloof from the climate debate."[14]

Having achieved cabinet status less than two years earlier, the Department of Energy was far from a passive funder of the event. The new department had inherited the old Atomic Energy Commission's role in developing nuclear power.[15] Because of its heritage, the department held nuclear physics and those who had been part of the Manhattan Project in particular esteem, including Alvin Weinberg. Weinberg earned his doctorate in mathematical biophysics from the University of Chicago in 1939 and then worked there with eminent scientists like Enrico Fermi, winning acclaim for his resolution of metallurgical issues related to building the atomic bomb. Weinberg later became the director of the Oak Ridge national laboratory (a part of the old AEC that ended up at DOE) and a science advisor to Presidents Eisenhower and Kennedy. In 1979, he served as the founding director of the Institute for Energy Analysis at Oak Ridge Associated Universities. He wielded considerable clout in energy circles.

An unabashed advocate for civilian nuclear power, Weinberg believed that nuclear and coal were the leading options for meeting the world's growing appetite for electricity. However, both had severe risks he viewed as "in a sense analogous." Nuclear plants that reprocessed fuel could lead to nuclear weapons proliferation. As for coal, carbon dioxide concentrations in the atmosphere might limit its ultimate use. Therefore, it was essential to learn more about climate change to evaluate the two technologies on a level playing field.[16] Because of this concern, Weinberg initiated studies at Oak Ridge on the environmental impacts of carbon dioxide and encouraged David Slade, director of DOE's carbon dioxide and climate division, to support the Annapolis conference.[17]

Workshop participants generally accepted the basics of climate science articulated in the recent *Energy and Climate* and JASON studies. Human use of fossil fuels was leading to an accumulation of carbon dioxide in the atmosphere that would increase the global mean temperature by 2 to 3 degrees Celsius when CO_2 doubled from preindustrial levels. However, Revelle warned that the increase in impacts might not be slow and continuous because of the possibility (though not the likelihood) that climate change would spin out of control and produce a "markedly different" climate from what humans were used to. Whether the change was steady or not, massive alterations might be "irreversible on a human time-scale, requiring several hundred to a thousand years before the added carbon dioxide is sequestered in deep ocean water."

The participants offered differing perspectives on the relative merits of altering the world's energy infrastructure to reduce carbon emissions versus adapting to a changing climate. Tom Wigley, director of climate research at the United Kingdom's University of East Anglia, concluded that climate change's social and economic impacts would be profound. However, he con-

sidered it "clear that mankind can cope if the decision is made to adapt."
He noted that the alternative route—preventing the buildup of CO_2 in the
atmosphere—was "especially difficult because of international ramifications." Jerry Olson, an expert on forest and plant ecology at the Oak Ridge
lab, reached a different conclusion on adaptation. He believed that climate
change might be "too abrupt for easy accommodation by natural or conventional social forces."

A primary focus during the week was the impact of climate change on
agriculture. The perspectives of the agricultural scientists depended on the
piece of the puzzle they examined. David Pimentel from the college of agriculture at Cornell University foresaw that an average 2 degree Celsius rise in
temperature would lead to a sharp decline in rainfall in the American farm
belt and result in a 25 percent reduction in corn production and a 12 percent
reduction in wheat. Further complications might arise if climate change led
to increases in insect pests. Others pointed out that American farmers had
proven they could update their practices to increase crop yields and would
do so in the future. Donald Baker of Mississippi State University and Jerry
Lambert from Clemson calculated that with the enhanced photosynthesis
expected from rising temperatures, cotton and corn production in the United
States could increase by 14 percent to 38 percent, assuming water availability and better insect management.[18] Overall, the agricultural panel found
the historical evidence that U.S. agriculture could cope with gradual global
warming to be reassuring.

The outlook for crop productivity looked different from an international
perspective. Shinichi Ichimura—an economist from Kyoto University in Japan with a doctorate from MIT—cautioned that dehydration of soils used
for food production in some, mostly poor countries without the resources
to adapt might lead to the migration of rural peasants into the cities. Such
migration would intensify the preexisting "social tension of urban life."

The final and most contentious panel in Annapolis met on the Friday before Palm Sunday. Chair Lester Lave, an environmental economist from
Carnegie-Mellon University, reported that the discussions on the "economic
and geopolitical consequences of a potential CO_2-induced climate change"
were "enlivened" by the diversity of participants and their "inherent opti-

mism or pessimism about our future." He cautioned that the lack of unanimity on panel 5 meant his summary "should not be taken to represent the
views of all participants."

The sixteen panelists included two from Oak Ridge (Weinberg, who
had to leave early, and Rotty). This panel was the only one with industry
representation—Henry Shaw from the research division of Exxon. Exxon
had been aware of the potential impacts of rising atmospheric carbon dioxide
since the 1950s, and Shaw recognized that climate change was an emerging
topic that might affect the company's business. He had worked with scientists at the Scripps Institution and the Lamont-Doherty Observatory, including Wally Broecker, to better understand the issue and establish Exxon as a
credible participant in advancing climate science. To this end, the company
outfitted a supertanker (the fourth largest ship in the world) to collect CO_2
data from the ocean, as it followed its dedicated route around the world
multiple times a year.[19]

The panel's summary, written with the assistance of National Academy
of Sciences senior staffer Jesse Ausubel, was likely reassuring for Exxon. For
starters, it did not consider climate change impacts more than fifty years
into the future, leaving out the later period when the buildup of atmospheric
carbon would have its most significant effects. Lave asserted that "it is impossible to forecast with confidence what the world fifty or one hundred years
from now will look like." To make his case, the economist asked people to
conceive of a hypothetical British Royal Commission in 1910 attempting to
forecast how much hay would be needed by the horses of London on the
basis of a forecast of human population in 1960.

Lave was also pessimistic about incorporating the cost of carbon emissions into policy decisions before 2030. He said that the evidence that carbon
accumulations were changing the climate was mainly based on the "theoretical predictions" of climate models. The evidence would be inconclusive until
physical measurements verified these predictions. He further argued that
there was not much reason for economists and social scientists to attempt
to assess the impacts of climate change until climate modelers could forecast
them at the regional level.[20]

The lack of such information didn't stop Lave from confidently concluding that the costs of reducing fossil fuels with expanded use of nuclear power

or energy conservation were "substantially larger than our present under-standing of the CO_2 problem can justify." He said it was premature to label CO_2 as a "problem" or a "threat" because it was "far from proven that CO_2 increases would prove a danger to the human race" in the next fifty years. Though skeptical about policies that might reduce carbon emissions, Lave said that "enough serious issues are raised by the potential climate change to make thoughtful consideration of adaptive strategies worthwhile." Like many economists accustomed to discount rates in cost–benefit studies, Lave favored adaptations, such as advancing agricultural sciences and building dikes, to cope with climate change.

Internal disagreements in panel 5 spilled over into June and July. Weinberg and Slade wrote Lave to complain about his draft conclusions, which effec-tively negated the need to reduce carbon emissions or even undertake the impact analysis DOE wanted to fund. But Lave dug in his heels. In a "Dear Alvin" letter, he asserted that the panel "could not think of a sequence of events that would make CO_2 a dominant issue in the next 50 years." In a sim-ilar "Dear Dave" reply, Lave said that "DOE's money wouldn't be well spent" attempting to "value the cost of a carbon dioxide build"—the purpose of the Annapolis workshop.[21] The adamant economist won the battle over the panel's report. According to Lave and Ausubel, the impacts of climate change should be assessed only if they did not occur before 2030 or after 2030.

The public and most conference participants did not know that the min-imization of climate change's effects had support at the highest level of the AAAS. Its president wrote the executive director on June 27 that, as he had told the board, he took exception to the workshop using the term "impact," as it indicated negativism. He warned "against a disproportionate emphasis on long-range climate concerns arising from activities of humankind"—the very message delivered by Lave and Ausubel.[22] The president of AAAS for 1979 was Edward David Jr., president of research and engineering at Exxon and Henry Shaw's boss.

There was no doubting David's heft as a scientist. As a Fortune 500 com-pany (lodged in second place between carmakers General Motors and Ford), Exxon employed a team of eminent scientists who mingled easily with their governmental and academic peers. Preeminent among these was David, who had earned his doctorate in electrical engineering at MIT, where his faculty lab mentor was Jerome Wiesner. He later followed Wiesner's path to become a chief presidential science advisor, in David's case for Richard Nixon.

Exxon was one of many companies adapting to the regulatory regime created during the 1970s. Environmental legislation, price controls, and federal allocation of supplies all added to the regulatory burden that the government imposed on oil companies. Exxon viewed climate change as an issue that might bring additional controls down the road.

In the summer and fall, another scientist working under David, Raymond Campion, participated in internal discussions about how the company's lobbyists in Washington should handle climate issues. In a July internal memo, he wrote that Exxon should take a low-profile approach to climate science. With the industry's credibility "not high," he argued that "should an API [American Petroleum Institute] study indicate no serious CO_2 problems, the results would be greeted with skepticism."

Campion identified the recent Annapolis conference as a credible source that might provide views compatible with industry's. According to his slanted version of the event, the symposium concluded that "no catastrophic hazards would be associated with the CO_2 buildup over the next 100 years and that society can cope readily with whatever problems ensue." Increases in ocean levels due to polar ice melting, because they were gradual, could be dealt with by "shoreline improvements." The movement of crop-growing regions northward, which would also be gradual, could be offset with better irrigation methods. A second Campion memo in September confirmed that API supported his recommendations.[23]

Thus did Exxon, in 1979, establish a long-term industry template for its climate strategy: Stress scientists' projections that observable climate change would be gradual but downplay the length of time carbon was expected to remain in the atmosphere. Identify ways to adapt to climate change that reduced the need to cut carbon emissions. Attempt to use technical experts to influence or interpret the work of those with credibility on the issue rather than speak in its own voice.

During his first two years, Carter had called on the country to expand its use of solar energy. In an April address at the National Academy of Sciences—the first presidential visit since Kennedy—Carter emphasized the role of science and technology in developing energy alternatives. "Over the next decade or so, we must rely mostly on existing technologies," he conceded, but his energy programs would lead to long-term changes in fuel choices. He predicted, "By the second quarter of the 21st century, we will have learned to rely on cleaner, essentially inexhaustible sources of energy." The principal candidates included nuclear power and solar photovoltaics.[24]

Despite the incentives for solar in the 1978 energy legislation and Carter's frequent references to it, many advocates thought progress fell short of the crash effort needed. After a detailed internal study and a productive discussion with solar proponents in Congress—including Albert Gore Jr., now a second-term member of the House—Carter was ready to announce a more aggressive program.

On June 20, 1979, the president unveiled his more robust solar initiative at the White House during the dedication of 611 square feet of rooftop thermal solar collectors, sparkling in the bright sunlight. The new panels would provide hot water for the White House mess and, in the words of an internal memo, "provide a highly visible demonstration of solar energy and the Carter administration's commitment to alternative energy resources."[25]

The assembled crowd included representatives of Friends of the Earth (Pomerance), the National Resources Defense Council, other environmental groups, the federal government, and solar businesses and organizations. The twenty members of Congress in attendance included Massachusetts Democrat Ed Markey, who twenty-eight years later would be appointed by House Speaker Nancy Pelosi to chair the Select Committee for Energy Independence and Global Warming.[26]

The event went well beyond a photo op. Carter announced an ambitious goal of generating 20 percent of the nation's energy from renewable resources by the end of the century.[27] He also proposed a variety of tax credits and the creation of a Solar Bank, initially funded at $100 million a year, enough to finance an estimated 100,000 solar units in the first year. Carter would go on to win the massive boosts in research and development for renewable technologies that solar advocates sought.

The solar focus of R&D was not the White House panels that heated water. Experts believed that the future of solar energy belonged to the photovoltaic cells that converted sunlight into electricity. To accelerate the effort,

Carter in July appointed Denis Hayes—who at the age of twenty-five had organized the 1970 Earth Day teach-ins—to run the rapidly expanding Solar Energy Research Institute. During the Carter presidency, a Department of Energy partnership with solar companies set a bold goal of producing photovoltaic cells with 15 percent efficiency (compared to the 9 percent Bell had achieved in the 1950s) and competitive prices by the mid-1980s.[28]

In June, the president flew to a G7 economic summit with the leaders of Britain, Canada, France, Italy, Japan, and West Germany. The Tokyo meeting began as long gasoline lines were peaking in the United States and the world reeled from another stiff price hike by OPEC. Nonetheless, the group found time to discuss, at least briefly, the specter of climate change.

Germany took the lead in raising the topic. During the opening session on the twenty-eighth, not open to the press, Chancellor Helmut Schmidt unexpectedly interjected that despite the current push to find more oil, the world in the next century might decide not to use hydrocarbons anymore. He envisaged that in one or two decades, "scientists will say we are heating up the outer atmosphere of the globe, when it will not be tolerable for nations to do this—when there will be too much heat and too little water, as in the Sahel." Schmidt called for "lots of money . . . for pure and applied research for renewable energy, which should come on stream by the middle 90s and by the end of the century enable us to use solar, geothermal, and nuclear energy more."[29]

The summit's final declaration briefly stated, "We need to expand alternative sources of energy, especially those which will help to prevent further pollution, particularly increases of carbon dioxide and sulfur dioxide in the atmosphere." However, the G7 placed the threat to the atmosphere in the context of the bigger energy picture. The declaration devoted more space to the need to expand coal use than to the risks of carbon dioxide.[30]

The week after the Tokyo summit, Speth sent the paper he had requested from MacDonald to Stuart Eizenstat, chief of the White House policy office. MacDonald's three coauthors were Revelle (now back full-time at the University of California), Dave Keeling, and George Woodwell, director of the Ecosystems Center of the Marine Biological Laboratory in Woods Hole, Massachusetts. Not surprisingly, the nine-page paper, *The Carbon Dioxide*

Problem, was in many respects a condensation of the 184-page JASON report, but at a length more suitable for busy policymakers.

JASON had issued the bluntest statement to date from prominent scientists. The paper delivered to the Council on Environmental Quality and circulated at the White House was even more direct in assessing human effects on climate. It began:

> Man is setting in motion a series of events that seem certain to cause a significant warming of world climates over the next decades unless mitigating steps are taken immediately. The cause is the accumulation of CO_2 and other heat-absorbing gases in the atmosphere. The result is expected to be . . . a warming that will probably be conspicuous within the next twenty years. If the trend is allowed to continue . . . a series of changes would have far-reaching implications for human welfare in an ever more crowded world, would threaten the stability of food supplies, and would present a further set of intractable problems to organized societies.[31]

After emphasizing that synthetic fuels would emit much more carbon than traditional fossil fuels, *The Carbon Dioxide Problem* recommended acknowledging the problem by making climate impacts part of national energy policy. The nation should, it said, reduce the use of fossil fuels, choose among the fossil fuels based on their release of CO_2, and grow forests that would absorb more carbon.[32] The authors were clear that sufficient information was available to adopt policies to limit future damages.

The CEQ's effort to slow down the synfuel train faced substantial opposition within the administration. With the pressure to develop alternatives to traditional oil and gas, the Department of Energy, the budget office, and the policy office supported a robust effort to produce oil and gas from coal.[33]

Unusual for a JASON often tasked with top-secret assignments, MacDonald went public with his views that were already circulating within government circles. He started doing press interviews about the conflict between synthetic fuels and the risks of climate change. On July 26, he even published a hard-hitting op-ed in the *Washington Post* headlined "The Synfuels Mistake."

7

More Reports from the Academy

Carter, July 1979 to January 1981

In July 1979, dozens of top scientists headed to yet another East Coast location, Woods Hole, Massachusetts, to consider the state of climate science. Located in the former whaling village of Cape Cod, the site had evolved into a coveted retreat for summer visitors and a world-class center for marine research. The week of July 16, the Climate Research Board—assigned to oversee the report Press had requested from the National Academy—spent six days reviewing previous climate change reports. Participants included leading contributors to earlier studies—Revelle, Smagorinsky, Suomi, Gates, and Woodwell.

The following week, the Ad Hoc Study Group on Carbon Dioxide and Climate convened to render its judgment on previous reports and "for the benefit of policymakers" assess what was known about "the carbon dioxide/climate issue." The Academy had selected Jule Charney of MIT to chair the group. The sixty-two-year-old mathematician was the son of Jewish parents who had emigrated from White Russia, a pioneer in using computers to predict the weather, and an expert in complex models. When the American Geophysical Union awarded him its William Bowie medal in 1976, it declared that he had "guided the postwar evolution of modern meteorology more than any other living figure." The report produced at Woods Hole, *Carbon Dioxide and Climate: A Scientific Assessment,* is generally remembered as "the Charney report."[1]

The study group also included a second member from MIT, two from the National Center for Atmospheric Research, and others from UCLA, the University of Washington, the University of Stockholm, Harvard, and Woods

Hole. Some participants brought spouses and children, giving the two weeks the aura of what writer Nathaniel Rich called "a family reunion."[2]

The Charney report examined the simple and complex models available and found them "basically consistent and mutually supporting." No model predicted "negligible warming." The report agreed that a doubling of carbon dioxide in the atmosphere from preindustrial levels would likely come sometime in the next century, depending on future emission rates. "We estimate," said the report, "the most probable global warming for a doubling of CO_2 to be near $3°C$ with a probability error of plus or minus 1.5." Increases would be more pronounced at high latitudes.[3]

Charney focused on the remaining uncertainties, including the role of oceans. More detailed modeling suggested that their ability to serve as a heat reservoir might delay thermal equilibrium for a few decades. This phenomenon would postpone when carbon added to the atmosphere affected air temperatures. Thus, the air temperatures expected from a doubling of atmospheric carbon might not be contemporary with the doubling but would remain unavoidable.[4]

Consistent with National Academy custom, the Charney report avoided specific policy recommendations. It also followed NAS practice by emphasizing the need for greater understanding, using the word "uncertainty" or its variants sixteen times. Still, the forward written by Climate Research Board chair Verner Suomi (recipient of the National Medal of Science in 1977) offered a blunt assessment of the climate challenge. It declared, "We now have incontrovertible evidence that the atmosphere is indeed changing and that we ourselves contribute to that change. Atmospheric concentrations of carbon dioxide are steadily increasing, and these changes are linked with man's use of fossil fuels and exploitation of the land. Since carbon dioxide plays a significant role in the heat budget of the atmosphere, it is reasonable to suppose that continued increases would affect climate." Agreeing with the report that anticipated changes would not be negligible, he warned public officials, "A wait-and-see policy may mean waiting until it is too late."

The four climate reports (JASON, AAAS, CEQ, and Charney) over several months, all intended more for use within government circles than for public consumption, recognized a close the tie between the burning of fossil fuels and a changing climate.

*   *   *

Discussions about climate change within the White House and Exxon corporate offices were usually private, becoming publicly visible only with the release of records decades later. But the U.S. Senate was asking in open proceedings whether concerns about global warming should affect the government's enthusiasm for coal-based synthetic fuels. On July 30, its Committee on Government Affairs invited nine scientists to a public symposium on the nexus between atmospheric carbon dioxide, synthetic fuels, and energy policy. Most of these scholars had joined the deliberations at Woods Hole a few days earlier.

The symposium's impetus came from committee chair Abraham Ribicoff. In June, the Connecticut Democrat had met in Washington with Chancellor Schmidt, who declared that the accumulation of CO_2 in the atmosphere "represented a major threat to the future of mankind." As a follow-up to the conversation, the German government dispatched one of its top climate scientists to appear before the committee. Ribicoff had held a climate hearing earlier in July that featured Gordon MacDonald, who briefed senators on the JASON and CEQ studies. "I was deeply disturbed by the testimony," Ribicoff recalled, "and I must confess that neither I, nor most of the people in this world, including most Senators, understand the implications of this phenomenon."[5] In its findings, the committee adopted many conclusions from JASON and CEQ.

Another motivation for the symposium was the committee's jurisdiction over synthetic fuels legislation. As a result, the questions focused on the ramifications of their displacing some traditional fossil fuels and how long the world could continue to rely on them.

The scientists generally accepted the administration's push for synthetic fuels due to concerns about energy scarcity. Smagorinsky, Rotty, Revelle, and even Broecker grudgingly agreed that the additional carbon emissions from the proposed level of synfuels over a few decades would not make a significant difference in the total accumulation in the atmosphere. Such investments would have justified their capital expense if phased out after that time. But the scientists worried that momentum for synfuels might last beyond several decades. Broecker feared that synfuels might become a "wedge" that would extend the life of coal production for as much as a hundred years.

Another worry was that investments in synfuels might distract from energy alternatives that produced no carbon emissions. Revelle testified, "I feel we need a sizable commitment to a major synthetic fuels program at the present time, but also we should devote considerable effort to developing alternative sources of energy which will not enhance the carbon dioxide effect on the atmosphere—I am thinking about the various forms of solar energy . . . I would argue also for nuclear."

Rotty laid out a long-term climate strategy with more precision than other witnesses. He calculated that if nonfossil energy began to grow at an annual rate of 9 percent, then by 2040, "you don't have to burn any fossil fuels anymore." If this happened, he said, atmospheric carbon would level off at 500 parts per million. As a result, the world would fall well short of doubling the preindustrial level, sharply reducing the chances that the average temperature would increase by 2 degrees Celsius.[6]

Not on the committee, Ed Muskie was invited to participate due to his status as, in Ribicoff's estimation, "the number 1 environmentalist in the Congress." Already known to senators like Muskie, Revelle got plenty of time to wrestle with some of the thornier challenges. His comments often reflected the deliberations at the recent AAAS conference in Annapolis. Because economists were increasingly joining physical scientists in these climate discussions, Revelle addressed issues largely skirted in previous work by the National Academy.

One topic raised by economists was the extent to which humans could adapt to climate change. Revelle expressed some optimism about this approach. He acknowledged that the land suitable for crops would shift as the planet warmed. However, he said that a panel of agricultural experts at Annapolis reported that their research programs at state experimental stations had already demonstrated their ability to adapt the genetics and agronomy of crops to temperature changes by about one-tenth of a degree per decade. Thus, U.S. farmers might be able to cope with slow global warming. Another leading concern about global warming, the fate of coastal cities threatened by rising sea levels, was also amenable to adaptation. Other than historical monuments, most of their infrastructure tended to be rebuilt in fifty-year cycles anyway. As with agriculture, the pace of global warming would affect the burden of adaptation.[7]

Revelle did not, however, regard adaptation as a complete answer to global warming. Rich nations had immense advantages in protecting themselves. He noted that the situation was quite different in less economically developed regions, where 80 percent of the world's people lived. He explained that poor countries lacked the social organizations, capital, technology, and educational resources to support the adaptations that would be necessary. The well-traveled scientist observed, "If the climate over large areas—if the Sahel region of Africa, for example, becomes dryer—probably there is no adaptation that those societies can make except to migrate. The people will have to move or die or depend on the rest of the world for their food supply."[8] The elder statesman of climate science was a rare voice warning that looking at the impacts of climate change on the United States alone ignored the damages that might result for the vast majority of the world's population.

The professor also questioned whether humans would have the foresight to plan for the needed reductions in emissions or adaptation. He observed:

Economists and politicians and all of us have a very high discount rate, as the economists say. We tend to best think about the immediate future, and as we go further and further into the future, things become less interesting, and we begin to take them much less into account. One of the reasons, of course, is that it is so hard to predict the future. But the CO_2 problem is very much like the inexorable advance of a glacier. From year to year, you cannot notice much difference, but in 50 years, you may be overwhelmed by the glacier.[9]

The difficulties of explaining the complexities of climate science became evident when some senators spoke of the need to protect coal jobs and did not understand that the scrubbers that reduced sulfur dioxide emissions did not capture carbon dioxide.

The occupant of the Oval Office did not ignore the concerns about synthetic fuels. Three days after the Ribicoff symposium, Carter sent an environmental message to Congress, acknowledging the risk of synfuels to the environment. He pledged to examine "the longer-term implications of increasing carbon dioxide concentrations in the atmosphere." Carter reiterated his strong support for solar energy and touted the hefty increases in his budget. Based on advice from the CEQ, he also noted that the global loss of forests and woodlands, nearly all occurring in or near the tropics, constituted a severe threat to the natural environment, including "the possibility that forest loss may adversely alter the global climate." He directed relevant federal agencies, including the State Department, "to place greater emphasis on world forest issues in their budget and program planning." The briefings by MacDonald, Pomerance, and others had elevated, perhaps faster than anticipated, awareness of the climate challenge at the highest levels of government.

October 1, 1979, constituted a critical marker in the saga of nuclear power—the release of a Carter commission report investigating the Three Mile Island nuclear accident, the industry's most severe in U.S. history. Dartmouth College president John Kemeny—a researcher on the Manhattan Project who later helped develop the BASIC programming language for computers—chaired the twelve-member body. The study would recommend future parameters for nuclear power.[10]

On a positive note, the commission found that the thick concrete walls of the nuclear plant's containment vessel had performed well. As a result, the radiation released into the atmosphere was relatively minor and had "no detectable adverse effects on public health in the vicinity."[11] However, the report chastised the Nuclear Regulatory Commission and the nuclear industry for the malfunctions in the cooling system that led to a core meltdown and the potential for more extensive damage.

The Kemeny report called for improved training, more transparent operational controls, and better emergency management for contingencies but didn't recommend a moratorium on nuclear power. There were clamors in Congress for such action. But Carter and sometime nuclear critic Congressman Morris (Mo) Udall—both with strong environmental credentials—helped block any ban on constructing new plants.

Popular support for nuclear power softened after the Three Mile Island accident, and no additional plants were ordered during the twentieth century (though many ordered earlier were completed). However, the accident was far from the industry's only problem. The faulty operation of existing plants, the previous cancellation of new orders, and the growing realization that nuclear power would prove more expensive than coal created substantial barriers to continued growth. Both nuclear and coal generators had to meet stringent environmental regulations; purchasers would base their decisions on economics. Senator Gary Hart, a Democrat from Colorado, remarked, "The jury is still out . . . The future of the [nuclear] industry is going to be determined as much on Wall Street as in Washington."[12]

Carter continued to articulate the need for light-water nuclear technology. "The steps I am taking today will help to assure that nuclear power plants are operated safely," he declared in his response to the report. "We cannot shut the door on nuclear power for the United States . . . Every domestic energy source, including nuclear power, is critical if we are to be free as a country from our present overdependence on unstable and uncertain sources of high-priced foreign oil."[13]

Carter's support for light-water reactors didn't extend to a giant leap in nuclear technology—the breeder reactor. Breeders were attractive to nuclear engineers who adhered to the "bigger is better" view of nuclear plants. But Carter balked at the massive spending required for an unproven capability that did not appear to be needed in the foreseeable future. However, he was willing to divert the money to research alternative nuclear strategies that might produce energy more safely and economically down the road. Negotiations with Congress to explore a longer list of nuclear options failed to get traction, though his appointees to the governing board heeded his wishes and refused to advance the Oak Ridge project.[14]

In November 1979, the National Academy discussed privately some reservations about the Charney report. The tight schedule imposed by the White House had deviated from the usual pattern of major studies taking several years. On November 12, Chairman Suomi convened a meeting of the Climate Research Board, which had overseen the Woods Hole project. Due to "hurried and irregular procedures" and concerns about too much focus on models, the board agreed to express no official view on the Charney findings. However, it did not challenge them and wrote him a letter of appreciation.[15]

Suomi asked Professor William Nierenberg to form a small group of board members and outside experts to explore options for following up on the Charney report and recommend a plan of action. Nierenberg was a logical choice to take on the assignment. The nuclear physicist with a doctorate from Columbia University had become an influential figure at the Academy. Like Revelle, he had governmental and academic experience, serving as Assistant Secretary General of the North Atlantic Treaty Organization (NATO) in the early 1960s. In 1965, he assumed Revelle's old position as director of the Scripps Institution of Oceanography. Nierenberg also served as a presidential science advisor under both Ford and Carter. More recently, he had helped prepare the JASON report on carbon dioxide.[16]

On a different track, Frank Press, in January 1980, formally asked the Academy to conduct "as promptly as possible" an assessment of "the likely foreseeable social and economic consequences of an increasing concentration of atmospheric carbon dioxide." Responding to the request, the Academy quickly set up an ad hoc study panel for another rushed study. The Academy chose Thomas Schelling, a Harvard economist considered one of the nation's top game theory and weapons policy experts, to chair the group. Schelling confessed he knew little about climate change at the time. But several people on his panel—Revelle, Nordhaus, Nierenberg, Smagorinsky, White, and Woodwell—were veterans of climate studies.[17] The panel's formation marked a turning point for the National Academy; economists like Schelling and Nordhaus would now play a more prominent role in how scientists framed the climate issue.

The Academy sent the "report" of the Schelling group to the White House on April 21. The document was an outlier in the annals of the National Academy. Delivered as an eleven-page letter, it contained no citations or documentation beyond a few references on its scientific assumptions. Given the diversity of perspectives among its coauthors, Schelling's epistle lent itself to various interpretations.

The group accepted the scientific conclusions of the two recent Academy studies on climate chaired by Revelle and Charney. At the same time, it increased the emphasis on the problems of uncertainty. The Schelling group acknowledged that increasing energy consumption using fossil fuels would have "increasingly undesirable climatic effects." But it also held out the prospect that the United States, the Soviet Union, and China might be "able to adapt to, or even benefit from, climate change."

A significant contribution of the Schelling report was its treatment of the international implications of climate change. The worldwide impacts of such change would be spread very unevenly due to national variations in hydrology and ability to adapt. Migration, it noted, had historically been the principal means of adapting to climate change. However, in the modern world, political barriers hampered long-distance relocations, and national boundaries were "not likely to be more open in the future." It suggested, "If migration out of climatically impoverished areas is not feasible, transfer payments and other technology and capital to aid local adaptation to climatic change may have to be considered."

Schelling recommended that near-term U.S. policy ensure that adaptation and prevention options remained open, place the carbon dioxide issue on the international agenda, and emphasize research needs. The government should undertake such efforts "with as low a political profile as possible" to avoid arousing public concerns about climate change.

The White House wanted quick access to the Schelling findings because Germany might raise the climate issue again at the next G7 meeting. On May 5, Press forwarded the Schelling letter to Carter. His cover memo emphasized that the Academy believed a better understanding of climate impacts would emerge in five to ten years. In the interim, he agreed with Schelling on the need for research to expand what was known, "with as low a political profile as possible until we reduce the uncertainty."

The Schelling report was another milestone in climate change history. An increasingly influential economist and the president's chief science advisor had agreed that the public should be protected from knowing too much about climate change.

In late June, Congress overwhelmingly approved the Energy Security Act of 1980. Despite congressional concerns about the impact on climate, the legislation authorized the construction of large synfuels projects using coal. But it also directed the White House science office to enter into an agreement with the National Academy for a comprehensive study of the projected impacts of fossil fuel combustion, synthetic fuels, and other sources on atmospheric carbon dioxide. The assignment reflected the views of Carter's science office, senators like Ribicoff, the Energy Department, and the Academy itself that a more comprehensive, less time-constrained study was needed.

Congress was specific about the kind of report it wanted to assess the "economic, physical, climatic, and social effects" of rising CO_2 levels. It called for international involvement to provide a "worldwide" perspective and protection from governmental influence on its content. The report should not, it said, be subject to White House preconditions or clearances.[18]

Nierenberg convened the Carbon Dioxide Assessment Committee's first meeting on October 3. Thirty-two days later, the U.S. electorate would decide whether Jimmy Carter or former California governor Ronald Reagan would receive its report. Revelle and Nordhaus were unsurprising picks for the committee. Other members included many involved in previous studies—Machta, Schelling, Smagorinsky, and Woodwell. Peter Brewer from the Woods Hole Oceanographic Institution and Paul Waggoner from the Connecticut Agricultural Experiment Station rounded out the group.[19]

What turned out to be Carter's last year as president flashed signs that climate change remained on the national agenda, though still on the periphery. In July, *The Global 2000 Report to the President*—a highly touted study on creating a sustainable future, led by the State Department and the Council on Environmental Quality—warned of atmospheric concentrations of carbon dioxide and ozone-depleting chemicals that were "expected to increase at rates that could alter the world's climate and upper atmosphere significantly by the year 2050."[20]

In August, the national Democratic Party platform added to the climate conversation. It declared, "To defend against environmental risks that cross national frontiers, international cooperation must be extended to new areas." This concern included the "buildup of carbon dioxide in the atmosphere." On matters that might also affect climate change, the document endorsed key elements of Carter's energy policy. These included increased coal use, enforcement of the fifty-five-mile-per-hour speed limit, efficiency standards for buildings, and a national goal to generate 20 percent of the nation's energy from solar and other renewable sources by the year 2000. It encouraged, as well, research and development for hydrogen or electric-powered vehicles. Overcoming White House opposition, nuclear opponents inserted language calling for "a national plan to coordinate an orderly phase-out of nuclear

THE PRESIDENTS AND THE PLANET

reactors [and] a moratorium on all licensing of new nuclear plants." As part of an agreement to remove the requirement for a "phase-out," the administration accepted a compromise that called for the moratorium, making the document still inconsistent with Carter's recommendation to Congress. The result was a platform that favored reducing carbon emissions, increasing energy production from coal and renewable resources, and blocking new nuclear construction.[21]

Ronald Reagan's victory in November made Carter a lame-duck president. Nonetheless, Carter continued to talk about the risks of rising carbon emissions. At a December 2, 1980, signing of legislation to protect Alaska's natural resources and in his January 1981 State of the Union message, he used identical language to defend his legacy. He asserted that his administration had "faced squarely such worldwide problems as deforestation, acid rain, toxic waste disposal, carbon dioxide buildup, and nuclear [weapons] proliferation." By the end of his term, Carter had mentioned the accumulation of CO_2 in the atmosphere more often than all previous presidents combined.

In mid-January 1981, the Council on Environmental Quality released *Global Energy Futures and the Carbon Dioxide Problem,* a report about what it called perhaps "the ultimate environmental dilemma."[22] The White House document laid out the scientific consensus on the subject in a style that made it accessible to nonscientists. With the fate of American hostages in Iran still unsettled and a new administration about to take power, the report got lost amid a string of heavy news days.

In his preface to the document, Gus Speth called on scholars to provide public leadership on climate. However, he challenged the growing influence of economists on policy. Leaders should not presume, Speth said, that "we can accurately assess the long-term costs and benefits of unprecedented changes in global climate" based solely on economic calculations. It was also necessary to "appreciate our ethical commitment to this planet and its inhabitants."

The CEQ foresaw that it might take twenty years for a clear "signal" of humans' effect on climate to emerge. At that point, the impacts would be evident enough to detect even in the presence of a large amount of "noise" due to natural variation. Some drastic actions could wait until more uncertainties were resolved. But postponing all action "would entail a risk of being unable to prevent long-term climate changes that prove serious and irreversible for centuries." The report recommended immediate efforts, on an urgent basis, to increase the global use of renewable energy and dramatically reduce the world's demand for energy by the end of the century, with U.S. leadership on both fronts.

The reports and hearings on climate change during the last two years of the Carter presidency differed somewhat in tone and sense of urgency. Still, they accepted what various models were saying about the fundamental science. There was also general agreement that the effects of emissions in the 1970s and 1980s might not be visible to the human eye until the twenty-first century. Given this time lag between cause and effect, climate change remained low on the list of national priorities. A scientist later recalled briefing a government official about a new study on climate change in 1979. The official—told that problems might become evident in fifty years—replied, "Get back to me in forty-nine."[23]

8

Changing Directions in the Reagan Era

The Reagan Presidency, January 1981 to September 1983

As the Republican governor of California, Ronald Reagan implemented pioneering pollution controls, positioning his state as a global leader in environmental protection. On the presidential campaign trail, however, the former actor cast himself as an ardent critic of federal spending and regulation. His campaign rhetoric often put him at odds with national environmental and energy initiatives adopted with bipartisan support during the 1970s.

After his inauguration in January 1981, Reagan showed that his pledged pruning of the federal government would not rely on gradualism. He quickly proposed, in the words of biographer Lou Cannon, "mild to fatal reductions in eighty-three federal programs."[1] Climate studies constituted a minuscule slice of overall spending. Still, based on early indications, programs to better understand climate change risked falling into the "fatal" category.

Dave Keeling's funding to monitor atmospheric carbon dioxide—the bedrock of empirical climate science—had been tenuous even before the change in presidents. As his measurements became more routinized, the agencies funding Keeling had suggested taking the Mauna Loa project away from a university research institution like Scripps. Arguing that he was still making essential enhancements, Keeling had obtained a series of extensions. But in the early weeks of the new administration, he received what appeared to be a firm no.[2]

Fortunately for Keeling, he had a powerful protector. His boss at Scripps, Bill Nierenberg, chaired the in-progress National Academy study on climate and had close ties to the Reagan team. In California, Nierenberg had been a vocal defender of nuclear power against the attacks of environmental organizations. More recently, he had served on Reagan's science advisory panel during the presidential transition. This influential group recommended priorities for science policy and evaluated the credentials of people under consideration for scientific positions in the new administration. Two former chief presidential science advisors also served on the panel—Nixon's Edward David, still president of Exxon research and engineering, and Ford's Guyford Stever, now director of the National Science Foundation. According to *New York Times* anonymous sources, Nierenberg was on the short list to become Reagan's top scientist.[3]

The day Keeling heard what he considered a final rejection, the key decision-makers at the Department of Energy reconvened after lunch, this time with Nierenberg presiding. To Keeling's amazement, the director of the carbon dioxide and climate division, David Slade, said quietly, without prodding, "I think DOE can pick up the tab for Keeling's program." Keeling later learned that Nierenberg had met with Edward Frieman, director of the department's powerful science office, who had worked with Nierenberg on the 1979 JASON climate report and later became his successor as the director at Scripps. It was Frieman who advised Slade that he considered the DOE funding justified. The sudden reversal preserved Keeling's ability to track atmospheric carbon with his characteristic precision.[4]

Other cost-cutting plans targeted the White House itself. The Council on Environmental Quality—the most active office on the climate issue after its creation in 1970—became another candidate for extinction. Malcolm Forbes Baldwin, a senior career acting director and self-professed "lifelong Republican," peppered Reagan advisors with pleas that the CEQ did not duplicate work at the Environmental Protection Agency and should continue its work. In a February memo, he noted recent reports that the administration was considering terminating CEQ and warned that "it would be a serious mistake to eliminate the Council or weaken it substantially."[5]

The President's Science Advisory Committee—active on the climate issue since the Kennedy presidency—also was on the chopping block. Since the committee oversaw funding for Nierenberg's National Academy study, changing priorities threatened the report's completion. However, like the CEQ, the science advisors and the Nierenberg study had statutory authorization and bipartisan support in Congress.

In mid-March, the White House staged a partial retreat, announcing it would keep its science advisors and the CEQ, though with draconian cuts to staffing and budgets.[6] The truncation of the two organizations hampered the recruitment of new members. The president did not announce until May his selection of forty-one-year-old George (Jay) Keyworth as chief science advisor. The head of the five-hundred-person physics division at the Los Alamos National Laboratory had earned his reputation as a leading expert on nuclear weapons.[7]

The following month, the administration told the National Academy that the Nierenberg study could proceed. But the White House ignored congressional mandates to give the Academy independence, reduced the budget for the project, and nixed several planned activities. It banned, for instance, the Academy's use of the scenario analysis that Revelle had suggested, based on concerns about how the press might cover any projections that appeared overly pessimistic.[8]

The administration also made a perfunctory effort to abolish the Department of Energy, which funded climate studies and—at much greater expense—research and development for alternative energy technologies. Facing broad congressional support for the still new department, Reagan had to settle for a massive restructuring of resources away from its civilian functions and toward nuclear weapons. Employees reeled from reductions that pushed those with seniority into lower ranks and others with less tenure out of the federal workforce.

In June, DOE officials traveled to the Solar Energy Research Institute. In a keynote address the previous year at the American Association for the Advancement of Science, SERI director Denis Hayes had touted recent advances in renewable energy and their potential for slowing climate change. Bolstered by an influx of dollars after the passage of Carter's windfall profits tax, the institute was making impressive strides in reducing the costs and increasing the efficiency of photovoltaic panels.

The visitors stunned their hosts by announcing the axing of about a third of the federal staff and all contract employees. The thousand people getting their walking papers included, Hayes later remembered, some who went on to win Nobel Prizes. The budget of $130 million—which he called "real money in those days"—was chopped to $30 million. Hayes called the event "the most horrible day of my life," harder than the day his parents died. He quickly resigned and spent much of the following year writing letters of recommendation for people he had lured to Golden and who now needed new jobs.[9] The massive cuts at SERI dealt a severe blow to U.S. global leadership on solar and other renewable technologies.

The White House also wanted to chop the federal climate studies recommended by the AAAS. To draw public attention to these cuts, three Democratic members of the House Committee on Science and Technology organized a July 31 hearing on "Carbon Dioxide and Climate: The Greenhouse Effect." Before the event—held in the smallish room 2325 of the Rayburn House Office Building—Representatives George E. Brown Jr. of California, James Scheuer of New York, and Albert Gore Jr. of Tennessee had devoted considerable attention to the climate issue. Scheuer welcomed a "very distinguished set of witnesses" to discuss "one of the truly massive environmental threats that we have detected so far."[10]

Two Republican members gave their blessing to the deliberations. Robert Walker of Pennsylvania declared that an examination of climate change was needed so that "our energy future is balanced by what we regard as acceptable and unacceptable risks." William Carney of New York warned that although CO_2 levels in the atmosphere might not reach critical levels until the twenty-first century, "the decisions that we take now regarding how much nuclear, how much fossil, and how much solar energy we will employ in future years . . . will determine the atmospheric levels of carbon dioxide in those years."

In his opening statement, Gore described his "disbelief" when he first learned about climate change. But the evidence continued to mount, he said, that increases in atmospheric carbon dioxide could lead to "a natural disaster of unprecedented proportions." Changes in the atmosphere had "the potential to radically alter our climate, upset our agriculture, and change the economic base of our society in fundamental ways." The young congressman expressed optimism that solutions could be found. "We are not helpless," he declared. "There is still time to understand the implications of this change in our atmosphere and still time to make the kinds of energy choices that could forestall" the threat.

Displaying a chart prepared by his staff, Gore conducted a tutorial on the Keeling Curve. The line on the accumulation of carbon concentrations had continued to rise. The monthly average of CO_2, which stood at 316 parts per million in 1958 (the first year of data from Mauna Loa), had reached 331 by 1974. The connection between Keeling's "very, very thorough" evidence and the burning of fossil fuels seemed to Gore "quite obvious."

Gore listed key questions needing further investigation. One was "What beneficial as well as what harmful effects might we see?" He cautioned that a mix of good and bad effects might be of "little comfort" if, for instance, climate change forced agricultural production to move from one region to another.

As the members of Congress were about to turn the microphone over to the witnesses, Gore intervened on "a point of personal privilege." Addressing the leadoff panelist, he declared, "I took your course at Harvard on population studies, and it was one of the best I have ever taken. You have been a pioneer in more than one field, and it is a great pleasure to see you here on this occasion." Representative Brown, who had chaired previous hearings on population and climate, heaped additional praise on the first witness for an earlier testimony that "took us to the mountaintop and gave us a world overview that was enormously thoughtful and stimulating." With expectations thus heightened, the floor now belonged to Roger Revelle.

Ignoring the kudos, Revelle explained carbon dioxide's vital role for plant and animal life and its atmospheric function in moderating the earth's temperature. More precisely, he said, "If there were no carbon dioxide in the air, it is very likely that all water would be frozen" because the earth's average temperature "would be well below the freezing point."

The professor continued, "There probably have been fluctuations in the carbon dioxide in the atmosphere over geologic time which have caused variations in climate." For example, one major factor contributing to temperatures during the Ice Age "was very likely that the carbon dioxide content of the air went down to about two-thirds of the value that it has now." Much earlier, "40 million years or so ago, the much higher temperatures at the high latitudes may also have been due to higher carbon dioxide in the air."

Revelle cited the work of early scientists who believed that burning coal might elevate the earth's temperature. In the late 1950s, to better understand any changes in the atmosphere's chemistry, the International Geophysical Year, he noted, supported a project to measure the atmospheric carbon dioxide. It was this effort that had allowed Keeling to set up monitoring stations on Mauna Loa and at the South Pole. The self-effacing Revelle neglected to mention his leadership at the IGY and Scripps that helped establish Keeling's project.

Revelle called Keeling "a remarkable person" who "except for his interest in music, has thought about nothing but carbon dioxide for the past 25 years." (Keeling played the piano to entertain friends and family.) Revelle lamented, with a dose of humor, that Scripps never had enough money to support Keeling's monitoring stations. It "always took twice as much money as he budgeted for. Nonetheless, he managed by sheer persistence and guts and making a nuisance of himself to make these extremely accurate measurements."

Revelle turned to the thornier question of when scientists could assert that they had "an early signal" from empirical observations that the climate was changing, as the models suggested, due to the emissions from fossil fuels. He cautiously observed that this point had not yet come. "The way scientists would put it is that the noise level is too high." The "noise" was the variation in world average air temperatures from year to year due to natural factors. This variability in the historical data, he said, "masks" whatever the carbon dioxide effect might be.

The long-range threat was that it could take a thousand years for atmospheric carbon to end up in the ocean. In addition to the persistence problem, carbon emissions were likely to rise faster as countries with smaller economies increased their energy use. Making synthetic fuels from coal would also increase the rate of emissions. As an alternative to synfuels, Revelle suggested, "The best fuel to use from the standpoint of the amount of energy you get per ton of carbon is natural gas." He said that experts believed there was "plenty of natural gas" and recommended utilizing more of it.

Scheuer interjected to cite the need to develop tidal and wind power. Gore jumped in: "Nuclear." Revelle, in response: "Yes. Nuclear is number one, perhaps, but biomass [from carefully managed forestry] is also a major possibility."

Future research, according to Revelle, should focus on the risks and benefits of a changing climate. The most worrisome danger was the disappearance of the Antarctic ice cap. This massive collapse might occur with a rise of 3 or 4 degrees Celsius in the average air temperature. In the part of the globe that would see the most significant temperature increases, Arctic ice might disappear much earlier and lead to a substantial sea level rise. He said that changes in the northernmost latitudes would have "a profound effect on climate" there and elsewhere. An additional risk was the release of methane (another greenhouse gas) from melting permafrost in northern latitudes.

On the plus side, carbon dioxide was a fertilizer, which might lead to higher-yielding crops. However, to take advantage of this phenomenon, Revelle said farmers would need to genetically modify seeds of staples such as wheat, corn, rice, and potatoes. Such enhancements in agricultural practices would help them deal with heat and water stress. He testified that the "effects of a changing climate might be much more drastic for the less developed countries than for the United States." He added that "their societies are so much less resilient, so much less able to cope with stresses of any kind."

Revelle's goal at the 1979 AAAS conference had been to expand the understanding of how a changing climate would affect various aspects of human life. But the results had not matched his aspirations. So Revelle had launched his own study of the potential impacts of climate change on water availability. He shared with the committee a preview of what he had found.

Revelle based his analysis on the work of the German climatologist Hermann Flohn. If atmospheric carbon reached 560 to 680 parts per million (roughly double or more preindustrial levels), Flohn estimated, the widespread impacts on climate would be extensive. Of particular importance, the changes in average surface temperature and precipitation would vary significantly by factors such as altitude and latitude. The German had argued that looking at climate as a global average was "neither informative nor complete." "We should think of climate," he asserted, "as the sum of weather variations." He argued that averaging divergent temperatures around the globe gave the appearance of less impact from climate change than was actually the case.[11]

Adopting Flohn's estimates of variability, Revelle focused on latitude 40 degrees north. If Flohn was correct, this latitude might experience greater temperature increases (plus 6 degrees Celsius, or 11 degrees Fahrenheit) and decreases in precipitation (minus 18 percent). This latitude, unlike those near the poles, was densely populated. Hotter temperatures would increase the rate of water evaporation and add to the environmental stress of lower rainfall. Revelle advised, "We westerners are most concerned about the Colorado River." Less precipitation and more evaporation "might cause as much as a 50-percent reduction" in its flow.

Similar effects might be felt at the 40th parallel around the globe. Revelle identified major world cities in this belt—including Lisbon, Naples, Athens, and Ankara—that might also see reduced river flows. He added that in China, the Huang He (Yellow) river would be subject to the same kind of change, as would the basins of the Amur Darya and the Syr Darya—sites of extensive irrigated farming in the Soviet Union.

In his written testimony, Revelle warned of possible population redistribution across international borders. He posed two related questions with significant political ramifications: "If climatic conditions for agriculture deteriorate in Mexico and Central American countries, will population pressure for migration to the United States be intensified? Or, if agro-climatic conditions should improve, will economic development accelerate enough to slow migration?"

Next in line was Joseph Smagorinsky, director of NOAA's Geophysical Fluid Dynamics Laboratory in Princeton, New Jersey. For decades, he had served as the federal government's chief weather modeler and was part of Revelle's team that drafted the climate change section in LBJ's 1965 environmental study. At the committee's request, "Smag" reviewed the Academy's previous findings on climate change, going back to 1979, and the Academy's ongoing work that would fulfill the mandate of the Energy Security Act of 1980.

Smagorinsky testified that the scientists working on these studies accepted the estimates of the World Meteorological Organization that atmospheric carbon dioxide in 2025 would reach 410 to 490 parts per million, with a most likely value of 450. Because these numbers depended on the levels of fossil fuel combustion, something the natural science models could not predict, climate scientists focused on what would happen when atmospheric CO_2 levels doubled from preindustrial times—that is, when they reached about 560 parts per million. He told the panel that the Charney report had estimated that a doubling would most likely lead to an average temperature increase of 3 degrees Celsius.

Climate modeling continued to confront challenges. For instance, Smagorinsky said, "The possible interactive role that clouds play in a shifting climate regime remains one of the major unsolved mysteries." Besides, modelers could not ignore the impacts of other gases, such as methane, fluorocarbons (Freon), and aerosols. However, model testing and knowledge from carbon dating about what happened during past natural changes in climate helped build confidence in model results. He testified, "Although questions have been raised about the magnitude of climate effects, none can deny that changes in atmospheric carbon dioxide concentrations have the potential to influence the heat balance of the earth and the atmosphere."

The third witness was Stephen Schneider, now with the climate project at the National Center for Atmospheric Research in Boulder, Colorado. The scientist (who no longer argued the earth was cooling) stressed that climate change was gradual and not very perceptible when viewed as just how one year differed from the previous one. This "creeping or incremental kind of issue," he said, was "one which our political systems have had a great deal of difficulty dealing with." But "if you step back and take a long look, you can see a big change." More than previous witnesses, Schneider addressed the choices facing policymakers. He cautioned against the idea that "there isn't very much we can do about" CO_2 buildup. This approach, he said, was not widely advocated "except by some special interests." At the least, he recommended increased study of the problem. He contended, "It is pretty hard to argue that more information would make rational policymaking any worse."

Not satisfied with just more studies, Schneider also wanted to "build resilience" and increase future options. As Revelle had suggested, a prime example was developing crop strains that could adapt to various environments. Such adaptability was a good idea in any case, but climate change provided an additional incentive to do so. Resilience also involved the "vigorous development of alternative energy systems stressing renewable energy and efficient technologies." Developing more fuel choices was a good idea for multiple reasons, but "carbon dioxide lends more impact to the need for doing that." The climatologist was skeptical that relying solely on private markets would lead to such strategic thinking. Developing alternatives would require "some degree of nonmarket intervention." He suggested that funding upfront costs for advanced technologies was like paying insurance premiums.

Lester Lave wrapped up the first panel. Like previous witnesses, the economist had played a significant role in the Annapolis conference. Like many in his discipline, he believed the world would have to adapt to a changing climate for a long time, given the attractiveness of fossil fuels. He added, however, "I think CO_2 is a serious problem. Some of the technologies we are pursuing with respect to synthetic fuels don't make any sense if this is the case."[12]

Gore then introduced several Reagan officials to discuss the future of climate studies. Though committee members expressed concerns about the plan to slash funding, the exchanges were only minimally contentious, with some verbal sparring but no fireworks. Administration witnesses, still new in their jobs, asserted that decisions about climate priorities remained under consideration.

The closest thing to a clash involved a question of semantics. Douglas Pewitt—who earned his doctorate in particle physics at Florida State University in 1974 and became acting director of energy research at the DOE in 1979—used some variation of the word "alarmist" five times during his oral testimony to describe unspecified statements about climate change. Comparing warnings about climate to those about the dangers of acid rain, Pewitt declared, "I absolutely refuse as an official in a responsible position to engage in the type of alarmism for the American public that I have seen in these areas time and time again, and I do not think that I can responsibly encourage that sort of alarmism."

Gore and Scheuer pushed back on the implication that they might be alarmists. Gore challenged Pewitt to identify any examples he had heard during the hearing. Gore stated his personal belief that it was too early to halt the synfuels program or dramatically move away from fossil fuels; however, the twenty-three years of data from Mauna Loa did portray a consistent pattern. Undeterred, Prewitt challenged the adequacy of Keeling's findings for drawing conclusions. He countered that those twenty-three years were insufficient to establish a pattern of increasing atmospheric CO_2. Furthermore, the available data did not show a consistent increase in global temperatures.

Pewitt asserted that decisions should be made only in the areas where scientists had achieved "certainty."

John Marcum from the White House Office of Science and Technology Policy was less combative. He testified that newly confirmed director Keyworth wanted to assure the committee that "he considers the atmospheric buildup of CO_2 an important issue, and it is one we think should be taken into account in formulating an overall approach on energy policy." The administration, he said, would be guided by the forthcoming Nierenberg report. The science office had corrected "some serious flaws" in the original proposal and warned the National Academy about the tendency for "low probability scenarios to take on a life of their own." Marcum expressed confidence in the project, which had put together, in his words, a "first-rate" panel that included Revelle, Smagorinsky, and Nierenberg.

Later in the year, as the administration's detailed plans became public, differences surfacing during the July hearing became increasingly divisive. On September 24, President Reagan announced his modified budget proposal, which included raising the science spending at the Department of Defense by almost $4 billion and chopping science at Energy by more than $1 billion. In December, Pewitt—recently elevated to assistant director at the White House science office—explained that the administration had reevaluated the role of the DOE labs because "we will no longer support things within the capability of private industry." Asked for examples of lab projects better handled by private companies, he cited improving the efficiency of solar photovoltaic cells and lightbulbs.[13] Schneider's argument for an active climate policy to "build resilience" that included "vigorous development of alternative energy systems stressing renewable energy and efficient technologies" had failed to sway the new administration.

Pewitt's skepticism about government efforts to improve the efficiency of solar panels and lighting was consistent with directives coming from Reagan's top advisors. Budget chief David Stockman—a former divinity student with an almost religious faith in the wisdom of markets unhindered by government concerns about the environment or national security—specifically mentioned wind power, photovoltaic cells, and synthetic fuels in his attacks on big government. They were part of the "great, malodorous garbage dump"

Carter had created at the Department of Energy, which the policy wonk saw as "experiments in high-cost, unproven energy technologies." He contended there was nothing wrong with such experimentation but it was "precisely the kind of thing Adam Smith had invented the free market to accomplish."[14]

Stockman's ambitions later suffered some setbacks from congressional opposition and the need for the administration to compromise at times. The tensions led to his eventual forced resignation. Nonetheless, his broadside confrontation with conventional wisdom in Washington achieved numerous successes in shrinking government and proved the adage that "elections have consequences."

As Reagan's 1982 budget cuts took effect, Keeling's continuous measurements of atmospheric CO_2 yet again faced a cutoff in federal support. And, once more, Nierenberg intervened to save the day. This time, Nierenberg found nongovernmental funding from the Electric Power Research Institute. The quasi-academic arm of the electric power industry provided support that allowed Keeling to continue his data collection and synchronize them with the results of other scientists studying the El Niño effect and carbon sinks.[15]

Nierenberg continued to champion Keeling's data collection. In a presentation the following year, he compared the projection of the Keeling Curve on his screen to "showing a picture of Moses and the Ten Commandments to a Bible class." He called the graph of the Mauna Loa data "unquestionably the single most important contribution to the clarification of the entire problem. It is also the only contribution that is indisputable."[16]

After the Gore/Scheuer hearing, all sides expected the National Academy's study to bring greater clarity to the understanding of climate sometime in 1983. But in the interim, additional statements by leading experts and another congressional hearing continued to probe the complex issues involved.

An August 1981 article by James Hansen and six colleagues at the Goddard Institute for Space Studies (part of NASA) addressed lingering questions about whether global temperatures were in fact rising in response to increasing levels of atmospheric CO_2 and whether they would do so in the future. The authors anticipated that by the end of the century anthropogenic carbon dioxide warming would emerge from "the noise level of natural climate variability" that Revelle had talked about. They also foresaw a high probability of warming during the 1980s. Hansen contended that the drop in temperature from 1940 to 1970 could be explained by dust from volcanic eruptions.[17]

Hansen, who worked out of an office near the Columbia University campus, seven floors above 112th Street and Broadway, soon got a chance to testify at a second hearing on "Carbon Dioxide and Climate: The Greenhouse Effect" in March 1982, also chaired by Scheuer and Gore. Unlike many climate scientists, Hansen relished opportunities to communicate with the public and lawmakers about what his models were showing.

Committee members expressed bipartisan support for tackling the threat of climate change, though with a few different slants. Scheuer warned, "If today's worst-case scenario becomes tomorrow's reality, it will be too late to reverse the atmospheric carbon dioxide buildup or to ameliorate the adverse human and environmental impact." Gore similarly declared, "There is now a broad consensus in the scientific community that the greenhouse effect is a reality." Both Democrats hammered the Reagan administration for drastic budget cuts to climate studies.[18]

Another Democrat, Bob Shamansky of Ohio, suggested that for lay audiences, the committee might want to use a term other than "greenhouse effect." "I always enjoyed going inside greenhouses, and everything seems to flourish there," he said. The "microwave oven" effect might describe the threat of warming "a little more accurately."

Republican Walker argued that it was time to talk more about solutions and less about more studies. "We have been told and told and told that there is a problem with increasing carbon dioxide in the atmosphere. We all accept that fact, and we realize that the potential consequences are certainly major in their impacts on mankind . . . How frequently must we confirm the evidence before we commence taking remedial steps?" Walker's preferred remedy was nuclear power, which could even produce hydrogen that might displace oil for fueling vehicles.

Hansen's testimony reported on recent temperature data from hundreds of stations worldwide. A black line in one of his Vu-Graph slides, illuminated for the screen with an overhead projector, showed a temperature rise of 0.4 degrees Celsius (0.7 degrees Fahrenheit) over the previous one hundred years—the data reported in his recent article. A red line showed a similar but somewhat higher increase expected by his climate model. He said that changes in the energy coming from the sun or so-called trace gases might explain some of the differences. Aerosols from volcanoes cooled the earth, while methane emissions (from energy and agriculture) and chlorofluorocarbons (such as those used in air-conditioning) had a warming effect. These other gases could not yet be measured with the same precision as carbon dioxide. Nonetheless, he concluded, they were, as a whole, "likely to substantially increase the magnitude and accelerate the rate of future climate change."

Hansen pointed to recent evidence that global sea levels were rising, as predicted by various models. His research suggested that levels over the previous hundred years—"a potentially sensitive indicator of climate change"—had risen ten centimeters (about four inches) due to climate change. Based on what had already occurred and model projections, he foresaw that the thermal expansion of seawater and some melting of ice sheets could lead to a sea level rise of thirty to forty centimeters (one foot or more) by 2050.[19]

In his written testimony, Hansen acknowledged that his climate model was more sensitive to atmospheric carbon than those at Princeton and elsewhere. But other models also showed significant impacts "outside the range of human experience" from a doubling of carbon. Asked by Gore whether the scientific community agreed that the greenhouse effect was in fact occurring, Hansen hedged his response. He answered that there was a consensus that the greenhouse effect was "real" and substantial climate change would accompany significant changes in atmospheric carbon. He cautioned, however, that factors other than carbon affected climate, so "it is perhaps premature to be very definitive about the greenhouse effect already occurring." Gore pressed on: "But the consequences of the effect, if it is real, are so incredibly unthinkable that we ought to get a handle on the problem. In that your view?" Hansen replied, "That certainly is my view, yes."

Administration witnesses attempted to disarm critics on the committee with their agreeability. They didn't, for instance, dispute the general parameters of climate science. Dr. James Kane, a longtime executive at the Department of Energy and its predecessors, caught Gore by surprise by asserting, "I would like to congratulate you on your choice of witnesses. There is almost nothing they said I would disagree with." He then praised Hansen for his handling of the uncertainties in the science. Gore questioned whether Kane's position was consistent with cutting the budget for climate science at the department by a third. Kane countered that within the resources of a smaller government, climate science was still getting priority treatment.

In a later House hearing, Gore and Scheuer hammered the administration for slashing Energy's research and development on alternative power sources, including photovoltaic cells.[20]

At work with his fellow panelists on the Nierenberg report, Bill Nordhaus updated his perspective on optimal climate policies in an address at an American Economic Association meeting, published in May 1982.[21] Building on his previous work, the influential professor was now ready to estimate the economic costs of climate change, a missing piece of previous cost–benefit analyses. He calculated that the impacts of a doubling of atmospheric carbon would range from a 12 percent loss of the global economy to a 5 percent gain. He said these effects were "clearly very large" but "also very uncertain."

Nordhaus asserted that the projected CO_2 buildup was not inevitable. Strategies to counter this trend included energy conservation, research and development, fuel switching, and building dikes. The costs of these responses would likely be "very substantial" because of the high volume of carbon emitted each year. He warned, in a rebuff to Stockman's worldview, that few of these responses were "likely to be efficiently provided by the invisible hand of competitive markets."

Digging deeper, Nordhaus identified the limitations of his economic projections. For one, models based solely on economic data excluded "ethical concerns about the sanctity of the climate." In addition, they concealed regional variations in impacts. Differential effects, particularly among nations, would create formidable hurdles for slowing climate change. As Schelling had suggested earlier, Russia and Canada, both prolific producers of fossil fuels and very cold, might benefit from global warming and refuse to join an international agreement to reduce carbon emissions. Their lack of participation in turn might deter other countries from making the sacrifices needed for an effective international treaty.

Nordhaus addressed head-on the idea that the inability to project the future impacts of climate change with anything approaching precision should delay taking action to slow the rise in carbon emissions. He advocated policies based on the "best guess" and adjusting them later based on the evolving science. His approach ruled out waiting for certainty to proceed but favored a middle ground. He wrote, "It is simply irrational to say we should wait for further information before we act, or to act purely because the consequences of inaction may be grave." For Nordhaus, certainty was not a true/false question. It was, instead, a progressive scale based on the level of evidence.

The professor used a discount rate of 5 percent (down from his earlier 10 percent) to calculate the most efficient approach for dealing with climate change. Considering the potential dangers and remaining uncertainties, his analysis supported some government action right away. One was a very modest tax on carbon emissions that would rise sharply in later years. Another was research and development to advance technologies that would reduce carbon emissions—the programs suffering deep budget cuts.

Another shoe dropped before the release of the Nierenberg report. An office at the Environmental Protection Agency issued its own analysis of the climate problem—*Can We Delay a Greenhouse Warming?*—just before the release of the more extensive NAS study.[22] The White House was not pleased.

The EPA publication, written by a contractor, synthesized previous work by the National Academy, the Department of Energy labs, and other leading climate scientists and economists. No federal agency disputed that carbon levels would continue to rise due to human activities. Moving into an area of greater controversy, the EPA assessed strategies to delay a 2 degree Celsius warming—what it called "a dramatic departure from historic trends." With no government intervention, EPA expected the global increase over pre-industrial levels to reach 2 degrees Celsius in 2040 and 5 degrees in 2100.[23]

One possible policy response was to adopt worldwide taxes that would eventually quadruple the market cost of fossil fuels—the favored method of leading economists unwilling to rely solely on adaptation. However, EPA models projected that these taxes would only delay a 2 degree warming by five years. The EPA foresaw much greater effectiveness with a policy banning all coal and shale oil, beginning in 2000. This option would delay the 2 degree increase by a more impressive twenty-five years. In addition, what happened with greenhouse gases other than carbon could make a sizeable difference in how fast the earth warmed.[24]

EPA viewed the high cost of alternative fuels and public opinion as formidable barriers to banning coal. This option appeared to be both "economically and politically infeasible," leaving no viable solution to the climate problem on the table. Nevertheless, EPA said the risk to the earth's climate was too enormous to delay action. It concluded, "Changes by the end of the 21st century could be catastrophic taken in the context of today's world. A soberness and sense of urgency should underlie our response to a greenhouse warming."[25]

Concepts like catastrophe, urgency, and bans on coal ran counter to the messaging the administration was preparing for the release of the Nierenberg study.

9

Discounting the Future

Reagan, October 1983 to February 1984

The National Academy of Sciences report of October 20, 1983, *Changing Climate,* was hardly a routine Academy publication. Congress had statutorily requested the study during Carter's last year in office. The Reagan White House, departing from precedent, helped script its unveiling. National television and front-page print coverage added to its luster.

On the eve of its release, the Academy underscored the report's stature with a celebratory dinner in its Great Hall for the eight panelists and seventy other guests. After cocktails, panel chair Bill Nierenberg joined White House Council on Environmental Quality leader Alan Hill and two Democratic members of the House of Representatives around table 1. Table 2 included Roger Revelle, another Democratic congressman, and National Academy of Engineering president Robert White. At table 3, panelist and Harvard economist Thomas Schelling supped with former CEQ chair Gus Speth (now president of the new World Resources Institute) and executives from the Peabody Coal Company and the Environmental Protection Agency. Other attendees included scientists from government agencies, congressional staff, private industry, universities, the National Academy, and foreign embassies. Table 8—with the director of Exxon's Theoretical and Math Sciences Laboratory seated next to an environmental advocate from the Conservation Foundation—also illustrated the broad audience the Academy sought for one of its most sensitive reports.[1]

The next day, Reagan science advisor Jay Keyworth compared the new study with the recent one from the EPA, branding the EPA release as "unwarranted and unnecessarily alarmist." He preferred the view of the Nierenberg panel, which he said found "no evidence to indicate that the gradual rise in carbon dioxide in the air would have environmental effects pronounced enough to require near-term corrective action." Nierenberg told the press, "We feel we have twenty years to examine options before we have to make drastic plans. In that twenty years, we can close critical gaps in our knowledge."

The national news media followed the lead of Keyworth and Nierenberg. *CBS Evening News* anchor Dan Rather reported, "The Academy wants more study before any major national energy policy changes are made." The *New York Times* headline the next day read, "Haste of Global Warming Trend Opposed."[2]

The characterization of *Changing Climate* and its media "hoopla" (as Academy staff described it) were music to the ears of an administration trying to deflect pressures to deal with the risks of climate change. Its goal was to limit climate action to scientific studies that pursued an elusive "certainty." The report's release portrayed a diminished urgency about climate change compared to reports released from 1977 to 1979, two from the Academy. After the *Times* story, Keyworth privately declared victory, telling White House counsel Ed Meese that the Academy "specifically advises against any action because of the lack of scientific knowledge regarding any potential effects of atmospheric CO_2 buildup." Although *Changing Climate* could always be misinterpreted, he said, it "should help defuse worst-case fears of the impacts of the greenhouse effect."[3]

Though it wasn't known at the time, the Nierenberg report fell short of the original intent of the panel responsible for it. At its first meeting, in October 1980, George Woodwell had argued against a wait-and-see attitude on action to limit climate change. The Woods Hole bioscientist believed that scientists would not be able to learn faster than the problem could develop. Thus, decisions should rely on what was already known. Nierenberg seemed to agree, opining that "in a complex society such as ours, many people are in fact making decisions incrementally on the basis of what we now know." Nordhaus expressed similar sentiments at the panel's December meeting, just weeks after Reagan's landslide victory. He recommended emphasizing low-cost approaches available right away to ease the problems caused by climate change.[4]

These early tilts by panelists toward some early incremental action on climate got lost in the shuffle by the time the report came off the presses in 1983.[5]

The substance of *Changing Climate* revealed a story with more internal tensions than initial impressions might have suggested. On some points, the report—containing a multitude of chapters written by the eight panel members with the help of eight coauthors—even seemed to contradict itself. For instance, *Changing Climate* agreed with the estimates of the Charney and Smagorinsky studies that a doubling of carbon accumulations would raise the average global temperature by 1.5 to 4.5 degrees Celsius. Thus, the scientists working on the report "expected" large-scale global warming due to increasing CO_2.[6] Elsewhere, however, the report downplayed these projections. Several authors, including influential modeler Syukuro Manabe, argued that numerous factors affected temperature change, not just carbon dioxide. In its scientific jargon, they said, "The failure [of earlier studies] to identify unequivocally a CO_2 signal in the noisy global temperature record suggests that attempts should be made to take into account other causal factors in order to reduce the residual variance, and to make a hypothesized CO_2 signal stand out more clearly."[7] Interpreters of the report were free to emphasize that climate change from human activities was expected or that the evidence was not yet unequivocal. Unsurprisingly, the White House pushed the latter view, despite what appeared to be a clear link between carbon accumulations and warming temperatures.

Agronomist Paul Waggoner bolstered the argument that action to slow global warming was not urgent. He concluded that the effects of climate change on farming in the year 2000 would be mixed—"modest, positive, and some negative." He based his forecast on the ability of humans to cope with climate change based on evidence that U.S. farmers in earlier times had been able to adapt crops and husbandry to changing environments.[8]

Waggoner's analysis tended to mask some severe impacts on agriculture. He dealt only with the United States, which provided farmers with the most advanced technical support. He also limited his analysis to the year 2000, even though there was general agreement among climate scientists that the impacts of climate change would become increasingly evident in the twenty-first century.

The synthesis near the front of the report—written by Nierenberg and senior National Academy staffer Jesse Ausubel—attempted to justify the exclusive national focus on agriculture by calling the American case "illustrative" of conditions worldwide. It did allow, however, that the report's views on agriculture would look different from a global perspective. It acknowledged that adaptation would be easier in countries like the United States that "span several climatic zones and have sufficient wealth, ingenious farmers, and capable scientists with a practical outlook."[9]

In contrast, Revelle's chapter on the impacts of atmospheric carbon on sea levels was not so reassuring. He projected that the transfer of water mass from glaciers to the oceans and the warming of water expected from increased radiation would raise sea levels by seventy centimeters (twenty-eight inches) over the next century. The melting of the West Antarctic ice sheet could (as he had noted in his 1965 report for the Johnson administration) have further "far-reaching consequences."

Changing Climate largely ignored impacts other than those on agricultural production and sea level, which bolstered its claims of a lack of urgency. However, panel member Smagorinsky briefly suggested another reason to worry about the carbon buildup. He wrote, "Although hurricanes are not explicitly treated at present in general-circulation models, one might infer that warmer ocean temperatures would increase their penetration into mid-latitudes; such inferences and the topic of climatic extremes deserve careful investigation." This risk did not go unnoticed in the synthesis. Thus, an early comment in the report read, "The frequency, severity, and track of hurricanes and other severe storms are likely to be affected by CO_2-induced climatic changes, such as the warming of ocean waters. Neither our current knowledge of storm genesis nor the current capabilities of climate models are great enough to allow convincing linkages at this time."[10]

At least one lack of focus on other impacts appears to have been a deliberate attempt to avoid sounding alarmist. Panelist Leonard Machta from NOAA warned privately that much discussion of human resettlement forced by climate change would "raise more questions with Congress than it would settle." The panel agreed it was "premature" to talk much about the potential forced movement of affected populations.[11]

It fell to Revelle to write the most provocative science papers, two of which he had previewed in his 1981 testimony to the House of Representatives. Both weakened the "wait-and-see" thrust of the report's unveiling.

His highly technical chapter "Methane Hydrates in Continental Drift Sediments and Increasing Atmospheric Carbon Dioxide" was hardly designed to attract a broad readership. Most Americans had no idea what methane hydrates (also known as clathrates) were. Nonetheless, Revelle was again on the cutting edge by identifying another challenge in unraveling the puzzles of climate change.[12]

Methane is natural gas and, like carbon dioxide, adds to the greenhouse effect when emitted into the atmosphere. Methane hydrates are stable because they are located in cold locations such as permafrost and ocean depths. Thus, this form of methane was unlikely to pollute the atmosphere unless exposed to higher temperatures. The understanding of methane's impact on the atmosphere was still at an early stage; there was, for instance, no functional equivalent of the Keeling Curve to track its atmospheric concentrations. However, Revelle built on recent work by other scientists, including Gordon MacDonald, and the Phillips Petroleum Company's collection of deep-sea sediments containing methane hydrates. Global warming, whether from natural or human causes, would destabilize these hydrates at some point, and their resulting release into the atmosphere would amplify whatever warming was already occurring. Revelle was not yet ready to make projections about the timing or extent of methane releases. But he did conclude, "The likelihood of the widespread occurrence of clathrates in continental slope sediments gives force to our argument that a systematic survey should be made in an attempt to determine their abundance and distribution."[13]

Revelle (with Waggoner as his coauthor) wrote another chapter of perhaps greater interest to an American audience. Their detailed calculations showed that "warmer air temperatures and a slight decrease in precipitation would probably severely reduce both the quantity and the quality of water resources in the western United States." They based their reasoning on the strong effect of the higher evaporation rates that would accompany temperature increases.

Revelle and Waggoner identified seven water regions that would suffer serious adverse effects. They were particularly interested in the Rio Grande and Colorado Rivers. California relied on the Colorado for 15 percent of its water imports, an amount that might fall to zero with global warming. Some areas with high rainfall would escape these effects, but many others would confront water supplies inadequate to meet current needs.

The authors concluded, "The probable serious economic and social consequences of a carbon dioxide-induced climatic change within the next 50 to 100 years warrant careful consideration by planners of ways to create more robust and resilient water-resource systems that will, insofar as possible, mitigate these effects." They suggested that the reengineering of major water-resource systems would take from three to five decades, implying that some such efforts should begin sooner than Nierenberg's twenty-year window before action became necessary.

Ivy League economists Nordhaus and Schelling exerted extensive influence on the overall tone of *Changing Climate.* Nordhaus called for adjusting the projected path of atmospheric CO_2 because the rising price of fossil fuels in recent years meant their use would increase more gradually than previously anticipated. Acknowledging the usual uncertainties, he now foresaw the midpoint estimate for atmospheric carbon in the year 2000 as 370 parts per million and the doubling of carbon from preindustrial levels as 2065. (Spoiler alert: the number for 2000 turned out to be 370, a bull's-eye for Nordhaus.) Consequently, there might be a few extra years to deal with the risks of climate change.[14]

One factor stood out in his examination of the sensitivities that affected his projections of carbon levels. "Ease of substitution between fossil and nonfossil fuels" outranked all other considerations. This factor was closely tied to research and development support from the federal government. The report, unlike Nordhaus's recent article, failed to link this finding to its implications for energy policy.[15]

Nordhaus still believed that the world's governments would eventually need to impose taxes on fossil fuels to slow carbon emissions. His new calculations of future emissions showed that a tax beginning in 2025 would be more effective than alternative strategies. Such conclusions required estimates of "elasticity" (how quickly energy demand would drop in response to taxes). With his usual candor, the professor acknowledged that "the empirical basis for this parameter [elasticity] is as weak as any we rely on." In effect, he was abandoning his position that some policy response was needed right away based on best guesses. He now proposed a delay of four decades for a significant policy response to climate change—and a policy without much evidence about how well it would work. His shift bolstered the White House argument for delay.[16]

Schelling contributed the final substantive chapter, "Climatic Change: Implications for Welfare and Policy." Unlike the other authors, he provided no survey of the relevant academic literature or list of references. The title, placement, and lack of documentation suggested to readers that his contribution was intended as a summary of the entire document.

Schelling delivered an economist's case for adapting to climate change rather than reducing greenhouse gas emissions. He foresaw that large nations—like the Soviet Union, Canada, and possibly the United States— would benefit more than they would lose if they avoided forced reductions in fossil fuels. Without full participation in any international agreement to slow greenhouse gas emissions, many countries taking action would give up more than they got. Schelling was not dismayed by the unlikelihood that carbon emissions could be reduced. He asserted, "It would be wrong to commit ourselves to the principle that if fossil fuels and carbon dioxide are where the problem arises, that must also be where the solution lies."[17]

Schelling's menu of adaptation strategies was wide-ranging. He believed "some control of hurricanes may be achievable" with weather modification. However, violent storms should not be suppressed everywhere because some regions might gain more from the additional rainfall than they lost from the damage. He also urged, without much evidence of careful reflection, consideration of injecting particles into the atmosphere to create a cooling effect.[18]

Schelling agreed that "a dramatic possible consequence of CO_2-induced climate change is a significant rise in sea level." The best approach was not to reduce greenhouse gas emissions. Instead, human populations could abandon coastal areas. Or they could "defend" coastal areas with dams, dikes, piers substantial enough for buildings, and landfills. He believed that defending coasts in densely populated areas would make economic sense "for a century or two." In support of his assertion, he cited the success of Dutch dikes in preventing flooding using existing technology. He said the Dutch approach was particularly suitable for protecting the Boston area, including his own Harvard campus. He acknowledged that nations' abilities to respond to sea level rise was not equal. According to Schelling, "The most severe dangers appear to be in areas like Bangladesh, where dense populations dependent on agriculture occupy low coastal plains already subject to freshwater and seawater flooding."[19]

Schelling foresaw little likelihood that the impact of climate on the overall global economy would be "of alarming magnitude." But, as illustrated by conditions in Bangladesh, it could severely affect some regions or countries. This situation would raise the question of "compensatory transfers of income, capital, and technical assistance" to poorer nations. He suggested that foreign aid packages might include such transfers.[20]

Schelling was the leading and most influential panelist telling readers that the specter of climate change was no cause for "alarm."

Consistent with common practice, the Academy had circulated a draft of *Changing Climate* to dozens of anonymous reviewers before its release. Many assessments offered general agreement and only minor suggestions for improvement. The most forceful critic was Alvin Weinberg. The former Eisenhower science advisor and current leader of Oak Ridge Associated Universities shed his anonymity to share with the authors his view that the draft report was "disingenuous" in its reasoning and "wishy-washy in proposing any actions."

In a "Dear Bill" letter, Weinberg asked Nierenberg not to regard his input as just "being the gripes of an old-time nuke!" Weinberg blistered the draft for relying on the "panaceas" of migration and irrigation.[21] "Does the Committee really believe that the United States or Western Europe or Canada would accept the huge influx of refugees from poor countries that have suffered a drastic shift in rainfall?" he asked. "I can't for the life of me see how historic migrations, which generally have taken place when political boundaries were far more permeable than they are now, can tell us anything about migrations 75 to 100 years from now."

As for water, Weinberg warned that the need for new irrigation on a warming planet could become "prodigious." Raising the prospect of U.S. dependence on Canadian waters, he said the needed infrastructure "would probably cost one-third of a year's gross national product." While these expenditures might be possible for the United States, "Such imaginings are entirely unrealistic in the underdeveloped world . . . For an official Academy report to imply that these are easily handled, business-as-usual developments simply makes no sense to me and certainly is not supported by analysis in the report."

Weinberg also took umbrage at the conclusion that "the evidence at hand does not support steps at this time to change fuel use patterns away from fossil fuels." He wondered whether "our Academy staff is afraid to hurt the sensibilities of some of its members or even sponsors, because any responsible assessment of strategies for reducing CO_2 emissions would require a forthright examination of nuclear and solar energy." He was "dismayed" at this omission, since the report had concluded that "the most important determinant of CO_2 accumulation is the rate of displacement of fossil fuel by nonfossil fuel." Despite Weinberg's withering critique of the draft report, the sentence favoring continued reliance on fossil fuels remained in the final version.[22]

Revelle, ever the team player, shared with Nierenberg a packed day briefing major media when *Changing Climate* was released, greatly enhancing its credibility. In many ways, it was an odd pairing. Nierenberg was known for his conservative views and support for President Reagan. Revelle was on the other end of the spectrum, even if he usually tried to keep his political views private. Reflecting on Reagan's prospects for reelection, he lamented in an archived interview, "I just shudder to think of having to live the next four years with this bastard as president." Revelle argued for scenario analysis in the report; Nierenberg worked with the administration to eliminate this approach. But the two were professional colleagues. Their families partied together on holidays. They both understood that individuals were sometimes constrained by the need for scientists to speak in one voice.[23]

Initially, there were only a few public attacks on *Changing Climate*. However, a couple of weeks after its release, the *New York Times* ran an opinion piece by an Environmental Defense Fund senior scientist strongly objecting to the "sanguine attitudes" of the climate studies released by the EPA and the National Academy.[24] Atmospheric physicist Michael Oppenheimer said both reports suggested that "a strange new world is in our future." Despite the threat, the EPA had stated, and the Academy implied, that little could "be done to forestall climate change and that Government efforts should focus on adaptation." Oppenheimer argued, "The United States currently uses about twice as much energy per capita as the average European nation and has barely scratched the surface of increased energy efficiency and alternative sources . . . The Academy report is skeptical about the desirability of fossil-fuel substitutions, but it presents no analysis of the cost of not preventing climate change."

The protest from an EDF scientist was notable because environmental advocacy groups had not yet played a substantial role in the climate debate. It may also have signaled the discomfort of one *Changing Climate* panel member. George Woodwell, who wrote the section on the biotic effects of atmospheric carbon dioxide, was a founder and trustee of the organization.

Changing Climate set a new direction in how many Americans viewed environmental and, more specifically, climate protection. The reversal in attitudes was massive enough to affect public policy for decades, and at the same time, subtle enough to escape adequate attention from historians.

In the 1960s and 1970s, policy discussions had often linked environmental protection to moral principles and a belief that nature had a value that could not be fully quantified. A notable example, the scientific team writing the 1965 White House report *Restoring the Quality of Our Environment* declared, "The responsibility of each polluter for all forms of damage caused by his pollution should be . . . generally accepted."[25] President Johnson was even more explicit about the ethical implications of environmental protection when he asserted that "the ultimate cost of pollution is incalculable" and "no person or company or government has a right in this day and age to pollute, to abuse resources, or to waste our common heritage."[26]

In tune with the times, Revelle argued in a January 1970 essay in the *New York Times* that human ecology and ethics were inseparable. He wrote, "Science can help in building the structure of concepts and natural laws that will enable man to understand his place in nature. Such understanding must be one basis of the moral values that should guide each human generation to exercise its stewardship over the earth." Like Johnson, he believed that not all the effects of pollution could be priced. He found it "easy to see that the economists' rational analysis must be supplemented by political action." Environmental improvement would have to rely on "a heightened sensitivity among young people" based on "moral and intellectual values."[27]

Standing before the combined houses of Congress ten days later for his State of the Union address, Nixon sang from the same hymnal. He proclaimed, "We can no longer afford to consider air and water common property, free to be abused by anyone without regard to the consequences. Instead, we should begin now to treat them as scarce resources, which we are no more free to contaminate than we are free to throw garbage into our neighbor's yard."[28]

Concerns over the ethical dimensions of environmental policy continued into the Carter years. In a March 1979 message to Congress, the president declared, "Man exists on this planet only with the consent of Nature."[29] In 1980, his Council on Environmental Quality argued, "To understand the full measure of the CO_2 problem, we must also appreciate our ethical commitment to this planet and its inhabitants."[30]

During Reagan's first term, the pendulum swung against the dominant environmental philosophy that had produced the Clean Air Act and the Environmental Protection Agency. Ethical concerns about the natural world were no longer in vogue at the White House. New policies had to prove their worth with transactional analyses of benefits and costs, with the effect on the gross domestic product being the ultimate judge of public policy. If environmental damages could not be calculated with precision, they did not, for all practical purposes, exist. For populous but low-income nations, damages to property and the loss of life attributable to climate change were of little account because the economic loss would be only a sliver of the world's total wealth.

The work of academic economists helped sanction the transition in environmental thinking. By incorporating discount rates into their projections, they created formidable barriers for any action requiring up-front expenditures to create long-term benefits. This problem was particularly daunting for a challenge like climate change that involved moving away from reliance on fossil fuels, whose carbon emissions would remain in the atmosphere for perhaps hundreds of years. The tools of economists skewed model results toward adapting to climate change over slowing it. In their transactional calculations, it was no longer the polluter who would pay for the cost of the pollution. Now it would be the victims of pollution who should foot the bill for adapting to a changing climate.

Among themselves, economists sometimes discussed the limitations of their transactional approach. They didn't hesitate, for instance, to critique the pros and cons of discount rates, sometimes under the auspices of Resources for the Future, a think tank organized in 1952 by the Ford Foundation to stimulate research and education on the economic aspects of natural resources and environmental policy. Its 1982 book *Discounting for Time and Risk in Energy Policy* provided a lively discussion of the role of discounting among its eighteen authors and commenters, all leading economists. Given the period, it is not surprising that the motivation for the book was making energy choices after the 1973 Arab oil embargo, not the still less conspicuous threat of climate change.[31]

In his brief preface, RFF president Emery Castle identified a significant limitation to economists' widespread use of discount rates—their compounded effect over long periods of time. He noted that "some major energy-related research and development projects are expected to produce benefits no sooner than thirty or forty years hence." Thus, "with almost any positive rate of discount, their benefits do not appear to justify their costs." According to Castle, this impact of cost–benefit analysis was relevant to a current inclination at the Department of Energy, the Office of Management and Budget, and the Council of Economic Advisers to apply a discount rate of 10 percent to federal investments in solar energy.[32] Jimmy Carter and his science advisor, Frank Press, had agreed to push solar research and development, even while acknowledging that photovoltaic cells might not compete with fossil fuels on price until the next century. Now economists, even some from Carter's administration, insisted on discount rates discouraging such investments.

One coauthor of the RFF book, Joseph Stiglitz of Princeton University, suggested another problem with discount rates. Government policy often addressed some "market failure," which led "market prices not to reflect social evaluations." In his view, there were many "important reasons that market rates of interest for private consumption goods might not be appropriate for evaluating public projects."[33] Stiglitz didn't want to abandon discount rates but believed they should not create an automatic ban on government investments with time-lagged benefits. He called for a case-by-case evaluation of their appropriateness and level.

A joint meeting of the American Economic Association and the American Association for the Advancement of Science in December 1981 also addressed how economists' cost–benefit analyses might influence the in-process *Changing Climate* study. Senior NAS staffer John Perry attended the discussion and reported to the Nierenberg panel on what he heard.[34]

Perry's memo summarized the views of Nordhaus that adopting a purely cost–benefit approach to climate eliminated ethical and moral questions that "might lead us to declare increases in carbon dioxide to be an anathema." Because of the persistence of CO_2 in the atmosphere, he said, action to deal with it should not be postponed because of the many uncertainties involved.

Nordhaus was not alone in questioning the universal applicability of discount rates. Glenn Loury of the University of Michigan noted several critical features of the CO_2 issue, including "questions of equity between societies and between generations." Because of these factors, he proposed using cost–benefit analyses less "as tools in formulating policy goals" and more "as aids in informing policymakers about the consequences of their actions." In this economist's view, "In considering the question of equity between generations, one might simply say that future generations will be richer than today and thus better able to cope with whatever damage we create. Alternatively, one might say that we have a moral obligation to the future. On such issues, economists can properly only advise and not dictate."

Lester Lave was blunter about the limitations of discounting future benefits. "For example," he said, "with normal discount rates, wiping out humanity in 200 years has virtually no present implications. This cannot be right."

The views of Nordhaus, Loury, Lave, and other economists offered a humbler view of their profession's tools than the unquestioned economic underpinnings of *Changing Climate*. The tilt of the Academy study toward postponing action on climate and favoring adaptation as the eventually preferred strategy relied on the largely hidden assumption of discount rates. The sections of the report on the physical sciences acknowledged differences of view within their disciplines. The economists' chapters, by contrast, lacked explicit discussion of discount rates or the views of some economists about their limitations.[35] Thus, the impacts of discount rates loomed large while remaining undetectable to most readers of National Academy and government reports.

After almost three decades of helping expand the understanding of climate change, Revelle was not about to pause his role as chief educator on the subject. By the time of the release of *Changing Climate,* he had already booked his seat on an Air India flight to Rome. The Vatican was paying the travel expenses for more than a hundred eminent scientists to serve as advisors at the biennial plenary session of the Pontifical Academy of Sciences, held November 7–11, 1983. The Vatican invited these experts without regard to religious affiliation.

Pope John Paul II addressed an enthusiastic audience of scientists, cardinals, bishops, priests, nuns, and church diplomats in the palatial stateroom Sala Regia. Built in the sixteenth century, its massive marbled walls sparkled with bold stuccos and frescoes by Italy's finest artists. He urged scientists to work for peace and "carry out a work of salvation analogous to that of a doctor who has sworn to use all his powers to heal the sick."[36]

The scientific rigor of the lengthy technical sessions impressed many attendees. The *Times* reporter covering the event, for one, observed, "Unlike the theologians of the 17th century who refused to even look through Galileo's telescope lest they see something to upset their faith, today's church leaders appear determined to keep abreast of the latest advances in science, to avoid unnecessary conflicts between religion and science, and to help guide science in directions beneficial to humanity."[37]

In one workshop, Revelle presented a paper, "Carbon Dioxide and Other Greenhouse Gases in Ocean, Atmosphere, and Biosphere, and Future Climate Impacts." As he had done so many times before, the professor provided a broad overview of the scientific consensus about matters ranging from the expected global temperature rise with a doubling of atmospheric carbon dioxide to the long-range threats of the melting of the Greenland and West Antarctic ice caps.[38] Revelle and other authors subsequently expanded their contributions for a book published by the Pontifical Academy.[39]

Climate change was far from the only topic at the Vatican gathering. Still, the dozen papers on the subject signaled its growing stature on the world stage.

*　　*　　*

In February 1984, Congressman Al Gore presided over a House hearing to review the recent reports from EPA and the National Academy.[40] The morning session focused mainly on a salient question raised by the Nierenberg panel: When would available evidence be sufficient for policymakers to initiate action?

Gore's witness list included climate scholars who had worked on the recent studies. It also included representatives from a rare environmental organization speaking publicly about climate change. The verbal jousting was congenial. Speakers on all sides mentioned their concerns about their grandchildren, current or expected, as reasons to take climate change seriously. A Republican congresswoman from Rhode Island, Claudine Schneider, a member of the Appropriations Committee, offered to help scientists beef up funding needed to plug gaps in knowledge. The scientists reported that the administration was now proposing more adequate funding for climate studies. Still, there was a big divide over the implications of the EPA and NAS studies.

The EPA's John Hoffman tried to soften his agency's pessimism about slowing climate change. Providing no written testimony due to lack of administration clearance, he spoke about three factors that explained his agency's time lags between climate policies and a significant slowing of climate change. First, the heat stored in oceans would linger long after carbon emissions were stabilized or even eliminated. Second, inaction on other greenhouse gases like methane would limit the impacts of slashing carbon emissions. Finally, the world's economy relied heavily on fossil fuels, and reversing long-term patterns would not happen quickly. More optimistically, according to Hoffman, "From 2040 to 2100 on, we found that almost all of the policies could significantly reduce the warming, some of them quite substantially."

Tom Malone—who headed the National Academy's Board of Atmospheric Sciences that oversaw the Nierenberg panel and presented its findings to the committee—won the descendants who might be affected by climate change contest by claiming nine and two-thirds grandchildren, "one of the main reasons I am here." The esteemed geophysicist strongly endorsed *Changing Climate*'s wait-and-see approach. "We recommend," he said, "expanding monitoring and continued research but no immediate change in energy policy." He quoted the judgment of the report: "The knowledge we can gain in the coming years should be more beneficial than a lack of action will be damaging." He hoped this concept would be "impressed deeply" on the members.

For an opposing view, the committee turned to Rafe Pomerance, who now served as the president of the Friends of the Earth—an organization with 30,000 members in the United States and affiliates in twenty-two countries. The group had recently adopted climate change as a priority. More politically attuned than other witnesses, he jested that he had to avoid calling Gore "Senator." (Gore was running for the Tennessee Senate seat Republican majority leader Howard Baker was vacating.) Pomerance warned the committee not to rely on scientists to tell them when or how to implement climate policies. "You are the ones that are going to have to make that decision," he asserted. "It's not their job." He called the National Academy's advice to wait and see "absolutely wrong." He cited the testimony from Cornell University's Carl Sagan and Columbia's Wally Broecker earlier in the morning that an average change in the world's temperature of just 1 degree Celsius could bring significant climate change. "So why wait?"

Pomerance challenged the Academy's approach to who had the burden of proof when determining climate policies. NAS acknowledged, he said, that the combustion of fossil fuels was an "unintentional experiment" whose precise effects remained unknown. *Changing Climate* argued that this lack of precision led to a recommendation of policy inaction. However, the basic knowledge was already available. Scientists should not wait for "the final increment of scientific certainty" before abandoning their business-as-usual position. Pomerance said they should, instead, have to demonstrate why the continuation of this grand experiment was safe.

Friends of the Earth argued that promoting efficient fuel use and investing in renewable energy offered "the opportunity to reduce fossil fuel consumption to a fraction of today's level." He cited evidence that these initiatives were compatible with continued economic growth. Challenging the economists' transactional view that nations should not act alone, he asserted that advanced technologies developed in America would likely spread to international markets and thus provide global, not just national, benefits.

Pomerance criticized the Reagan administration, which had, he said, "dismantled or attacked every federal energy program in the area of solar and conservation." Though it was unmentioned, Reagan had also chopped and eventually eliminated funding for synthetic fuels—the program Pomerance and MacDonald had warned against.

The highlight of Gore's hearing was his skeptical but respectful probing of Malone's defense of the National Academy's wait-and-see strategy. The chair started by asking why it was premature to be concerned about climate change. The sixty-six-year-old scientist replied that, at his age, he had experienced "strident voices" calling for action on the depletion of stratospheric ozone to prevent a "pending catastrophe." He said Academy reports had significantly lowered the estimate of missing ozone over the past decade. With this additional knowledge, he questioned whether ozone was actually a "formidable problem." Gore: "Really?" Malone did not attempt to assess whether scientists might have overestimated some environmental threats when first discovered and underestimated others.

Gore asked whether the Academy's tilt toward adaptation, which might involve abandoning 40 percent of Florida, presented "an overly benign view of how catastrophic these changes could be." Malone said the Academy took the issue seriously, which was why it brought together the best minds it could get for *Changing Climate,* a "review process which [it had] never seen the likes of." He asserted that the Academy did not "feel that abrupt change" was needed "when we have not yet been able to identify clearly a climatic change associated with increased carbon." The purpose of careful monitoring was to detect an "unmistakable" signal, and then "we can press the start button."

The EPA's Hoffman agreed that scientists were not yet "statistically sure" of the relationship between carbon emissions and climate change. But he clarified how high the bar was for such certainty. "Scientists don't operate on the basis of it's 3 to 1; they don't operate on the basis that it's 9 to 1. Most scientists operate on the basis that it's 19 to 1 [95 percent] that this couldn't happen by chance."

Eager to interject, Pomerance declared, "The longer we wait, the more CO_2 we are locked into because we are sitting on a very powerful engine, world coal use. If you say, 'Stop,' it's going to take a long, long, long time to stop."

Despite Gore's evident unease with a wait-and-see approach, the Reagan administration, for all practical purposes, had won the battle over *Changing*

Climate. With its stringent standards of proof, the National Academy had found no reason to alter the existing energy system based on fossil fuels. Without a declaration of certainty, public officials shied away from pushing for such changes. Early in the questioning, Gore had acknowledged the political realities of confronting the risks of a changing climate. He conceded, "Our natural inclination in political systems is to side with those who say it is too early to act. That's just the political inertia that is a fact of life around here, and I assume in other countries as well."

10

Bipartisan Rebound

Reagan, October 1985 to December 1987

Changing Climate discouraged, in a very visible way, actions to delay global warming. Nonetheless, international cooperation on climate protection took a leap forward in the fall of 1985 and suggested that the wait-and-see approach of the National Academy of Sciences and the Reagan administration might not prevail.

The United Nations Environmental Programme, the World Meteorological Organization, and the International Council of Scientific Unions convened scientists from twenty-nine nations in Villach, Austria, to assess the state of climate science. The October conference, located at the eastern foot of the Alps, welcomed eighty-nine participants, including twenty from the United States. The U.S. delegation included Tom Malone, Roger Revelle, Syukuro Manabe, and others who had worked on the 1983 NAS report, as well as representatives from the Department of Energy and its national labs. Scientists from Africa, Europe, China, India, Japan, and the Soviet Union contributed their expertise.[1]

The consensus report from Villach accepted the *Changing Climate* projections on expected global warming. It agreed with the modeling of Manabe and others that doubling atmospheric greenhouse gases would likely increase mean global temperature by 3 degrees Celsius, plus or minus 1.5 degrees. However, Villach diverged from a dominant theme of the Academy report. While identifying scientific questions needing further investigation, the new report did not regard these unknowns as barriers to collective action.

Instead, the summary recommendations declared, "The understanding of the greenhouse question is sufficiently developed that scientists and policymakers should begin active collaboration to explore the effectiveness of alternative policies and adjustments."[2] The report didn't provide a detailed policy road map. But its tone justified a statement from one EPA participant that the scientists had agreed that governments should go beyond research and turn toward economic and social policies.[3]

Revelle served on the panel examining potential socioeconomic impacts. Not surprisingly, the group highlighted his section of *Changing Climate* asserting that scientists would underestimate climate stress if they did not combine effects on rainfall with those on evaporation. It also called for economists to analyze the "efficient use of resources" but added that they should also address the "equity of their distribution." The equity reference recognized that the adaptation strategy advocated by Thomas Schelling might favor wealthy nations that emitted much more carbon dioxide per person and had more resources to adapt over poorer countries that emitted much less and lacked the ability to adjust to climate change. Revelle's group, notably, contributed the sentence in the report's summary advocating going beyond just more research.[4]

Malone's comments at Villach reflected a stunning reversal from his spirited defense of *Changing Climate* before the Gore subcommittee the previous year. New information had convinced him that the impacts of climate change might be "earlier by several decades than previous estimates." Furthermore, he now praised scientists for recognizing the risks of ozone depletion. In contrast to his congressional testimony, he declared at Villach that scientists should not stop with pleas for more research. They should also work to illuminate "the policy issues and options."[5]

With six Americans playing prominent roles in both the *Changing Climate* and Villach reports, the conflict between the two documents cried out for explanation. Did American scientists take different positions abroad than they did at home? At Villach, Malone elaborated on his changed outlook. "As a reversal of a position I held a year or so ago," he said, "I believe it is timely to *start* on the long, tedious, and sensitive task of framing a convention on greenhouse gases, climate change, and energy. The task will likely take longer than the decade of discussion needed to draft an ozone convention." Malone had come to believe that when scientists spoke with clarity on climate change, policymakers would have to do more than, as he had told Congress, just "press the start button."[6] (In 1991, the Franciscan Center of Environmental Studies in Rome awarded Malone—a devout Catholic who had attended the 1983 meeting of scientists with the pope—its international Saint Francis Prize for the environment, Canticle of All Creatures.)

The U.S. State Department would base future climate negotiations on the Villach statement. But back in Washington it was not clear which of the dueling reports more accurately mirrored Reagan's position.

In December, a Senate subcommittee examined the implications of what climate scientists were saying. Previous congressional hearings on the subject had taken place in the House under Democratic leadership. But an international treaty on climate would require Senate ratification, giving its views greater weight. With a Republican majority in the Senate, this hearing would also test the issue's bipartisan appeal.

The event turned out to be remarkably collegial. GOP chair David Durenberger slated Democratic freshman senator Al Gore, a leading voice on the threat of climate change during his days in the House, as the first witness. When Durenberger had to leave, he stretched comity even further by designating Gore, who wasn't even on the committee, as the temporary chair. Both senators and witnesses cited the importance of the Villach report, whose authors Durenberger called, as near as he could tell, "a collection of the finest scientific minds in the world on this topic." The senator from Minnesota observed, "Depending on who you ask, the consequences [of climate change] could be a disaster of biblical proportions or just maybe nothing. Unfortunately, more and more of the bets are on biblical and fewer and fewer on nothing."[7]

Quentin Burdick—a seventy-seven-year-old Democrat from North Dakota—was sparse in his comments, asking only, "How long have we got?" No clear answer to Burdick's question emerged. Gore worried that equivocation by scientists would block efforts to do anything about changing climate in the short and mid terms. "The caveats and qualifications," he said, "diminish the political will [to act] out of all proportion to the effect they should have."

Manabe, the government's chief weather and climate modeler, testified that scientists would reduce but not eliminate all uncertainties. Gordon Mac-

Al Gore, surrounded by family, celebrating his 1984 election to the U.S. Senate, where he continued to push for hearings on climate change. Courtesy Hulton Archive/Getty Images.

Donald agreed: "The uncertainties will always be there." The lead author of the 1979 JASON report cautioned against waiting very long to act because it would take many decades to transition to new energy systems. Celebrity scientist Carl Sagan asserted the time to act was "now" and called a focus on international and intergenerational equity and Chinese involvement. "We are all in this greenhouse together," he declared.

Gore proposed transitioning to the action stage with a bill for an "international year on climate science and policy options," patterned after the International Geophysical Year during the 1950s. It was the IGY, he noted, that had supported the careful measurement of atmospheric carbon dioxide at Mauna Loa.

Witness Dean Abrahamson from the University of Minnesota (in Durenberger's home state) called for moving swiftly to policy implementation. The professor said the optimal time for action depended on how much damage one wanted to avoid. He and Gordon MacDonald argued that efficiency standards for buildings, appliances, and vehicles would reduce carbon emissions and were justified even without a climate argument. They also suggested incorporating climate change into environmental impact statements. They complained that the Reagan administration was moving in the wrong direction by weakening the auto efficiency standards established during the Ford and Carter years. Abrahamson urged the country "to avoid coal at all costs" and move to natural gas as a transition fuel.

The rapport among senators and witnesses didn't eliminate the sticky problems lurking ahead. Only Durenberger and Gore attended the hearing for more than a few minutes. Moreover, some eyes glazed over during the explanation of topics like the infrared spectrum. Durenberger acknowledged the issue's complexity by citing one Energy official's comment that "those who link global warming to CO_2 buildup are blowing smoke." Durenberger separated himself from that skepticism but conceded, "One of the frustrations of dealing with this issue is that it's virtually beyond the grasp of human imagination."

Despite the sparse attendance, the calls for action by witnesses did attract some national press attention the next day.[8]

Additional Senate hearings in June 1986 displayed solid bipartisan resistance to the administration's wait-and-see approach on climate.

The subcommittee on environmental pollution, chaired by Republican John Chafee of Rhode Island, began by praising the work of the Villach conference and the Durenberger subcommittee. Again, the first witness was Gore, who repeated his call for an "international year" on the greenhouse effect.[9] Over two days, Chafee's hearing manifested increased participation by senators and stronger probing about when to adopt policies that could slow climate change.[10] The agenda interspersed the discussion of climate with the thinning of the earth's ozone layer and, to a lesser extent, with acid rain—the big three environmental issues of the decade. But the senators appeared well versed in what scientists were telling them, avoiding potential confusion about the topics.

Several senators pressed the sixteen witnesses on whether they agreed with the stance of the National Academy of Sciences and the Reagan administration in 1983 that climate action could wait until all pertinent scientific questions could be fully resolved. The answers provided a diversity of views and even some daylight between the witnesses from the administration.

Among the scientists, James Hansen projected the most dramatic temperature increases. However, he cautioned that the evidence was insufficient to justify policy responses. After a Chafee question about what he would do if he were "a king," Hansen responded, "I would like to understand the problem better before I order any dramatic actions . . . I think that what I would like to see most of all . . . is global observations of the atmosphere, of the oceans, of the land surfaces, which allow us to see what is happening better and allow us to develop and test the models to represent what is happening." But another climate expert, George Woodwell, countered that it was already time to act. He testified, "I think we know enough about this topic, and we have known enough about it for at least a decade to move toward alleviating the problem." The Woods Hole scientist's role in the Nierenberg report gave special significance to his disavowing its advocacy of delay.

Former Democratic congressman Andrew Maguire—representing the World Resources Institute, where Gus Speth and Rafe Pomerance were now colleagues—was blunter in distancing himself from Hansen's view on timing. He argued, "We must act before all the dimensions of the problem are fully known, or we will risk irreversible catastrophic changes . . . The timing of action will also have a considerable impact on how precipitous the actions must ultimately be and, therefore, how difficult and costly." Maguire offered an extensive agenda for climate action. He called on the United States to consult with the Soviet Union and the People's Republic of China, adopt efficiency standards for appliances patterned after those for automobiles, reduce deforestation, and reassess existing federal curbs on natural gas to encourage a transition to less-carbon-intensive fuels. Lastly, he testified, "We should begin now . . . to fashion a political consensus for more aggressive policies such as a carbon tax on fossil fuels."

On the second day, Chafee put administration witnesses in the hot seat by probing their approach to the climate problem. William Graham—a PhD in electrical engineering from Stanford University, current deputy administrator of NASA, and soon to be confirmed as the next science advisor to the president—was the senior Reagan appointee at the witness table. When Chafee tried to pin him down on whether some action was needed to lower emissions, Graham echoed the administration's stance: "I would be more comfortable with taking additional specific action when I found the basis of understanding for what is going on now to be more sound."

However, two Reagan agencies distanced themselves from Graham on the issue of certainty. Richard Benedick, an assistant secretary at the State Department, volunteered that complete foresight on the complex alterations of the atmosphere was unattainable. He told the committee, "Policymakers may be called on to act even while there is still scientific uncertainty over the causes or the extent of potential threats to environment and health." Similarly, the new administrator of the Environmental Protection Agency, Lee Thomas, bristled at Democratic senator George Mitchell's suggestion that administration policy required that "everything must be known before anything can be done." He retorted, "I am the guy who has to make the regulatory decisions, and frankly, I make most of them, if not all of them, where everything is not known."

At the Chafee hearings, historian Theodore (Ted) Rabb rebuked scientists insisting on more time to understand the climate puzzle before governments decided what to do about it. He testified that political leaders had to recognize "a congenital hesitancy always in the scientific community in making unequivocal statements." Born in Czechoslovakia and raised in London, the champion of the humanities and quantitative methods at Princeton had examined the impacts of natural climate change in earlier times and coedited a 1981 book on the subject.[11] Drawing on his expertise in European history, he argued that the origins of modern science as a discipline began at a time when "vicious religious and ideological conflicts . . . threatened to tear apart the very fabric of European civilization." In response, scientists wanted their work to be "an oasis in the wilderness of hate and emotion" and adopted the language of "unadorned and objective" neutrality. In current times, scientists continued to treat advocacy as "a mortal sin."

The professor testified that the significant cooling during the seventeenth century demonstrated the dramatic impacts of natural climate change. The results were "dreadful, dreadful hardship—starvation, plague." He opined that none of these upheavals involved "a climatic break of a magnitude, in the short term, that even approaches what the studies now tell us is likely to happen to us in the 21st century." In response to Rabb's scolding, Chafee defended the scientists testifying before the committee who "spoke right out and said, we have a disaster on our hands here, unless we do something."

Chafee—an enlisted Marine who fought at the battle of Guadalcanal during World War II, a popular former governor, and part of a pro-environment bloc of Republicans in the Senate—left no doubt where he stood on the need for action. He declared, "By not making policy choices today, we may, in fact, be making a passive choice. By allowing these gases to continue to build up in the atmosphere, this generation may be committing all of us to severe economic and environmental disruption without ever having decided that the value of 'business as usual' is worth the risks." He compared the current path to a "form of planetary Russian roulette."

The chair advocated several measures to deal with the climate challenge. First, he would ask the Environmental Protection Agency and the Office of Technological Assessment "to launch immediate and separate studies setting forth policy options" to stabilize the levels of atmospheric gases. At the international level, he called on the State Department to begin discussions with the Soviet Union and China (the other two nations with vast coal reserves) and on the United Nations "to convene a meeting to negotiate a convention on climate change in the near future." He also said he would request the Council on Environmental Quality to direct "all federal agencies to recognize ozone depletion, the greenhouse effect, and climate change as environmental impacts that must be considered in the [environmental impact review] process."

The news media sometimes confused the ozone hole with the greenhouse effect. But climate activists like Pomerance were pleased that the Senate had given both topics an elevated sense of urgency.[12] Moreover, after the hearings, Chafee's requests to the targeted agencies were signed by three Republicans and three Democrats. The Republicans included Vermont's Robert Stafford, chair of the full Environment Committee. The Democrats included Maine's Mitchell, a former staffer for Senator Muskie, who had inherited his mantle as a leader on environmental issues.

By the time of the Chafee hearings, the Reagan administration had softened its stand on land, air, and water protection. During the president's first term, harsh attacks from environmental organizations had taken a political toll, leading to the replacement of appointees who had provoked the severest ire. Because of these changes, Chafee's initiatives found parts of the Reagan administration responsive to his concerns. For instance, the official *EPA Journal* of December 1986, on the theme "Our Fragile Atmosphere: The Greenhouse Effect and Ozone Depletion," opened with an article by Administrator Thomas that identified stratospheric ozone depletion and climate change as "global challenges" facing the agency. He warned, "There will always be scientific uncertainty associated with these complex problems. We will have to be prepared to act despite these uncertainties." Moreover, the issue contained articles by Chafee and Pomerance, both prominent critics of the administration's caution on climate.

The need for international cooperation on climate gave State a pivotal role. The department's leader added to its clout. Secretary George Shultz, an economist by training, had been a colleague of Milton Friedman at the University of Chicago, run three federal departments for Richard Nixon, and joined the Reagan cabinet in 1982 after heading the international behemoth Bechtel Corporation. Among the living policy heavyweights of the Republican Party, only Henry Kissinger (no longer in government) outranked Shultz in stature.

Despite flashes of environmental moderation within the administration, Reagan took a combative stance against calls from Senate witnesses for energy efficiency standards. California's regulation of household appliances such as refrigerators to promote energy conservation during the 1970s had produced fuel savings, reduced pollution, and become the model for federal standards authorized by the 1978 Energy Policy and Conservation Act. Carter ran out of time to finalize the regulations needed to implement the law, and Reagan quickly scrapped the effort. In 1982, the administration dug in its heels even deeper, ruling that federal standards offered no benefit over incentives provided by "the market." It argued further that its failure to adopt a standard was, in fact, a standard that preempted states like California and New York from adopting their own regulations. In response, the National Resource Defense Council launched litigation challenging the "no-standard standard." The federal DC circuit court agreed with the plaintiff and ruled in 1985 that the administration's position was unsupported by substantial evidence and contrary to law.

Subsequently, momentum grew in Congress for strict federal mandates on appliance efficiency, with backing from the NRDC, the American Council for an Energy-Efficient Economy, state energy offices, and eventually even appliance manufacturers. Advocates argued that since the savings of reduced energy bills more than offset the higher costs of more efficient appliances, this form of energy conservation paid for itself. In addition, many appliance manufacturers came to believe they could benefit from selling to a national market with a uniform standard rather than dealing with different standards in different states.

The Democratically controlled House of Representatives brought a new appliance efficiency bill to the floor in September 1986. Its standards were so detailed that they could take effect without rulemaking by the executive branch. Floor manager Ed Markey, a Democrat from Massachusetts, predicted that national standards for eleven major appliances would reduce residential energy use by 6 percent, and manufacturers could produce better products at lower costs if they did not face diverse state regulations. He praised Republicans for helping put the bill on a fast track and asked for passage with unanimous consent. Markey's GOP counterpart, Carlos Morehead of California, returned the compliment and strongly endorsed the bill. No one spoke against the legislation, which passed on a voice vote, a strong signal the bill lacked controversy.[13]

In October, the efficiency bill moved to the Republican-led Senate and received an equally friendly reception. GOP floor manager and former Washington State governor Dan Evans estimated that the standards would save an average household $300 per year and commended the "spirit of cooperation" across party lines in crafting the bill. Minority floor manager Bennett Johnston of Louisiana declared, "As a nation, we are once again headed toward an energy crisis similar to that which we experienced in the early 1970s. I believe that energy conservation must become and remain an integral part of our national energy strategy." He noted the support of more than forty organizations, "running from the National Association of Manufacturers to the Sierra Club and the Methodist church." He added that the bill would provide savings that the free market would not. As in the House, the bill passed without objection.

Despite the unanimous congressional backing for efficiency standards, Reagan, on November 1, announced a pocket veto, not subject to an override. The president contended, "The bill intrudes unduly on the free market, limits the freedom of choice available to consumers who would be denied the opportunity to purchase lower-cost appliances, and constitutes a substantial intrusion into traditional state responsibilities and prerogatives."[14] Markey called the veto "a triumph of extreme right-wing ideology over practicality and common sense."[15] Moreover, with his endorsement of state-by-state prerogatives, Reagan had overplayed his hand and further weakened his position with appliance manufacturers. He had also failed to quash the enthusiasm in Congress for appliance efficiency.

In January 1987, two environmental subcommittees in the Senate held yet another hearing on climate change. Because the November elections had flipped control of the body, Democrats George Mitchell and Max Baucus would now wield the gavel. Mitchell praised the outgoing GOP chairs for running their subcommittees like "one big family," across party lines, and predicted Democrats would do the same.[16]

The new hearing called on two stalwarts of climate science. Wally Broecker suggested that climate models might lull people into complacency with their simulations showing gradual warming over a hundred years or so. The Columbia scientist's historical data from polar ice and ocean sediments painted a different picture. He testified, "What these records tell us is that the Earth's climate does not respond in a smooth and gradual way; rather it responds in sharp jumps . . . Coping with this type of change is clearly a far more serious matter than coping with a gradual warming."[17]

Gordon MacDonald found assumptions that societies could just adapt incrementally to climate change "unconvincing." He asserted that market economics could not handle the job. "The intangibles of climate are not readily quantified in conventional market terms; they are one of the externalities with which markets find so hard to deal." He advocated preparing the way for "alternative energy sources, including nuclear and renewable sources, as well as increasing overall energy efficiency."[18]

The Senate hearings on climate and the bipartisan collaboration calling for new approaches to energy constituted the strongest evidence to date that, even in the Reagan era, the climate issue had political legs. Still, the House and the Senate held hundreds of hearings a year on various topics. Moreover, even among environmental issues, the thinning of the ozone layer and the acid rain resulting from sulfur emissions outranked climate in congressional priorities.

MacDonald's call for energy efficiency faced a turning point the next month, when the Senate tried to resurrect the appliance standards bill Reagan had vetoed. The spectacle proved the adage that laws were like sausage: you enjoy them more if you don't watch them being made.

On a busy Thursday, with senators trying to escape to the airport, Republican Phil Gramm of Texas was determined to block the bill until his amendment was accepted. The first-term senator—a former economics professor at Texas A&M University and member of the House of Representatives, where he frequently teamed with other antiregulation stalwarts—began by opposing the traditional unanimous consent approval of the body's official journal. The Texan then demanded several quorum calls that forced members to rush back to the chamber while the clerk conducted time-consuming roll calls to certify their presence. Robert Byrd of West Virginia, the new majority leader, believed that the rules did not allow such "dilatory" interruptions.

But the White House, which had announced it would veto the bill again, was playing hardball. It dispatched Vice President George H. W. Bush to perform his constitutional but rarely exercised duty as presiding officer. Bush's rulings from the chair supported Gramm's right to gum up the works. Byrd called for suspending the rules to allow the presiding officer to defend his actions, which provoked an extended sharp but polite discussion about the disruption. When the majority leader offered the vice president a chance to have the last word on points of order, Bush demurred and responded, "I just want to get out of here."[19]

The assault on the appliance efficiency bill seemed futile since it had previously passed the Senate on a voice vote and now appeared veto-proof with sixty-eight cosponsors. But the White House wanted to avoid the embarrassment of successful legislation it had previously blocked.

After Bush gave up the gavel, senators tried to explain what had just happened. Bennett Johnston, floor manager for the bill, complained, "If there was ever a bill that did not merit this kind of constitutional challenge, . . . it is this simple, bipartisan, nonpartisan wide array of interests coming together in one bill . . . Lord help us when we get to something that might be controversial."

Many Republicans found themselves caught between their support for the bill and the need to back a party colleague's right to object. William Armstrong of Colorado "defended" Gramm's obstruction on the floor, saying, "I think, by gosh, it is the tradition of this body that if a senator wants to rise to his feet even to make a fool of himself, he is entitled to do that." Dan Evans also backed the Texan's right to intervene but pled that the bill deserved to be "judged on its own merit and not tied down to nongermane amendments and dilatory tactics that are designed to kill the bill rather than to help it."

Following more back-and-forth, Byrd lamented, "We could go on and on and on ad infinitum. I think it has already reached the point of being ad nauseum." After Gramm demanded another quorum call without enough senators still on-site to meet the threshold, the Senate moved on to other business.

While the Senate debated aid to the Nicaraguan Contras, Johnston and Gramm huddled privately about a possible compromise. They finally agreed that the Senate could begin discussing the bill but would delay action until its next working session twelve days hence. At that time, Gramm could offer his amendment.

On February 17, the Senate resumed its consideration of the appliance efficiency bill with a deluge of decorum and members praising each other for their mutual courtesies. Gramm had given up all avenues of delay in exchange for Democratic and Republican sponsors and the groups supporting the bill accepting his amendment. Senator Johnston told senators, "The legislative process requires compromise. I think this is a case where all parties have been reasonable." With comity back in fashion, the amendment carried by unanimous consent.

Gramm's amendment left intact the bill's mandates for stiff reductions in the energy consumption of refrigerators, freezers, furnaces, room and central air conditioners, water heaters, dishwashers, washers and dryers, televisions, kitchen ranges, and ovens. But it did weaken the requirement for the Department of Energy to issue periodic updates to the standards based on advances in technology. Gramm said the compromise assured that the president's senior advisors would recommend that he sign the bill. On final passage, the new version garnered 89 votes from both sides of the aisle, though Gramm joined five of his GOP colleagues in opposition.

From left, Senate Majority Leader Robert Byrd (D-WV) with Senators John Chafee (R-RI), Robert Stafford (R-VT), George Mitchell (D-ME), and Quentin Burdick (D-ND), preparing to discuss their upcoming bipartisan override of President Reagan's veto of a clean water bill, 1987. Courtesy AP Photo/Lana Harris.

Nine days later, the White House kept its side of the bargain. President Reagan sent a terse, one-sentence message to Congress: "The administration has no objection to the enactment of H.R. 87."[20] Given his veto of a similar bill just months earlier, his acceptance of the slightly pared-down National Appliance Energy Conservation Act of 1987 was a big deal, even if the message avoided any mention of the topic.

On March 3, with major hurdles out of the way, the House floor managers again praised their colleagues across the aisle for their collaboration on efficiency standards.[21] The bill sailed to overwhelming voice approval. But this time two second-term Republicans from Texas voiced their opposition. Joe Barton did not accept that Congress "must protect the consumer from himself." Tom DeLay complained, "We will take away another consumer choice, which, when left to the free market, would have accomplished the same things." Reagan signed the bill without fanfare two weeks later.

Advocates justified the new efficiency standards based on their ability to delay the depletion of fuels and the need for expensive new power plants. They did not cite its implications for climate change. Nonetheless, many who supported the bill (such as Gore, Mitchell, Baucus, Stafford, Chafee, and Durenberger) had attended climate hearings where witnesses suggested energy efficiency as one way to deal with the climate problem and knew that the new law was a small, early step toward slowing global warming.

On May 21, Reagan signed another bill with implications for climate change, this one without any reservations. Earlier in the month, Congress had voted to amend the Fuel Use Act of 1978, which had forced electric utilities and other industrial users to burn coal instead of oil or natural gas. By 1987, the goal of spurring coal-generated electricity had been largely achieved. At the same time, the potential supplies of natural gas appeared more substantial than foreseen during the Carter years, reducing the need to limit its use.

Reagan's January State of the Union address had called for an end to the limits on natural gas as part of his plan for "regulatory relief." With Democratic heavyweights like Bennett Johnston in the Senate and John Dingell in the House on board, the measure to eliminate restrictions on the industrial use of gas and ease them for electric power generation won swift congressional approval.[22]

Expert testimony from Revelle, MacDonald, and the World Resources Institute had suggested substituting gas for coal to slow global warming. As with energy efficiency legislation, climate was not a primary objective for lowering the barriers to gas. However, allowing its expanded use, if price competitive, was another step in adopting the advice of several climate scientists.

On September 16, 1987, representatives from twenty-four nations (including the United States) plus the Commission of European Communities approved a historic breakthrough in global environmental protection. The signatories to the Montreal Protocol agreed to phase out chemicals expected to damage the earth's stratospheric ozone layer. Because the ozone layer blocks harmful ultraviolet radiation from reaching the earth's surface, its thinning posed risks to human health, including increased rates of skin cancer and cataracts. Another likely effect of these pollutants was climate change, since the chemicals involved—several types of chlorofluorocarbons and halons—remain in the atmosphere, where they trap heat. Despite the compromises involved, the diplomatic triumph in Canada was a cause for celebration among environmental advocates. In addition to protecting against ultraviolet rays, the new protocol would help slow the pace of global warming and demonstrate that nations with diverse interests could hammer out complex solutions requiring international action.

Compared to climate change, the science supporting the deliberations on the ozone layer was quite recent, with the first scholarly articles appearing in the mid-1970s. Like the climate science of the 1970s, during the "global cooling detour," the evidence on the ozone layer was sometimes contradictory. Nonetheless, U.S. federal agencies and international organizations quickly began efforts to understand the environmental risks. President Carter mentioned it on several occasions as one of the nation's long-term challenges. Curtailing the use of these chemicals would not be easy. The proposed caps on the pollutants damaging the ozone layer would have worldwide impacts on aerosol sprays, the chemical industry, and air conditioning that relied on certain refrigerants.

The State Department—with the assistance of NOAA, the EPA, and NASA—led international efforts to advance the science and consider appropriate policy responses. Richard Benedick, a fifty-two-year-old career foreign service officer whom scientists regarded as "a very capable and forceful guy," coordinated the U.S. team.[23] Assessing their national economic interests, the United Kingdom and France initially resisted the strict regulations favored by the United States. However, West Germany sided with the Americans, helping offset opposition within the European Community (later the European Union). Other nations, particularly the less affluent ones, had less interest in the issue, having neither chemical manufacturing nor much air-conditioning. In these cases, the leadership of the United Nations Environment Programme's Mostafa Tolba, a former professor of biology at the University of Cairo, helped guide undecided countries toward ratifying a forceful agreement.

For a time, the United States was in the awkward position of calling for an eventual ban on ozone-depleting chemicals amid doubts that its negotiators had the president's full backing. Any substantial disagreements within the administration had to be resolved by the Domestic Policy Council—created to ensure policy consistency and chaired by Reagan confidant and now attorney general Ed Meese. On May 20, 1987, Benedick defended State's position in the international negotiations at a DPC meeting attended by more than two dozen administration heavyweights. Interior secretary Donald Hodel criticized the U.S. stance, arguing that it was too early to determine whether natural phenomena or manufactured chemicals caused ozone depletion. If the administration was under pressure to do something, it should urge other countries to adopt restrictions on the aerosols used for spray cans, something the U.S. had already done. Reagan science advisor Graham backed up Hodel, saying the science was "very uncertain" and more studies were needed before mandating stiff reductions.[24]

Hodel went public with his reservations about the ozone negotiations the following week. He told the *Wall Street Journal* that increases in ultraviolet radiation were not a problem if people didn't stand out in the sun.[25] In Benedick's view, Hodel was part of the "antiregulatory forces in the Reagan administration [who] mounted a rearguard action . . . to undermine the U.S. position on protecting the ozone layer." The diplomat complained that "no persuasion appeared possible" to overcome the libertarian argument that "skin cancer was a 'self-inflicted disease' attributable to personal lifestyle preferences, and therefore protection against excessive radiation was the responsibility of the individual, not the government."[26]

In the first week of June, Secretary Shultz sent a "Dear Ed" letter to Meese expressing his "strong personal interest in the early and successful completion of an effective international treaty to protect the stratospheric ozone layer." He was, he said, concerned that a few agencies were pushing to reopen the entire international negotiations scheduled for completion in September. He concurred with EPA administrator Thomas that the current U.S. position reflected "a prudent approach to risk management . . . Although scientific certitude is probably unattainable, I am impressed by the growing international consensus on the threat to the ozone layer, largely due to research by our own NASA and NOAA." Based on his contacts with the chemical industry, Shultz believed it could develop the substitute products required by aggressive international standards. He also argued that U.S. companies would compete at a disadvantage without a global agreement because EPA had some regulations in place, authorized under existing U.S. law, and might be forced to adopt additional controls because of pending court cases. Furthermore, a U.S. reversal of position at this late stage would prove embarrassing. Not waiting for permission, Shultz said he would continue his current course unless the Domestic Policy Council objected. "In that case, I propose that we, together with Lee Thomas, take this matter to the President without further delay."[27]

Three days later, the Senate provided strong backing for the Shultz position. Senators Baucus and Chafee cosponsored a bipartisan resolution endorsing the State Department's negotiating posture—a prompt reduction of not less than 50 percent of ozone-depleting chemicals and their virtual elimination over time. Eighty senators voted aye. There were only two nays.[28]

The Domestic Policy Council met for its second discussion of the ozone negotiations on June 11. Thomas and Hodel continued to joust over State's push for an aggressive agreement. Graham, again allying himself with Hodel, criticized EPA's predictive models, noting they extended up to two hundred years into the future and did not reflect likely changes in skin-care protection.

The animated debate provided evidence that the State/EPA position might prevail. Beryl Sprinkel, chair of the Council of Economic Advisers and a noted advocate of Milton Friedman's defense of free markets, said an international agreement should be regarded as an insurance policy with a specified rate of return. He reported that cost–benefit analysis based on

EPA's models justified sharp reductions in CFCs and halons. He added, "If we wait too long, international and congressional actions will have passed us by." Meese echoed part of Spinkel's argument. He suggested it was important how the administration dealt with the scientific aspects of the issue but even more important how it handled the political situation.[29]

On the afternoon of June 18, Reagan personally chaired the final DPC meeting on ozone. Attendees included Vice President Bush and Chief of Staff Baker. The president asked several basic questions and gave Thomas and Hodel plenty of time to restate their arguments. However, the objections of the Interior secretary were more muted than before. Baker—former Senate majority leader and a voice for moderation after arriving at the White House—made perhaps the most telling argument. Just thirteen days after the overwhelming vote in the Senate, he observed, "While the science is in dispute, there is pressure in Congress for a strong protocol." The president announced he would consider the comments and convey his decision later.[30]

Reagan set his parameters for the negotiations in a confidential memorandum of June 25. It concluded, "It is the U.S. position that the ultimate objective is protecting the ozone layer by eventual elimination of realistic threats from man-made chemicals and that we support actions determined to be necessary based on regularly scheduled scientific assessments." In effect, his decision dismissed the objections from Interior and his chief science advisor and endorsed the negotiating strategy of State and EPA.

Benedick speculated about the influence of Reagan's removal of skin cancers in 1985 and 1987 on his eventual decision. The behind-the-scenes influence of Shultz and Baker—neither part of the Reagan team in 1981 that declared war on previous environmental policies—clearly played a role. In addition, the record reveals the significant impact of the solid, bipartisan support for the treaty in the Senate.

The chief U.S. negotiator also acknowledged the assistance of the Council of Economic Advisers' cost–benefit analysis based on EPA models. Using EPA's risk management methodology avoided wiping out the long-term benefits of action with discount rates.[31] Benedick further warned that "the risks of waiting for more complex evidence were finally deemed to be too great." As he told an early negotiating conference, we do not demand certainty that a bridge will collapse as a justification for strengthening it.[32]

The Senate voted 83–0 for ratification in March 1988. Later, numerous additional nations added their signatures to the agreement, the science became increasingly convincing, and substitute chemicals entered the market in a timely fashion.

The UN's Tolba predicted the next year, "The mechanisms we design for the Protocol will—very likely—become the blueprint for the institutional apparatus designed to control greenhouse gases and adaptation to climate change."[33]

Other observers were less ready to view the Montreal agreement as a precursor of international action on climate. They noted that the economic scale of the CFC industry was a few tens of billions of dollars. In contrast, that of energy industries and associated sectors amounted to a more daunting trillions.[34] *Time* magazine concluded, in a lengthy cover story on threats to the environment:

> Any similar attempt to ease the greenhouse effect by imposing limits on CO_2 and other emissions is unlikely . . . Obviously, the most far-reaching step would be to cut back on the use of fossil fuels, a measure hard to accomplish in industrialized countries without a wholesale turn to energy conservation or alternative forms of power. In developing countries, such reductions might be technologically feasible but would be all but impossible to carry out politically and economically.[35]

But there was another difference between regulating ozone-depleting chemicals and greenhouse gases. Early U.S. controls on aerosol sprays legally required that they could "reasonably be anticipated to affect the atmosphere, especially ozone in the stratosphere, if such effects may be reasonably be anticipated to endanger public health or welfare."[36] Similar language in the 1970 Clean Air Act pertained to climate protection. However, the greenhouse gas standard had for many people become reduction of the remaining uncertainties, with no specific indication of when this quest might end. There was a dramatic difference between reasonable anticipation and certainty. No environmental challenge had ever passed (or probably ever would) the test of absolute certainty.

*   *   *

In November 1987, two days of hearings on greenhouse gases by the Senate Committee on Energy and Natural Resources appeared to justify Tolba's optimism about the future of climate negotiations. The Environment Committee, which had conducted previous Senate investigations of the issue, tended to attract members inclined to support environmental protection. Now a committee with more representation from energy-producing states would take over.[37]

Unlike earlier climate hearings during the Reagan years, this one did not feature Al Gore. The first-term senator was busy running for president. Stepping in to drive the climate debate was another first-termer, forty-eight-year-old Tim Wirth of Colorado. When serving in the House, the Democrat had earned a reputation as a long-term thinker and fan of the 1970 classic *Future Shock* by Alvin Toffler. His home state traditionally took nuanced stances on energy and environmental issues since it had both a substantial fossil fuels industry and popular support for conserving its scenic beauty. The state was also home to the Solar Energy Research Institute and the National Center for Atmospheric Research.

In the absence of more senior members, Wirth chaired the proceedings and aggressively pressed witnesses for solutions to the climate problem. His opening statement predicted, "These hearings may show that in the process of developing a national energy policy, it will not be enough to merely consider our dependency on foreign oil resources . . . We must take into account global climate change as well."

John Firor, a physicist from NCAR and the leadoff witness, testified, "The kind of climate change that is being projected now as a result of the increases of greenhouse gases is many times faster, five or ten times faster than climate changes that we have studied in the past." Heeding Wirth's call for solutions, Firor advocated cutting U.S. fossil fuel use in half over the next thirty years, mainly through conservation encouraged by subsidies, taxes, or regulation. Such action, he said, would have the side effect of reducing urban pollution and help the country in many other ways. "For the longer term," he added, "we must examine the difficult problems of how to develop major sources of renewable energy" as well as stabilize the earth's population and help developing countries "move to the prosperous future they desire without repeating the fossil fuel evolution we went through."

Another panel member, James Hansen, was hardly a stranger to congressional testimony, but this statement came with an unusual twist. He noted that he directed NASA's Goddard Institute for Space Studies but was appearing as a private citizen. His written testimony listed his home in Ridgewood, New Jersey, rather than his institutional affiliation. This change in procedure resulted from the refusal of the budget office to clear his written testimony without alterations that he angrily refused to accept. Such reviews of congressional testimony were common practice but usually uneventful. But they could facilitate the censorship of scientific findings. And that was Hansen's interpretation.

Hansen displayed a greater intensity about climate science than some other giants in the field, like Revelle and MacDonald, who sometimes responded to grave questions with a trace of humor. Hansen's grimness might have traced back to his doctoral dissertation on the atmosphere of Venus, a dystopian example of extreme heat. This planet's extremely high concentrations of atmospheric carbon were associated with scorching temperatures that made it hostile to human life. Before the committee, he used his knowledge of the inner planets to advocate for the validity of his and other climate models. Considering factors such as distance from the sun, the temperatures of Venus, Earth, and Mars seemed to correlate well with their levels of atmospheric carbon.

Hansen believed that human impacts on global warming would soon surpass natural variation. However, this was not the only way to look at the evidence. He wanted to describe the effects of climate change for "the man in the street" who wasn't concerned about annual mean global temperatures or familiar with the metric system. To this end, he provided the estimated days a year that temperatures would exceed 100 degrees Fahrenheit for eight American cities. Memphis, for instance, averaged four such events from 1950 to 1983. With a doubling of atmospheric carbon from preindustrial levels, the number would likely rise to forty-two.

The following day, Wirth reconvened the committee with a new panel, led off by James Decker, acting director of the Department of Energy's office of science research. The well-liked physicist had earned his doctorate at Yale, worked at Bell Labs and the Atomic Energy Commission, and joined DOE as a career senior executive during the Carter administration. Hewing to the administration's position, he conceded, "Man's activities are modifying the composition of the world's atmosphere." But he also cautioned, "We should not panic at this point and make policy decisions in the United States that may be unsound."

Decker's first concern affecting policy was the emergence of China. He noted that in 1950 North America contributed 43 percent of the world's carbon dioxide production but "only" 25 percent in 1984. By contrast, Chinese emissions had climbed from less than 2 percent to 10 percent. With the likelihood that the emerging Asian power would rapidly increase its use of coal, "Clearly a change in U.S. energy policy alone would not have a significant effect on CO_2 buildup."

Though not a political appointee, Decker became the target of Wirth's cross-examination on his department's failure to develop climate policies. With some prodding from the senator, Decker acknowledged that global warming was likely, carbon dioxide contributed significantly to that, and part of the CO_2 came from natural or agricultural processes and some came from fossil fuels.

Wirth: There is some disagreement about how rapidly that [warming] is going to occur and its intensity, but it is certainly going to be there. Correct?

Decker: Yes.

Wirth: Is it not logical to then that we have to begin to think what the consequences of that might be?

Decker: Surely.

Wirth: And that we probably ought to begin now to take some preventive action.

Decker: Sure.

Wirth: What efforts are under way at the Department of Energy to transmit this understanding of the potential warming, the significant contribution of fossil fuels to CO2 . . . to the energy policy people?

Decker: The greenhouse effect . . . has not perhaps been incorporated in our policies to a large extent at this point because of the fairly large uncertainties that exist in prediction and the—

Wirth (interrupting): Did we not just establish the fact that there is broad agreement? [He inferred from the Department's testimony] . . . a tendency, as I read it, to perhaps hide a little bit behind the uncertainty rather than to come out and say, "Hey, we have some very significant problems here, and we ought to change the policies of the Department of Energy accordingly." Do you think that is a fair conclusion?

Decker: We do not want to rush into some policy decisions too early.

Wirth: One policy that is pretty obvious is that we ought to be really encouraging conservation?

Decker: I would agree that certainly conservation is important.

Wirth: If the policy implication is that we ought to be encouraging conservation, for example, then why would it be rational to, say, [weaken] miles-per-gallon standards for new automobiles? Or why would it be a logical conclusion to dramatically cut funding for solar energy?

Decker: The fact of the matter is, with energy prices being relatively low, conservation has not been on the minds of the public nearly as much as it was a number of years ago.

Wirth: Has there been an advocacy of conservation from the Department of Energy?

Decker: I would say it has not been a major push.

Wirth then turned to Gordon MacDonald, who, after reviewing the likely impacts of climate change, delivered the discourse on policy the senator was looking for. MacDonald's priority was "to reestablish nuclear energy as a viable energy source." This goal would require the availability of waste disposal, which, as a geologist, he believed was feasible. He said the government also needed to help the industry move beyond its "1950s technology" toward safer and smaller reactors. His other major thrust was "a better pricing scheme so that the cost to society of the disposal of carbon dioxide into the atmosphere is reflected in the price of the fuel itself." Proper incentives would encourage conservation, nuclear power, and solar energy. They would also favor natural gas over coal.

The next witness, Gus Speth of the World Resources Institute, called the accumulation of greenhouse gases "the most serious environmental issue of them all." He declared that the case for action was "getting steadily stronger . . . Far from having the luxury of additional time, we are late getting started." Going well beyond the conclusions of leading climate scientists, he foresaw the possibility of as much as 5 degree Fahrenheit increase in global average temperatures by the year 2000. As Carter's CEQ chair, Speth had been tutored by MacDonald on climate change, and he endorsed several of the geophysicist's policy recommendations, including a tax on carbon fuels. He also reported that he had written Alan Hill, Reagan's CEQ chair, urging that federal environmental impact statements take carbon emissions into account. However, he separated himself from MacDonald's emphasis on nuclear power. He feared that expenditures for nuclear and "so-called clean coal" (which filters out several pollutants but not carbon dioxide) might eclipse "better opportunities." Speth was optimistic that the nation would develop an appropriate response to climate change because of the multiple side benefits of dealing with the greenhouse effect. Moreover, the recently signed ozone convention was "another straw in the wind" demonstrating the ability to do something about climate.

The final witness was George Hidy from the Electric Power Research Institute. Hidy argued, "The consensus on climate change and heating is not really quite as strong as one might assume, I think, from your witnesses so far." He acknowledged the possibility of increasing temperatures but said it was too early to tell if carbon dioxide was leading to global warming. He cited the evidence that from roughly 1940 to 1980 global temperatures had dropped, despite increases in atmospheric carbon. The cooling effect of particulate matter, also emitted from the combustion of fossil fuels, created additional doubt about whether the planet would warm in the future. The Electric Power Research Institute stated that the U.S. electric industry contributed "only a few percent of the total global carbon dioxide emissions." Nonetheless, the organization funded climate researchers, including Keeling at Scripps. As for the timing of policy initiatives, the industry's position was "Because the phenomenon is so complex and is global in character, it is premature to consider an action beyond intensified research on the potential environmental and economic impacts of climate change."

Debriefing with Pomerance after the hearing, Hansen was in a grumpy mood. Preparing testimony, he complained, took a lot of work. Attendance had been sparse. Press coverage was much lighter than for the Environment Committee hearings the previous year, and no one had asked about the censorship of his testimony. The two agreed it might be better to schedule such hearings in the summer since lawmakers might pay more attention to climate change with sweltering weather outside.

Just before Christmas, Congress passed a bill incorporating several ideas from congressional climate hearings. Senator Joe Biden had introduced his Global Climate Protection Act of 1987 earlier in the year. Its inclusion in the much larger foreign relations authorization passed in December made the climate provisions easy to miss.

When Biden presented his bill on the floor (after ending his brief presidential campaign three months earlier), he had lamented that "our public policies—even with responsible democratic government—seldom take serious account of consequences on a time horizon stretching beyond the next election." He warned that foresight was needed to address the "greenhouse effect" since "even a small rise in temperature could disrupt the entire complicated environment that has nurtured life as we know it." He credited the Chafee hearing during the summer of 1986 for helping establish an agenda for climate action.[38]

The Biden bill provided direction for several federal agencies. It tasked the EPA with proposing to Congress "a coordinated national policy on global climate change." It assigned the State Department "to coordinate such policy in the international arena." The bill also directed State to promote an International Year of Global Climate Protection (Al Gore's idea) and urged the president "to accord the problem of climate protection a high priority on the agenda of U.S.–Soviet relations."[39] The legislation signaled that the climate hearings were having some impact.

In the 100th Congress, the forces calling for action to slow global warming were winning more battles than those insisting on "wait and see."

11

The Hottest Summer

Reagan, January to August 1988

A State Department directive to the U.S. permanent representative to the World Meteorological Organization in January 1988 provided additional evidence of active U.S. engagement in global climate negotiations. It confirmed America's official support for the establishment of the Intergovernmental Panel on Climate Change. The IPCC, it said, should "identify for the scientific community information needed by policymakers to evaluate possible response strategies and communicate to policymakers what is known—and what is not known—about climate change, in order to ensure that considerations of climate-related policies are based on solid scientific information."[1]

U.S. endorsement of the new organization checked a significant box in the list of steps recommended by congressional witnesses who opposed a wait-and-see approach to climate policy. This one had the support of the president.

In March, Tim Wirth, John Chafee, and forty other senators sent a letter to President Reagan commending him for discussing threats to the climate during a recent summit meeting with Soviet General Secretary Mikhail Gorbachev. The senators noted that the two countries together accounted for almost half of the world's carbon dioxide emissions. As a next step, the letter urged the president to "call upon all the nations of the world to begin the negotiation of a convention to protect the global climate" modeled after the one to protect the ozone layer.

The 1986 elections had produced a Senate made up of fifty-five Democrats and forty-five Republicans. The signatories to the Wirth/Chafee letter were 55 percent Democratic and 45 percent Republican, a precise reflection of the latest election results.[2] It provided more evidence that the bipartisan success in curbing ozone-destroying chemicals might lead to similar action on greenhouse gases.

Al Gore, who had signed Wirth's letter, would soon resume his full duties in the Senate. Earlier in the month, he had finished in a virtual three-way tie for first in the new southern presidential primary, not good enough in a region where he had run as a southerner tough on national defense and concentrated most of his campaign resources. In April, he suspended his run for president.

While seeking higher office, Gore had tried to talk about the health of the ozone layer and the greenhouse effect but found little interest in either topic. He later told reporters that the environment was "heavily discounted" by the press and politicians and not the kind of issue on which one runs for president.[3] The failure of climate to attract attention during his hunt for convention delegates suggested that the momentum on climate in Congress might lack appeal at the grass roots.

Gore's disappointing plunge into presidential politics didn't change the view that he remained a leading contender for future national office. A few weeks after the November general election, *Newsweek* included him in a list of five potential presidential candidates with improved prospects for 1992. The magazine's assessment of the senator was "Old CW [conventional wisdom]: miserable campaign. New CW: valuable campaign experience." A forty-year-old real estate developer was also on the short list of rising stars. *Newsweek* said of Donald Trump, "Good NY–DC Trump [airline] Shuttle service could sway Eastern media elite."[4]

In May and June 1988, an expansion of the previous year's appliance efficiency legislation sped through the Senate and the House, despite presidential opposition. A broad coalition, including electric utilities, supported new national standards for fluorescent lighting, and Reagan finally agreed to sign the bill on June 17. A unique aspect of this legislation was the argument that it would reduce atmospheric carbon dioxide. On the Senate floor, Wirth identified "worldwide emissions of carbon dioxide from combustion of fossil fuels" as the cause of rising temperatures. He added, "Scientists are very concerned that this global warming is proceeding at a rapid pace and will cause changes in the earth's climate that far exceed not only human experience but also the ability of plants and animals to adapt to a changing environment. It is clear from these scientific findings that energy efficiency is an essential component of our efforts to protect the environment."[5] The statement was another milestone—an explicit and ultimately successful appeal to pass a law to reduce carbon emissions rather than just study them.

Twelve days in late June featured two international meetings in Toronto sandwiched around a highly publicized Senate hearing—each elevating the prominence of the climate issue.

When the G7 nations met on June 19–21, under tight security provided by the Royal Canadian Mounted Police, the leaders of the richest industrialized countries took positions on a wide variety of global issues, including a call for

British prime minister Margaret Thatcher (*left*) sitting next to President Reagan at a G7 economic summit in Toronto that endorsed the creation of an international panel on climate change, June 19, 1988. Courtesy AP Photo/Martin Cleaver.

"sustainable development." The G7's environmental agenda included climate. Most notably, it urged "the establishment of an inter-governmental panel on global climate change under the auspices of UNEP [United Nations Environment Programme] and the World Meteorological Organization."[6]

This G7 summit wasn't the first body to raise the topic of damage to the atmosphere from greenhouse gases. However, its endorsement of the inter-governmental panel was particularly timely since negotiations to form such a body were in progress. Moreover, the support of Prime Minister Thatcher and President Reagan—both skeptics of big government—broadened the support for global cooperation on climate.

The Senate hearing scheduled for June 23 on "Greenhouse Effect and Global Climate Change" was in some respects a rehash of previous congressional hearings.[7] Many witnesses were making repeat appearances. Twenty-eight years after Keeling published his first data on atmospheric carbon, the world was already moving to establish an international panel devoted to climate change. A bipartisan group of forty-two senators had just endorsed the effort. Yet this was not just another climate hearing.

Most leading climate scientists refrained from connecting changes in the daily weather to what their climate models were showing. Their adherence to their discipline's rigorous standards of proof frustrated efforts to explain the risks to nonscientists and ran counter to the proclivities of the daily news cycle. Against the backdrop of record-breaking heat waves around the country in the summer of 1988, this hearing would not ignore the connection between local weather reports and the predictions of more significant global warming ahead.

With three television cameras on-site, committee chair Bennett Johnston opened the proceedings on a dire note. The Louisiana Democrat observed that the temperature that day of 101 degrees Fahrenheit in Washington, and dried-out soils around the country destroying soybean, corn, and cotton crops, were consistent with the testimony of previous witnesses about the greenhouse effect. He warned, "We only have one planet. If we screw it up, we have no place to go. The possibility, indeed, the fact of our mistreating this planet by burning too much fossil fuels and putting too much CO_2 in the atmosphere and thereby causing the greenhouse effect is now a major concern of Members of Congress and of people everywhere." He predicted, "The problem is only going to get worse" and that the solutions "are going to be both expensive as we find alternative fuels and are going to involve massive international efforts." Johnston represented a state with substantial oil and gas interests and was the Senate's most influential member on energy, giving added weight to his declaration.

Wirth, who wielded the gavel after Johnston departed for other business, also stressed extreme weather events around the country. "Meteorologists are already recording this as the worst drought we have experienced since the Dust Bowl days of the 1930s," he reported. "On Tuesday, the Mississippi River sank to its lowest point since at least 1872 when the U.S. Navy first began measurements. And in my home state of Colorado, peak [water] flows are among the lowest on record." On his weekend trip back to North Dakota, Democratic senator Kent Conrad had observed the drought's devastation, which had given the farmland there "the appearance of a moonscape."

James Hansen would be the leadoff and most anticipated witness. Wirth spotlighted Hansen's testimony by predicting that it would clarify the relationship between current weather conditions and climate change. The extra attention weighed heavily on the scientist. This time, the administration had cleared his testimony. However, he was still sensitive to the reality that most of his peers, including those at NASA, did not believe historical data had detected the "signal" that would separate a persistent pattern of global warming from the "noise" of natural fluctuations. On the other hand, he did not want scientific caution to confuse the public about what experts had concluded. He had told the committee in advance that this testimony would be his most potent.

Hansen opened with an assertion, supported by a graph, that the earth was warmer in 1988 than it had been at any time during the hundred-year history of recorded measurements. This record was not an aberration. The four warmest years were all in the 1980s.

His second point broke new ground in what scientists were telling Congress and the public about causality. He displayed another graph with surface air temperatures from 1958, when Keeling started collecting data, to 1987. The single line had plenty of ups and downs, but to the naked eye the overall trend appeared upward. Hansen calculated that global warming was almost 0.4 degrees Celsius relative to climatology—defined as the thirty-year mean from 1950 to 1980. In a pathbreaking finding, he concluded, "The probability of a chance warming of that magnitude is about 1 percent. So, with 99 percent confidence, we can state that the warming during this time period is a real warming trend." Hansen added the usual scientists' caveat that they needed more data but was clear that a line had been crossed. He declared, "The greenhouse effect has been detected, and it is changing our climate now."[8]

His third point was that the greenhouse effect was significant enough to affect the probability of extreme weather events, such as summer heat waves. He was careful to distinguish that the effect did not "cause" a particular event, only that it increased the likelihood that such events would occur.

Hansen's presentation dominated the front page of the next morning's *New York Times.* The stark top headline read, "Global Warming Has Begun, Expert Tells Senate." Directly below was the two-column copy of Hansen's graph showing the historical rise of global temperatures. One story, "Drought Raising Food Prices," sat just beneath the chart. Another, by science writer Philip Shabecoff and devoted to the hearing—with the subheadline "Sharp Cut in Burning of Fossil Fuels Urged to Battle Shift in Climate"—occupied the column to the right of the graph.[9] Shabecoff's account led with the significance of Hansen's new calculations: "Until now, scientists have been cautious about attributing rising global temperatures of recent years to the predicted global warming caused by pollutants in the atmosphere known as 'the greenhouse effect.' But today, Dr. James Hansen . . . told a Congressional committee that it was 99 percent certain that the warming trend was not a natural variation but was caused by a buildup of carbon dioxide and other artificial gases in the atmosphere."

A comment not in Hansen's testimony also garnered attention. In an interview after the hearing, he proclaimed, "It is time to stop waffling so much and say that the evidence is pretty strong that the greenhouse effect is here." Not wanting to antagonize his bosses at NASA any more than necessary, he had omitted the comment from his formal remarks but had jotted it down the night before (while listening to a Yankees' baseball game), hoping he could work it into an answer to questions from the committee or the press.[10] His sense of what would grab public attention proved prescient. An insert on page 2 identified his post-testimony comment as the paper's "Quotation of the Day."

Other witnesses provided greater specificity than Hansen on the need for new policies to deal with climate change. For instance, George Woodwell called for a massive reduction in the world's use of fossil fuels, "the sooner, the better." The Woods Hole scientist also recommended rebuilding forests over a vast area, which could "store carbon at the rate of approximately a billion tons a year for 40 to 50 years as that forest grows toward maturity." Woodwell and other witnesses did not engage in the argument over whether global warming had been proven at a sufficient confidence level. For them, plenty of evidence linking CO_2 and warming already justified the need for action.

No committee member challenged the climate science they heard, but Senator Frank Murkowski questioned how quickly the nation could reduce its carbon emissions. During the examination of witnesses, Woodwell called for "a policy immediately of reducing our emissions of fossil fuels by 50 percent over the course of the next years, perhaps a decade or so." When he argued this strategy would be "*totally* consistent with economic strength," the skeptical Alaska Republican interjected, "Are you prepared to recommend, how, doctor?" Woodwell responded, "Simply through changing standards, for instance, of efficiency for automobiles, by super-insulating houses, by building houses that don't require as much energy. And there are many, many ways of doing that."

After the hearing, several Hansen colleagues pushed back on his announcement of a climate "signal." To quell potential backlash, he sent out a letter to influential climate scientists, including Revelle, "to clarify what I actually said, which differs in some cases from what was reported." He attached a copy of his written testimony.[11]

Hansen's appearance quickly gained fame as a pivotal point in climate science. Two months later, the *New York Times* quoted the glowing observation of another witness at the Senate hearing, Michael Oppenheimer. The Environmental Defense Fund scientist declared, "I've never seen an environmental issue mature so quickly, shifting from science to the policy realm overnight. It took a Government forum during a drought and heat wave and one scientist with guts to say, 'Yes, it looks like it has begun, and we've detected it.'" The paper found Hansen's statement all the more remarkable because he realized that "he was risking his reputation as a cautious and careful scientist."[12]

Hansen's success relied in part on his acceptance of a scientist's responsibility—as described by LBJ science advisor Donald Honig—to communicate in terms nonscientists could understand. To do so, he became the rare scholar who discussed temperature changes in degrees Fahrenheit, the unit familiar to most Americans. He used his model to forecast the likelihood of a greater number of extreme heat waves in specific cities—something people could easily visualize. With the help of eager press coverage, he broke through the scholarly barriers that had often confused the broader populous.

Despite the significance of Hansen's testimony, later accounts of it sometimes failed to match reality. This mythology included the assertion that Senate Energy in 1988 "was the first committee to explore the problem of global warming, an issue that until then was confined mainly to academic journals." Similarly, another version claimed that the unusual weather and Hansen's testimony should be remembered because "for the first time, global climate change became a significant public and policy concern in the United States."[13] These views, though widely shared, were not supported by the historical record and were far from isolated.

Congressional hearings had in fact discussed the threat of climate change as early as the 1950s and heard numerous days of testimony focused on the topic well before 1988. At these earlier hearings, many witnesses had suggested specific policies to slow global warming. Moreover, Presidents Johnson, Nixon, and Carter had issued substantive reports on climate change. The Reagan administration was already working to establish an intergovernmental panel on climate change, with bipartisan support in the Senate. The national press had already carried plenty of stories on the risks of atmospheric carbon dioxide, occasionally on the front page. Climate change had in fact escaped its confinement to academic journals very quickly after Keeling first published early data from Mauna Loa in 1960.

Another embellishment came from the legend about the energy committee's efforts to accentuate the connection between weather and climate. According to an oft-told tale, the committee opened hearing room windows the previous night so that sweltering room temperatures would reinforce the expected testimony. Unfortunately for those who enjoy an entertaining story, no contemporary evidence supports this narrative. Moreover, the committee staff and Hansen insisted, years later, that this intentional indoor heating did not and could not have happened, partly because senators would not have tolerated it. For additional perspective, the ABC television news report and archival photographs show no beads of sweat on the brows of the witnesses. In 2015, the *Washington Post* reexamined the open windows myth, got some spreaders to recant their version of events, and gave what had become conventional wisdom: four Pinocchios for falsification.[14]

Inaccurate memories of the June 1988 hearing distorted, for a long time, a history that is interesting enough without the fictional garnishes. The idea that this event was the first to bring what scientists knew to the attention of policymakers erases much of the history of how climate science arose in the academic world and made its way into the broader public discourse. It ignores the longtime work of scientists like Revelle and MacDonald, who tried to push the findings of climate science into the world of climate policy in the 1960s and 1970s. Climate history looks very different if the first warnings occurred well before 1988.

The overdramatization of the June 1988 hearing had another unintended consequence: acceptance of the idea that certainty about climate science was a necessary precursor to climate action. Policymakers had not required a 95 percent statistical confidence level for other environmental issues. Environment policies were traditionally based on probabilities. Standards of certainty seemed especially inappropriate in the case of CO_2, given its prolonged persistence in the atmosphere.

The National Academy of Sciences and Reagan's science advisors had pushed the idea that climate policies required scientific certainty. Hansen now argued that the warming trend had already emerged from the "noise" of natural variation. But there were additional questions related to the role of other gases and the effects of global warming on natural systems. The challenges of achieving certainty on all these matters were immense and likely insurmountable. Even if the difficulties could be addressed, there was little prospect that computers would be powerful enough in the foreseeable future to capture the complex interactions of nature. Scientists had made significant progress in predicting weather and climate based on probabilities rather than certainties. A broad-based quest for certainty in 1988 was a recipe for perpetual delay.

Roger Revelle—still collecting prestigious awards and honorary degrees and by now taking time to play with his great-grandchildren—was no longer the go-to witness for congressional hearings on climate change. But privately Wirth did seek his views on the Hansen testimony. Like many prominent climate scientists, Revelle did not agree with Hansen that the "signal" had yet been identified and he cautioned, "We must be careful not to arouse too much alarm until the rate and amount of warming becomes clearer." But he maintained his position that because of the long residence of carbon in the atmosphere, it was unwise to do nothing about it. He recommended taking actions that made sense whether or not the greenhouse effect materialized— such as the expanded use of nuclear power and planting massive new forests to capture carbon. He told the senator that such actions "could reduce carbon dioxide emissions very drastically, to quite a safe level."[15]

Four days after Hansen's testimony, Canadian Prime Minister Brian Mulroney welcomed the month's second international meeting in Toronto, this one organized by the World Meteorological Organization and the United Nations, with the topic "Changing Atmosphere: Implications for Global Security." Its three hundred scientists, government officials, and representatives of nongovernmental organizations brought additional attention to the international implications of acid rain, ozone depletion, and climate change.

The conference described the scientific basis for concern about global warming in terms that resembled those of the Villach report. The breakthrough in Toronto was the specificity of its policy recommendations. The report set a "challenging" (some participants thought unrealistic) target of reducing annual global CO_2 emissions by 20 percent by the year 2005 "through improved energy efficiency, altered energy supply, and energy conservation." It called for research and development for renewable energy, fuels from plants (biomass), and nuclear reactors with reduced risks of accidents, waste, and arms proliferation. It also favored reducing deforestation, accelerating reforestation, and imposing taxes on pollution to reflect its external costs.[16]

Failure to confront threats to the atmosphere, the report warned, would "imperil human health and well-being." Included in the long list of threats were diminished global food security, increased political instability, and the more rapid extinction of animal and plant species "upon which human survival depends."[17]

The report said that international cooperation to slow global warming should reflect the differentiated resources of the world's nations. It declared, "The countries of the industrially developed world are the main source of greenhouse gases and, therefore, bear the main responsibility to the world community for ensuring that measures are implemented to address issues posed by climate change." Since developing nations with high rates of poverty would need to increase their energy use to improve standards of living, industrialized countries would need to make "compensating reductions."[18]

Wirth's keynote address in Toronto stressed the connections between the ozone treaty, the G7 call for action, and what he called Hansen's "stunning and broadly reported testimony" with its "ominous new warning . . . *that the greenhouse effect is upon us.*"[19] He urged all nations to use energy efficiency to reduce carbon emissions. This approach, he said, was "good environmental policy, good energy policy, and good economic policy, and for the U.S., good for our national security." The senator argued that alternative energy sources would also have multiple benefits. Solar photovoltaics were environmentally benign and could provide flexible sources of energy for developing nations. He said the United States would probably have to "start all over again with nuclear, strengthening our research efforts if we can develop a new generation of passively-safe, economical nuclear power plants."

The flurry of congressional hearings and international conferences on climate change took place amid the contest to replace term-limited Reagan. In mid-July, the Democratic faithful gathered in Atlanta to confirm Massachusetts Governor Michael Dukakis as their presidential nominee and adopt their party platform. The document briefly mentioned climate change, with a call for an international summit to address the "greenhouse effect." However, the party's energy proposals differed from suggestions by Wirth and others at congressional hearings about how to deal with the threats of climate change. The platform called for "encouraging the use of our vast natural gas and coal reserves," as well as new oil and gas drilling, renewable sources of energy, and energy conservation. These measures, it hoped, would allow the country to "reduce its reliance on nuclear power."[20] As with earlier party platforms, the Democrats' robust support for coal over nuclear raised questions about U.S. capabilities to respond to international agreements calling on nations to reduce carbon emissions.

At the New Orleans Superdome a month later, Republicans chose Vice President George H. W. Bush to head their ticket and adopted their vision of U.S. priorities. The Republican language on climate resembled that of the Democrats. It cited the successes in protecting the ozone layer and called for similar "international agreements to solve complex global problems such as tropical forest destruction, ocean dumping, climate change, and earthquakes." The platform's rhetoric on energy did not sound very different from the Democrats', except on one matter. The GOP wanted to increase rather than reduce reliance on nuclear power. However, elsewhere in the document, the Republicans' strident attacks on government and taxes of any kind (amplified by Bush's call for "no new taxes") created their own obstacles for climate initiatives recommended in congressional hearings.[21]

Several days later, Reagan reaffirmed his support for nuclear power and for the first time mentioned, albeit briefly, the climate implications of fuel choices. At his signing of amendments to the Price-Anderson Act that guaranteed government compensation to victims in the event of a nuclear accident, he touted the benefits of nuclear power, which now furnished close to a fifth of U.S. electricity. "To replace this energy with electricity produced by oil would require two million barrels of oil per day, pump 350 million tons of carbon dioxide into the atmosphere each year, and, if the demand is not met from domestic reserves, worsen our trade deficit by more than 1 billion dollars per month to purchase foreign oil."[22] It was a rare and easy-to-miss Reagan assertion that reducing carbon emissions was a worthy goal.

12

Maggie Thatcher's Science

Reagan, September 1988 to January 1989

On September 27, 1988, the Royal Society welcomed as its dinner speaker a longtime member—British prime minister Margaret (Maggie) Thatcher. In London's historic Fishmongers' Hall, the former chemistry major at Oxford's Somerville College quipped, "I am quite pleased that I didn't continue my work on glyceride monolayers in the 1950s, or I might never have got here!"[1]

The Conservative Party stalwart heaped praise on the 343-year-old institution and mentioned that many portraits of the country's eminent scientists—including Michael Faraday and Isaac Newton—adorned the walls at 10 Downing Street. She underscored her support for science by quoting Alfred North Whitehead's dictum that "a nation which does not value trained intelligence is doomed."

Acting on a careful plan, the "iron lady" emphasized the complicated interplay between scientific progress and the environment. For instance: "Research on agriculture has developed seeds and fertilizers sufficient to sustain [a] rising population, contrary to the gloomy prophesies of two or three decades ago. But we are left with pollution from nitrates and an enormous increase in methane which is causing problems." Similarly: "Engineering and scientific advances have given us transport by land and air, the capacity and need to exploit fossil fuels which had lain unused for millions of years. One result is a vast increase in carbon dioxide. And this has happened just when great tracts of forests which help to absorb it have been cut down."

Echoing Roger Revelle's 1957 article, she warned, "For generations, we have assumed that the efforts of mankind would leave the fundamental equilibrium of the world's systems and atmosphere stable. But it is possible that with all these enormous changes (population, agricultural, use of fossil fuels) concentrated into such a short period of time, we have unwittingly begun a massive experiment with the system of this planet itself." Sounding like James Hansen, she noted that "the five warmest years in a century of records have all been in the 1980s—though we may not have seen much evidence in Britain!"

It was time, she said, to consider the broader implications of climate change for energy production, fuel efficiency, and reforestation, which she considered "no small task." To achieve the needed results, she said, "We must ensure that what we do is founded on good science to establish cause and effect." She considered economic growth and environmental protection as mutually compatible. "Stable prosperity can be achieved throughout the world provided the environment is nurtured and safeguarded. Protecting this balance of nature is, therefore, one of the great challenges of the late twentieth century."

Thatcher had worked with her advisors on the Royal Society talk for two weeks because she believed confronting global warming using hardheaded scientific principles was consistent with her conservative views of government. Reflecting on the speech years later, she saw her approach as a balance to the "environmental lobby," which "used the concern about global warming to attack capitalism, growth, and industry." She also complained about the environmentalists' antipathy to nuclear power, even though "it was a far cleaner source of power than coal."[2]

Thatcher was disappointed with "the extraordinary lack of media interest" in her ground-breaking statement. Contrary to her expectations, no film crews covered the occasion. Still, the call for climate action from an influential leader with solid conservative credentials was another critical step in building an international coalition working toward slowing global warming. Her stance was even more remarkable because the United Kingdom generated three-quarters of its electricity from domestically produced coal. Some analysts had predicted that coal-dependent nations might block effective international agreements to cut CO_2 emissions. But Thatcher appeared willing to abandon coal—yet another milestone in climate policy history.

*　*　*

With additional international climate meetings scheduled for late 1988, the U.S. State Department's bureau of oceans and international environmental and scientific affairs asked its external advisory committee to prepare a discussion paper on the "U.S. International Posture on Climate Change."[3] The document wasn't intended for public distribution but still mattered, partly because of who wrote it.

The advisors included three top foreign policy advisors plus three of the country's most influential scientists—Robert White, Gordon MacDonald, and William Nierenberg. This committee was the first collaboration between MacDonald and Nierenberg since they worked on the 1979 JASON report that recommended changes in energy policy because of the risks of climate change. After that, their views appeared to diverge. Before congressional committees and others who would listen, MacDonald patiently spelled out why the scientific evidence justified fear of climate change and advocated specific policies to slow the expected warming. Nierenberg had chaired *Changing Climate,* the bible of those arguing it was too early to adopt climate policies. Both were veteran advisors at the highest levels of government who understood the expectation of a united front. There would be no minority report.

The committee's October 31 conclusions used the Villach report (not the unmentioned *Changing Climate*) as its starting point. As for the science, the committee cited the works of Murray Mitchell, Tom Wigley, and James Hansen supporting the hypothesis that concentrations of atmospheric carbon dioxide, rather than natural fluctuations, were causing increases in global temperature. While not totally abandoning the view that it still might be the "noise," the report tilted in that direction. The advisors urged a review of the data and analyses under international auspices, perhaps with the technical assistance of the National Center for Atmospheric Research. They also recommended establishing new monitoring systems, addressing deficiencies in mathematical models, and supporting *"continued studies of potential effects of climate change on social activities* such as agriculture, water resources, coastal habitability, etc."

The new report adopted a nuanced but pro-activist approach to government policymaking. It recommended that *"the U.S. government advocate a prudent set of energy options that: (1) favor lower CO_2 emissions, (2) would be* *desirable for other environmental reasons, and (3) would not foreclose future options that might be desirable in the light of improved understanding.*" With greater specificity, it called for adopting energy efficiency technologies, using fossil fuels with the least carbon dioxide emissions, and ultimately increasing the fraction of world energy production and consumption coming from non-fossil sources. On the latter, it opined that the nuclear energy option needed to be revisited and noted the improving efficiency of solar photovoltaic cells.

The advisors discussed the recent Toronto Conference advocacy for a 20 percent reduction in CO_2 emissions by the year 2005. But they suggested that "*it is premature to set quantitative goals for CO_2 reduction at the present time because of present scientific uncertainties and a lack of understanding of the full economic and social consequences and how such a goal would be achieved.*" Nonetheless, they urged U.S. advocacy that every country explore how it could achieve feasible goals. Despite several notes of caution, the overall paper sounded more like John Chafee or Gordon MacDonald than the White House science office.

The committee also urged the United States to exert leadership, with the president issuing a strong statement or appearing at the United Nations. The presidential general election the following week would make Reagan a lame-duck president. However, the report noted that "presidential candidates of both parties have declared that they will call an international meeting to address the climate issue."

In November, Roger Revelle (with coauthor David Burns from the American Association for the Advancement of Science) contributed a brief paper to the World Congress on Climate and Development in Hamburg, Germany.[4] The University of California professor was still trying to resolve the protracted puzzles of climate change. No person had a longer perspective on the inter-play between climate science and politics.

In "Energy Options and Climatic Effects," Revelle displayed considerable disappointment with the state of both the science and the politics. Despite the rosy optimism (bordering on hubris) of the 1950s and 1960s, scientists still did not understand many fundamental aspects of climate. He lamented the limitations of climate modeling after all these years, saying much work remained before it would be useful for regional impact studies.

Revelle was nonetheless clear that the long-term risks were grave. Previewing the article in an interview with the *Los Angeles Times,* conducted at his home overlooking the Pacific Ocean, Revelle foresaw a sea level rise of about one meter in the next hundred years. "That doesn't seem very spectacular, but that's quite serious in countries like Bangladesh, where about one-third of the country is within one meter of sea level. Florida is the same way. And, of course, Louisiana."[5]

Revelle's quest for greater certainty derived in part from his frequent interactions with government officials. Reagan's science office had argued against climate action until all was known. Moreover, the professor reported that Senator Gore had told him that his political colleagues were uncomfortable with "approximations." Revelle urged reducing the unknowns about climate change. However, because of the long persistence of greenhouse gases in the atmosphere, he stuck with his position that protective action should not wait on scientific certainty. For him, the main question was "Can we design a long-term energy strategy that slows down or reduces the emission of greenhouse gases, *without penalizing human welfare?*" The answer would depend on the "feasibility and cost of a wide array of improvements in energy technology."

For many countries, he said, nuclear energy was the most promising way to achieve long-run reductions in fossil fuel combustion. Despite the unpopularity of nuclear in the United States and some European countries, he regarded increasing its use as "a cause worth arguing for." But his advocacy came with caveats—the necessity of coming up "with persuasive solutions to cost and safety questions, and convincing and reassuring ways of dealing with potential dangers from waste disposal and proliferation."

Energy conservation could "buy time" and save money. Massive reductions in U.S. energy use came after the "extraordinary price increases in the cost of oil imports in the 1970s." But fifteen years later, "the message seems to have been forgotten. Many of the laws and policies are no longer being applied or have been watered down or slowed down to a point of ineffectiveness. Where energy is concerned, we may need OPEC to rescue us from ourselves!"

Revelle concluded that price was "the most effective mechanism ever devised for conserving a scarce resource," and a tax on carbon taxes might be "the only mechanism that is truly effective over time." Ever the realist, who consulted frequently with members of Congress, Revelle admitted that despite economists' calls for such levies, they would probably never "get past the talking stage."

Unwilling to give up all hope of dealing with "undesirable global warming," Revelle urged the consideration of another possible solution that had not received much attention. A changing climate would move the boundaries of boreal forests (made up mainly of spruce, pine, and fir trees) further north due to warming. This change would create the opportunity for massive forestation, particularly in Canada and Russia. These trees could be excellent carbon sinks, though they would be insufficient, absent other actions, to deal with the expected increase in CO_2 emissions.

If a younger scholar had written the article, it could be viewed as a politically realistic overview of climate change options. But this was Roger Revelle. For more than three decades, the pioneer of climate science had worked tirelessly to alert scientists and nonscientists worldwide about a threat to the planet and suggest possible solutions. It is tempting to read this long-forgotten contribution from late in his life as a call to persist in the effort to find solutions, yet permeated by a sense of pessimism about the future and a foreboding that humans might not be up to the challenge.

The first meeting of the Intergovernmental Panel on Climate Change, November 9 to 11, 1988, brought together twenty-nine nations. In Geneva, the body established its structure for future proceedings and selected its leadership. The United States was pleased with the results.[6]

The IPCC set up three groups to conduct its work. The United Kingdom would chair the one on climate science. The Union of Soviet Socialist Republics would head a second, on the socioeconomic impacts of climate change. U.S. assistant secretary of state Fred Bernthal would lead the panel on policy responses for limiting or adapting to climate change—the working group with perhaps the most significant political repercussions. Bert Bolin (pronounced "bo-LEEN"), a highly regarded meteorologist at the University of Stockholm, would chair the overall effort. The U.S. delegation viewed the Swede as "very competent" and "sympathetic to the U.S. interests." Top appointments also went to Saudi Arabia and Nigeria. With five of the six top leadership positions going to the world's leading oil producers, the organization was designed to deflect threats to oil interests.

Senator Rudy Boschwitz struck an upbeat but noncommittal tone when he delivered the official American position. The Minnesota Republican told his audience, "While we do not yet have conclusive evidence of climate change, the United States believes that it is sensible to begin to examine our potential options for response." The United States, he said, welcomed the formation of the IPCC and stood "ready to work." Boschwitz was particularly bullish on a strategy of forestation. He spotlighted an ongoing project to plant 52 million trees in Guatemala to offset carbon emissions from a new coal-burning plant in Connecticut.

On December 6, the UN General Assembly adopted a resolution proposed by the government of Malta on "Protection of the Global Climate for Present and Future Generations of Mankind." The statement noted "with concern that the emerging evidence indicates that continued growth in atmospheric concentrations of 'greenhouse gases' could produce global warming with an eventual rise in sea levels, the effects of which could be disastrous for mankind if timely steps are not taken." It strongly endorsed the IPCC and urged full support for its activities. The UN resolution represented a substantial victory for those pushing to put the climate issue on the international agenda.

Alan Hill's appointment to head the White House Council on Environmental Quality in 1981 was followed by a long period when he largely flew under the political radar, with a staff slashed from previous levels. Despite service under Governor Reagan in California, he had little clout at the White House. Unlike the presidential appointees at the EPA and the Department of the Interior, his name rarely appeared in the news. Yet in the last weeks of the Reagan presidency, it was Hill who was left standing—a rare environmental appointee who survived both presidential terms.

In April 1987, the CEQ had begun a series of meetings with government scientists about atmospheric ozone depletion and global warming. It also discussed the subject with the National Resources Defense Council, a frequent litigant on environmental issues. The CEQ believed it was legally compelled to act under the National Environmental Policy Act if these newly identified threats were "reasonably foreseeable" and federal programs would affect them.

After examining the evidence, the CEQ concluded that there was indeed sufficient cause for serious concern. Rather than wait for the resolution of all scientific uncertainties, it prepared draft guidance directing federal agencies to examine how their actions might contribute to, and could be affected by, global climate change.[7] With Senator Chafee and former CEQ chair Speth lobbying for such action, Hill was quietly moving forward to use NEPA authority to expand the scope of environmental impact statements and, ultimately, government climate policies.

The final decision lingered into the last weeks of the Reagan presidency. Finding out the proposal was headed to the Oval Office, the Domestic Policy Council and the White House legal office complained about a lack of consultation. They also warned that new requirements would lead to extensive delays from legal challenges because of the uncertainties of climate science. Hill's proposal hit Reagan's desk on January 11, 1989—nine days before the inauguration of Vice President Bush as the forty-first president of the United States.

Reagan tersely penned his decision into his Wednesday diary: "Nancy Risque [secretary to the cabinet] has ordered 'Council on Environmental Control' to kill a proposal re the greenhouse effort. It would have made practically everything illegal."[8]

Reagan's mislabeling of the CEQ and abrupt dismissal of an early attempt to deal with the climate problem raised questions that would carry into the next administration. Prime Minister Thatcher had strongly supported international cooperation on climate and accepted the idea that this would ultimately force her country to alter its energy infrastructure. Under Reagan, the United States had joined other nations to put climate change on the international agenda. But the White House's antiregulatory zeal stymied efforts to constrain greenhouse gas emissions in the United States.

At a November 1987 climate hearing, Senator Wirth had raised the possibility that the U.S. government might be better "coordinating and pursuing our efforts internationally than we may be domestically." The rejection of the CEQ proposal added credence to Wirth's speculation. What if international collaboration produced calls to reduce greenhouse emissions and the world's greatest emitter proved unwilling to do so?

13

Bush in the Middle

The Bush Presidency, 1989

George Herbert Walker Bush of Texas began his presidency in January 1989 with interest in the climate issue on the rise. The former oil executive and U.S. vice president had stirred expectations of an active approach during his campaign. On the trail, he proclaimed, "Those who think we are powerless to do anything about the greenhouse effect forget about the 'White House effect'; as president, I intend to do something about it."[1]

Moreover, the State Department was moving briskly to support the new Intergovernmental Panel on Climate Change. Secretary of State James Baker endorsed global efforts to confront climate change in remarks to the IPCC's response strategies working group on January 30, just ten days after the inauguration. Baker's comments carried particular weight due to his stature in the administration. The savvy Houston attorney with a stellar résumé had served as Reagan's chief of staff and Treasury secretary. He had also managed Bush's losing and winning presidential campaigns and been a close Bush friend for a quarter-century.[2]

Baker warmly welcomed Fred Bernthal (remaining as the working group's chair) and Bert Bolin (still heading the IPCC). He also noted the presence of Bush's nominee to head the Environmental Protection Agency, Bill Reilly—a Harvard law grad, thirtyish senior staffer at Nixon's Council on Environmental Quality, president of the World Wildlife Fund U.S., and a popular choice for environmental advocates. Baker appeared to distance himself from the previous administration's wait-and-see approach by declaring, "We can probably not afford to wait until all of the [scientific] uncertainties have been resolved before we act. Time will not make the problem go away." While scientists refined the state of knowledge, he told attendees, policies should focus "right away" on steps justified on grounds other than climate change. These measures included reducing emissions that harmed the ozone layer, greater energy efficiency, and reforestation.[3]

Baker's ties to the president didn't guarantee the White House was on board with his remarks. Indeed, they provoked a skeptical reaction when a transcript reached the desk of Chief of Staff John Sununu. Sununu had earned a doctorate in mechanical engineering at MIT and taught at Tufts University. He left academia to win election as the pugnacious governor of New Hampshire, where he provided critical assistance to Bush's 1988 presidential campaign. Former colleagues regarded Bush's top aide as extraordinarily confident in his views on many topics, including climate change. Gordon MacDonald remembered him as an external advisor to Nixon's CEQ and described him as someone who was "totally self-assured" and "believed that having an IQ of 180 . . . set him apart from everybody else."[4] Sununu revealed his opinion on the need for climate action with annotations on the text of Baker's remarks. In one sentence, he underlined "probably cannot afford to wait until" and wrote in the margin, "Not true."

The fault line between waiting and acting on climate change had divided Reagan's advisors. The rift would continue, with Bush now the man in the middle.

The State Department wasn't the only agency working on climate change. Reilly fired an early salvo on the subject in an April memorandum to the president proposing a "state of the environment" message.[5] The rarely reticent administrator with media connections outlined a robust agenda for a potential presidential address giving climate change a prominent role. He congratulated Bush for his decision to "phase out ozone-destroying CFCs by the end of the century." He also endorsed higher corporate average fuel economy standards for new vehicles. Legislation signed by Gerald Ford had mandated 27.5 miles per gallon by 1985, but the Reagan administration had, by rule, delayed the final increment into the Bush presidency. The EPA's leader predicted that a decision to implement the last step would reduce carbon dioxide emissions by 2 million tons a year and rank as a "significant plus" for the administration.

Reilly also advocated for the president's personal leadership backing an international framework convention on climate change. He warned, "Scientists agree that a change of even 1–2 degrees Celsius could produce significant flooding of our coastal population centers" and droughts that "could plague many currently fertile agricultural areas." He acknowledged that some Bush advisors were wary of an international agreement because other countries might force the United States to go beyond what it could accept. Reilly countered that all the industrialized nations would face hurdles in reducing emissions, so reasonable compromises should be attainable.

Three days later, Bush forwarded Reilly's memorandum to Sununu, saying, "It makes me realize that this is a full agenda we need to move on." He told the chief of staff to get back to him about the possibilities of an environmental speech and "how fast we can move on the other issues."[6]

Even as Bush pushed for action, delaying tactics within the administration threatened to derail Reilly's climate agenda. A week earlier, budget analysts had raised objections to the Transportation Department's draft rule to allow the 27.5-mile-per-gallon standard.[7] The Office of Management and Budget didn't oppose letting the 1975 law go into effect but wanted to remove the description of its benefits, lest favorable comments undermine any future attempts to repeal the standards. The budget office specifically objected to Transportation's language that its decision reflected its "recognition of the role of oil conservation in reducing carbon dioxide emissions which contribute to the greenhouse effect." The OMB complained that the department should not highlight problems "that we do not agree exist."

State, the EPA, and Transportation appeared headed in one direction on climate change, the president's chief of staff and the OMB in another. It was too early to discern which side would prevail.

International talks scheduled for May forced several early decisions on the administration's intentions. The day that Bush told Sununu to get moving on Reilly's suggestions, the Domestic Policy Council was working to limit U.S. commitments at upcoming meetings. DPC staff briefed director Roger Porter for a call to Robert Zoellick, counselor to Secretary Baker. The call's objective was to deflect National Security Council pressure for U.S. leadership in setting a global climate policy agenda. The NSC believed a U.S. delay in agreeing to emissions reductions would cede leadership to Europe and Japan, making the U.S. negotiating team the "caboose" in leading the world on climate. The DPC rebutted: "No matter how limited, the process could be uncontrollable and could result in international pressure for the U.S. to sign a protocol which is not based on sound science and could have a serious impact on the U.S. economy."[8]

The DPC won the debate. The United States went into long-scheduled international meetings opposing goals and timelines for emissions reductions. Previous secretary of state Shultz had told Reagan's DPC that losing U.S. momentum on limiting ozone-depleting gases would embarrass the country and he would continue on his current course until the president told him to stop. The new secretary would not join the fight.

The legislative branch of government was also grappling with the threat of climate change. In the first year of the Bush presidency, members of Congress filed dozens of bills on the subject. Many limited their focus to enhancing the understanding of climate science—a goal without much controversy.

In February, Representative Claudine Schneider went a step further and filed one of the most comprehensive and widely supported bills—the Global Warming Prevention Act of 1989. She argued that Congress often adopted national policies while uncertainties remained. Based on the current understanding of the issue, she proposed that society now "phase out greenhouse gases with nonpolluting or less polluting methods that provide the same economic services." The "gentlelady from Rhode Island" was encouraged that improvements in the efficiency of buildings, appliances, vehicles, and industrial equipment over the previous fifteen years had cut energy consumption by one-third and carbon dioxide emissions from what they would have been otherwise. This trend had trimmed U.S. energy bills "by a phenomenal $160 billion per year." She concluded, "Success breeds success, and this is undeniably true about remaining opportunities to save even more money and cut greenhouse emissions through additional efficiency improvements."[9]

In this Republican's view, one attractive opportunity was the compact fluorescent light bulb, which could screw into existing sockets. This "technological marvel" provided the same light as the conventional incandescent bulb advanced by Thomas Edison while using only a fourth of the electricity and lasting ten times longer. Its adoption would reduce CO_2 emissions.

The nation's more than 130 million cars and trucks—"responsible for about one-third of U.S. greenhouse gas emissions"—provided even more potential for energy savings. She declared that mass production of highly efficient vehicles would lower their price and significantly reduce greenhouse gases. To increase their market penetration, the bill mandated a 65 percent increase in the energy performance of the federal government's fleets and provided incentives of as much as $2,000 per car to encourage the general public to buy ones with better mileage.[10]

The legislation offered a wide range of solutions, from reforestation to expanding the use of solar energy. It was not highly prescriptive, largely calling on federal agencies to prepare "plans." Still, it recognized that there was no single answer to the problem. The battle would have to be fought on multiple fronts.

Over a third of the House signed on to Schneider's bill. Despite its GOP authorship, the bulk of cosponsors came from the Democratic side of the aisle. But the twenty-five Republican cosponsors included the party's new minority whip, Newt Gingrich of Georgia.[11]

Later that year, Vermont's Patrick Leahy introduced a similar bill in the Senate. Nine other Democrats (including Tim Wirth and Al Gore) and six Republicans (including Rudy Boschwitz of Minnesota and John Heinz of Pennsylvania) joined as cosponsors.[12]

Gore's Senate subcommittee held its second climate hearing of the year in early May. The seven scientists testifying were making repeat appearances, and senators continued to stress the seriousness of the issue. Full committee chair Fritz Hollings of South Carolina declared, "We must ensure that scientific facts are not dismissed merely because they force us to make uncomfortable decisions. For the sake of our children and our grandchildren, we must face the consequences of our actions now." Senior Republican Larry Pressler called global warming "very likely the most serious problem facing this world in the near future." He foresaw that it would affect everyone, adding, "In my home state of South Dakota, for instance, rising sea levels are not likely to be a great concern, but drought and destruction of forests can be devastating."[13]

Despite the evidence of bipartisanship, the Bush administration was in for a rough morning. Gore excoriated the budget office's altering the written testimony of James Hansen—more evidence of the tension between being a scientist and a government employee. An indignant Gore called the administration's actions part of "a dangerous pattern of environmental fraud that is becoming all too routine." When Hansen's slides failed to display, Gore jested: "We saw an OMB official near the projector."

Adding to the tension, senators had just learned of the Domestic Policy Council's success in reining in U.S. commitments at the IPCC meeting. Gore exclaimed, "Today in Geneva, our negotiators on the global warming problem have been instructed to argue against their better judgment that no actions are necessary because we need more study . . . President Bush, only months ago, told this nation that he was an environmentalist. And yet, in the past few days alone, we have seen his administration back away from a critical diplomatic initiative on global warming."

The hearing attracted front-page coverage in the *New York Times*. Tipped off in advance to the administration's influence on Hansen's statement, the daily published a story, "Scientist Says Budget Office Altered His Testimony," just hours before the hearing began. The next day, the paper's lead editorial asserted, "Many countries are now eager for President Bush to take the lead on . . . the feared warming of the earth's atmosphere by pollutant gases like carbon dioxide. But despite Mr. Bush's ringing campaign pledge to do just that, his Administration flounders in confusion and timidity." It opined, "Leadership on the issue has thus fallen to Europe. Last month, Prime Minister Thatcher made her cabinet sit through a daylong briefing on the greenhouse effect from climatologists. And yesterday, the British delegate to the United Nations called for a new international convention to deal with global warming." The paper believed, "It is far too soon to advocate the most direct and drastic remedy, which is to stop burning coal. But it makes eminent sense to buy insurance against global warming with steps that are worth taking in their own right, from raising auto efficiency to protecting tropical forests."[14]

The clamor over Hansen's testimony and the U.S. retreat in Geneva overshadowed some of Gore's witnesses' nerdier but consequential points. They described, for instance, some progress in providing more detailed, geograph-ically specific projections of climate impacts. The scientists considered their models generally reliable but far from perfect.

Leadoff witness Stephen Schneider described a climate model as "a math-ematical representation of the atmosphere, oceans, and the important com-ponents of the climate system, where we describe them by basic physical laws." Solving the equations in models required dividing the planet into man-ageable grids. The resolution possible with the computer power available in the late 1980s could analyze a grid as small as the state of Colorado. This size was good enough to identify some regional trends but not adequate for analyzing climate change impacts on, for instance, clouds or violent storms. He testified that predicting changes in cloudiness "was an essential part of reliable simulation. Yet no global climate model now available or likely to be available in the next few decades has a grid fine enough to resolve individual clouds, which tend to be a few kilometers rather than a few hundred kilo-meters in size." Schneider asserted that "the greenhouse effect" was, none-theless, more than "a speculative theory." It had been validated by literally thousands of laboratory experiments and millions and millions of satellite and balloon observations.

Schneider concluded, "Predictions about something as complex as the global climate system response to forcings controlled by human behavior will always be somewhat uncertain." He cautioned, "Of course, while scien-tists study and debate, the world becomes committed to a growing dose of greenhouse gases and their impacts."

Later in the hearing, the committee called on Jerry Mahlman, director of NOAA's geophysical fluid dynamics laboratory, who outlined more clearly than witnesses at dozens of previous congressional hearings how policy-makers should consider the findings of science. Like Hansen a government employee, Mahlman had convinced his overseers that he could express his personal assessment of the subject as a scientist rather than hew to a gov-ernment line.

Mahlman testified that, when discussing climate change, scientists and policymakers often divided predictions into "certain" and "uncertain." Since scientists strenuously avoided claims of "certainty," even in the face of overwhelming evidence, all findings ended up in the big bucket of "uncertainty." This simplistic dichotomy had provided the foundation for the wait-and-see approach of the 1983 National Academy study and the Reagan science office.

Mahlman described another way to sort out the scientific evidence. He proposed four levels of confidence and probability. He eschewed a category for "certainty" because scientists were rarely sure about anything. Even his level of "virtually certain" climate model indications was nearly empty. He urged more attention to matters that were "very probable" (betting odds of 9 out of 10), "probable" (around 2 out of 3), and "uncertain" (defined as hypothesized effects that the climate modeling community had not yet been able to evaluate properly).

Even with a high level of scientific skepticism, Mahlman considered "global surface warming over the next century" and "arctic winter surface warming of a substantial magnitude greater than the global average accompanied by a reduction of sea ice in high northern latitudes" as "very probable." The difficulty of putting more impacts in this category was the resolution problem and the limitations of computer power that Schneider and Hansen had discussed. Mahlman's list of "probable" trends included "a marked decrease of the soil moisture over some midlatitude interior continental regions during the summer" and a "rise in global mean sea level."

Mahlman also suggested a different way of looking at uncertainty. Even if theories had not been rigorously evaluated—and perhaps could not be, due to the inability of computers to fully reproduce complex natural phenomena—they should not be ignored. Mahlman pointed to one particular risk, one that had attracted the attention of Revelle and his colleagues in 1965. In his model, the Antarctic region was "refusing to warm up for virtually a century." He hypothesized, with "some corroborating evidence from the real world," this might be attributed to an upwelling of colder water in the Antarctic ocean "that effectively provides a thermal buffer against the greenhouse effect." This explanation was far from a "prediction" of the eventual massive melting of ice when the effects of the buffer waned. But it identified a potential "big surprise" whose complexities taxed the capabilities of computers and likely would for a long time.

Mahlman's schematic resembled that used in much of government policymaking, from health to economics to the environment. Climate change was a rare field held to a standard requiring a complete understanding of the natural world that could predict the future with precision at the local level. The NOAA modeler offered a viable approach to apply the rigorous standards of science and communicate them in a way the public could understand.

Reflecting on what the hearing had accomplished, Hansen regretted that the finer points of his testimony (such as the complex relationship between greater average rainfall and more intense droughts in some regions) had gotten "lost in the brouhaha." He resolved to avoid future hearings and "go back to pure science and leave media interactions to people such as Steve Schneider and Michael Oppenheimer, people who were more articulate and seemed to enjoy the process."[15]

Stung by questions about his commitment to climate action, the president issued a statement four days after the Gore hearing. He boasted of the U.S. role at the Geneva meeting, "establishing a process for considering how to respond to climate change." He said the response strategies group would hold a workshop in the fall hosted by the United States. He looked "forward, personally, to reviewing its results."[16]

On June 8, Bush made another attempt to shore up relations with environmentalists by hosting a White House "listening session" with seventeen of the nation's leading environmental organizations in the Roosevelt Room. Invitees included Jay Hair (president of the National Wildlife Federation), Kathryn Fuller (president of the World Wildlife Fund), Peter Berle (president of the National Audubon Society), Michael Fischer (executive director of the Sierra Club), and Fred Krupp (executive director of the Environmental Defense Fund). Climate was on the agenda, and Bush reiterated his support for the IPCC process. The president assured his visitors that he was pleased with Reilly's performance at the EPA. In addition, he mentioned his intent to nominate Mike Deland, an EPA regional director also popular with advocacy groups, as the head of the Council on Environmental Quality (replacing Alan Hill, held over from the Reagan administration).[17]

Fifteen days later, Reilly launched two memoranda to Sununu on how to deal with climate change. One on "Reforesting America" presented an ambitious plan to plant 10 billion trees on private, state, local, and federal lands. He noted, "Trees will cool houses, pull carbon out of the atmosphere to improve air quality, reduce energy demand in urban areas, and provide an initially small but alternative energy source from tree plantations." The proposal was compatible with the idea that early climate policies should have multiple benefits. His second communication called for "An International 'Mission to Planet Earth.'" Reilly urged the president to work with other nations to launch new Earth-monitoring spacecraft "to provide essential information on global climate change caused by increasing atmospheric levels of 'greenhouse gases.'"[18]

The CEQ was even more ready to push the envelope in dealing with climate change. On June 21, Hill distributed draft guidance to federal agencies on incorporating climate impacts in environmental impact statements. In mid-February, the lame-duck chair had told the *Los Angeles Times* he was optimistic that the Bush administration was more positive about the idea than the previous one. If finalized, the directive would affect federal decisions on vehicle mileage standards, oil and gas development on federal lands, and forest management.[19]

While Reilly and Hill urged action, other forces within the administration again tried to slow things down. One hurdle at the White House for climate advocates was an influential insider—Boyden Gray, counsel to the president. The North Carolina–born attorney and former clerk to U.S. Supreme Court Chief Justice Earl Warren had the president's ear. During the Reagan years, Gray had served as counsel to the Presidential Task Force on Regulatory Relief, headed by then vice president Bush. The task force insisted on rigorous cost–benefit analysis, rarely found a regulatory solution to a national problem it liked, and advocated repealing energy efficiency legislation already in place.

On June 22, Gray warned national security advisor Brent Scowcroft to be wary of European interest in climate change. For instance, England and France "would benefit from using the global warming issue to limit Germany's industrial prowess." In addition, the president's chief lawyer adamantly opposed new U.S. standards for vehicle efficiency. "No focus on the automobile . . . can stand scrutiny if *all* factors of greenhouse gas production *and extraction* are taken into consideration." He recommended that international global warming initiatives be limited to continued scientific studies and reforestation.[20]

The White House science chief agreed with its top lawyer. On July 7, while awaiting Senate confirmation, Allan Bromley summarized for the president the state of climate research. The Canadian-born scientist—hard to miss because of his striking gray hair and colorful bow ties—had a stellar résumé. His previous positions included Henry Ford II Professor of Physics at Yale (with a specialization in the dynamics of atomic nuclei), president of the American Association for the Advancement of Science (1982), and recipient of the National Medal of Science (1988). Bromley told Bush, "I remain convinced that at the present time, our predictive capabilities and the available data are not adequate to justify any large-scale projects aimed at ameliorating global warming."[21]

Additional evidence that the enthusiasm of the EPA and the CEQ for climate action would encounter roadblocks at the White House came in mid-July. The Domestic Policy Council informed agency heads that the CEQ's draft guidance on environmental impact statements had been "circulated without proper clearance" and did "not reflect administration policy at this point." The intent of the DPC memo was, in its drafters' internal comments, "to undo Hill's mischief."[22]

In late July, Bush signed the Natural Gas Wellhead Decontrol Act of 1989, completing the move to market pricing of gas begun by Carter and expanded by Reagan. In previous decades, natural gas price controls had been one of the most divisive issues in American politics. This time, price reform breezed through Congress with broad support. During remarks at the ceremony, the president predicted the legislation would help eliminate the gas shortages of the past. In addition, "Natural gas burns much more cleanly than other fossil fuels," he said, "and can play a larger role in our efforts to clean up our air and our water." He did not specifically mention climate. However, scientists like Revelle and MacDonald had suggested to Congress over the years that an initial approach to slowing global warming could be to displace some coal with gas, now a greater possibility with the removal of price controls. Bush cited estimates that U.S. natural gas reserves would "take us to the year 2025 and beyond."[23]

*　*　*

During the first week of November, the differing views of the United States and other nations on climate responses came to a head in Noordwijk, a small former fishing village on the Dutch coast of the North Sea. Sixty-seven countries sent representatives, usually their top environmental official, and Reilly led the U.S. delegation. The result, intended for use in the IPCC process, was a call for action.[24]

The Noordwijk declaration began, "Based on our current understanding, society is being threatened by man-made changes to the global climate . . . Predictions available today indicate potentially severe economic and social dislocations for future generations. Assuming these predictions, delay in action may endanger the future of the planet as we know it."

The declaration cited the special responsibilities of the industrialized nations due to their contributions to greenhouse gas concentrations and their capabilities for responding. It urged these countries to set an example with domestic action to reduce emissions. They should also, it said, support nations with lower incomes, where "the protection of the atmosphere would prove to be an excessive burden" and promote sustainability in the nations establishing industrial facilities for the first time, where there was "a unique opportunity to include up-to-date technologies for controlling the emissions of greenhouse gases."

The participants found it "timely to investigate quantitative emissions targets to limit or reduce CO_2 emissions" and encouraged the IPCC to analyze such options. They also endorsed the concept of "CO_2 equivalence" as a single parameter to describe the radiative effects of the various greenhouse gases, an idea the United States had encouraged. This approach would allow countries to select the most cost-effective strategies when setting their targets.

Although the declaration appeared to break new ground, it drew sharp criticism from environmental advocates in the United States. They believed that targets and timelines for emissions reductions were essential ingredients of an effective international strategy, and many participants at Noordwijk were ready to do more than "investigate" them. The Dutch, for instance, announced plans to unilaterally hold carbon emissions to their 1988 level by 2000, whatever the conference results. Several European nations had supported targets and timelines, but the Americans insisted on a vaguer commitment.

The wrangling at Noordwijk put Reilly in a tough spot. The EPA chief had a long meeting there with German chancellor Helmut Kohl just days before the fall of the Berlin Wall. At the end of the session, Kohl stood up, looked Reilly in the eye, and said, "On the shoulders of your country and my country rests whatever hope the planet has to avoid a planetary catastrophe. And on your shoulders, Mr. Reilly, is the responsibility to bring your president to understand the significance of this issue." Years later, Reilly still got chills when he recalled Kohl's plea.[25]

The U.S. position at Noordwijk took flak back home. Both parties in the Senate—including now majority leader George Mitchell and John Heinz, a Republican from Pennsylvania—criticized the American delegation for resisting the European push to move faster. The headline in the *New York Times* read, "U.S., Japan, and Soviets Prevent Accord to Limit Carbon Dioxide." The article quoted the Audubon Society, which charged, "The White House has sabotaged the first international effort to make good on the President's word. This is not the way to lead the world."

Back in DC, Reilly had an hour-and-a-half lunch with the president, during which he discussed the economics and politics of international cooperation. But in his words, "It wasn't enough." In public, however, Reilly pushed back against the negative press by suggesting that U.S. hesitance to endorse targets and deadlines was more a matter of timing than policy.

Reilly's interpretation of events received some credence when Margaret Thatcher addressed the United Nations on climate change, just as the Noordwijk conference ended. Evoking Charles Darwin, John Milton, and the Garden of Eden, the prime minister depicted the issue in a broad context: "As we travel through space, as we pass one dead planet after another, we look back on our earth, a speck of life in an infinite void. It is life itself, incomparably precious, that distinguishes us from the other planets. It is life itself—human life, the innumerable species of our planet—that we wantonly destroy. It is life itself that we must battle to preserve."

She dismissed the idea that complete scientific understanding had to precede any remedial action. She declared, "The evidence is there. The damage is being done. What do we, the International Community, do about it?" She also argued that because global climate change affected everyone, "It is no good squabbling over who is responsible or who should pay." Advances in climate science should guide policy decisions. "We must use science," the former chemistry major said, "to cast a light ahead so that we can move step by step in the right direction."

The Iron Lady urged climate advocates to resist blaming "modern multinational industry for the damage which is being done to the environment. Far from being villains, it is on them that we rely to do the research and find the solutions." She also offered some advice for the leaders of industry: "The multinationals have to take the long view. There will be no profit or satisfaction for anyone if pollution continues to destroy our planet."[26]

Some of the prime minister's address might be dismissed as more rhetoric than substance. But behind the literary flourishes, she provided a general road map for the United Kingdom's response to climate change. The country would work to revamp its approach to industrial pollution, agricultural practices, and transportation. On the latter, the UK would look for ways to "strengthen controls over vehicle emissions" that "produced better solutions than current technology for carbon dioxide emissions and the greenhouse effect."

Wading into an issue that might further separate her from the U.S. negotiating position on climate, Thatcher declared, "Each country has to contribute, and those countries who are industrialized must contribute more to help those who are not." She also insisted that future protocols "must be binding, and there must be effective regimes to supervise and monitor their application. Otherwise, those nations which accept and abide by environmental agreements, thus adding to their industrial costs, will lose out competitively to those who do not."

Thatcher accepted that setting targets and deadlines could wait until the following year, partially supporting Reilly's view that the critics of the U.S. position at Noordwijk might have overdramatized the situation.

Behind-the-scenes maneuvering at the White House, however, provided evidence that the administration wanted more than a delay in targets and timetables.

Bromley had alerted Sununu in October about "rumors that EPA was moving toward support of CO_2 reduction programs of a substantial nature."

He reported one German official's surprise at the White House science office's position that "the currently available science did not support such programs at the present time" when the EPA "had just indicated strong support for programs directed toward major CO_2 emissions reduction prior to 2000." Bromley told the chief of staff, "It is important for us to settle this matter internally so that we can present a united coherent posture in our discussions with foreign groups and leaders."[27]

In response, Sununu quickly established a working group on global climate change within the Domestic Policy Council. It would include the heads of agencies with climate responsibilities, such as the Departments of State, Energy, and Agriculture, and the EPA. From the White House, the Council on Environmental Quality and the Council of Economic Advisers also joined. Bromley chaired the new group, solidifying his central role in briefing Sununu and Bush on the subject.

As part of his assignment to keep Reilly in line, Bromley had attended the Noordwijk conference. In mid-November, he sent the president his meeting summary with the complaint that many nations and organizations were pushing for climate change responses "carried along by considerable emotion and a degree of mob psychology . . . Scientific facts are largely irrelevant to current discussions." For example, he said, these groups focused on historical, thousands-of-years ice core records showing a striking correlation between atmospheric CO_2 content and temperature, without discussing "the still open question as to which is cause and which is effect."

Bromley issued one more warning. The United Kingdom, he claimed, was trying "to wrest world leadership from the U.S. in matters of the environment" and "should hereafter be viewed as a less than reliable partner."[28]

Establishing the economic context for the climate working group fell to the Council of Economic Advisers (like many at the White House, producing papers on their WordPerfect-readable diskettes). The CEA utilized orthodox economic theory in its analysis, and the shadow of Bill Nordhaus loomed large. At forty-eight, the Yale professor was still actively refining his models and publishing his latest results. The economists working on the matter for the Bush White House often cited him and other economists he had influenced as authoritative sources on the economics of climate change.

Consistent with Nordhaus's views expressed initially in the 1970s, the cost–benefit analysis would use a discount rate (now set at 5 percent) to reflect the real value of money.[29] Discussions among academic economists had previously identified a fundamental flaw in using discount rates for assessing atmospheric carbon where the benefits of cutting emissions were very long term. As if none of these concerns had ever existed, discount rates went unchallenged at the White House, even though Reagan's economists had used different analytic tools to analyze the threat to the ozone layer.

Bush's economists also believed that "the economic effects of living with global warming" before 2050 should not be part of their analysis. This approach was more or less consistent with the Nordhaus view that increases in average global temperature were not worth worrying about until they reached 2 degrees Celsius (3.6 Fahrenheit)—an assumption Nordhaus once called "deeply unsatisfactory." Over the years, the view of tolerable climate change became embedded in the work of leading climate economists and scientists. This view was so standard that it rarely surfaced in public discussions and never merited careful analysis. In economists' professional jargon, they assumed that greenhouse gas emissions had insignificant "external costs" until the year 2050.

Bromley's enhanced role as climate science briefer in chief affected the flow of information to the Oval Office. However, Sununu remained the highest White House official in history to delve deeply into climate science, and his views on the subject were already well established. In contrast, the president claimed no expertise on the subject. He found the ideas fascinating, he said in a handwritten September 1989 note to Bromley, but admitted to "not understanding all the science."[30]

Bromley didn't adopt a traditional briefer's role of presenting the general state of knowledge with his personal views included at the end. The physicist was on a mission: convincing Bush that there was insufficient scientific understanding to justify any action on emissions reductions. He revealed his strategy in an October memo alerting Sununu to an article by an Australian scientist in a recent issue of *Nature*. The piece identified cloud phenomena that would offset the impacts of carbon dioxide on global warming.[31] According to Bromley, the two-page research note underscored "the uncertainties in present predictive capabilities in this area and the difficulties and dangers involved in basing major greenhouse amelioratory programs on current predictions and understanding." In fact, leading climate scientists agreed that they needed to know more about clouds and that some expected phenomena would amplify global warming while others would have the opposite effect. In the big picture, Bromley's "discovery" was hardly something that would move the needle of understanding very far.

Bromley wasn't the only prominent scientist who doubted that the current understanding of climate justified preventive measures. Most notably, Professor Richard Lindzen, a respected atmospheric physicist at MIT, claimed that for a doubling of carbon concentration, correcting the flaws in climate models "could easily reduce the predictions for warming to well under a degree" Celsius.[32]

Bromley's most striking commentary on climate science and technology came in a mid-December memorandum to Sununu in which he disputed suggestions on climate policies from the World Resources Institute.[33] WRI had written Bush earlier in the year urging him to place global climate change *"among the top issues to be addressed by your new administration."* The organization stressed the need for U.S. research and development on energy supply technologies that didn't release carbon dioxide and other pollutants into the atmosphere. Its preferred options included "solar, wind, biofuels, geothermal and other renewables, hydrogen fuel and electric vehicles." It conceded that nuclear power might also contribute, but only with the resolution of the waste problem and the availability "of a new generation of safe, cost-competitive reactors and proliferation-resistant fuel cycles that will be acceptable to investors and the public."

In rebuttal, Bromley argued that "currently the only option for the production of large blocks of electric energy—thus replacing fossil fuel combustion—is the nuclear fission one." He agreed with the WRI that there had to be a "sea change" in nuclear technology, personally favoring not yet available modular high-temperature gas-cooled reactors with standardized units of about 350 megawatts. Bush's science advisor dismissed WRI's renewables list as "inherently niche sources."

There was little argument with the idea that in 1989 most renewables remained niche energy options, though some were more market-ready than modular nuclear reactors. Photovoltaic cells, for instance, were competitive in remote locations far removed from electricity transmission lines, but they weren't yet competitive as substantial energy suppliers. But Bromley's use of the word "inherently" implied that there was something about renewables that prevented them from ever becoming viable alternatives to fossil fuels or nuclear reactors. This position put him on very shaky ground.

Bromley ignored substantial and readily available evidence that some renewables were on the path mapped by the Energy Department during the Carter presidency. The cost of photovoltaic electricity, for example, declined rapidly from $60 per kilowatt-hour in 1970 to one dollar in 1980. By the early years of the Bush administration, the price had dropped even further, to twenty to thirty cents.[34] These lower prices were not yet sufficient to make PV competitive beyond niche markets. But they did support projections that photovoltaics (and other forms of renewable energy) would be able to supply more massive amounts of energy sometime in the early twenty-first century. After years of scientists like Roger Revelle and Frank Press identifying renewable energy as a long-term solution to the carbon dioxide problem, Bromley's blunt dismissal of the renewables option was a notable low point in the history of the White House science office.

Bromley's views aligned well with those of the president's chief of staff. Sununu later recounted that he had advocated for specific positions with the president when he had "very strong feelings about" them.[35] His "personal agenda" included two big environmental controversies of the time. As governor of New Hampshire he had spoken out about the threat of acid rain, and at the White House he had pushed for amendments to the Clean Air Act to ratchet down sulfur dioxide emissions. But it was climate change that stood out as the most significant issue on which Sununu pushed his own views.

The MIT-trained engineer boasted that the Bush White House tried to know "more about every issue than anybody we had to deal with, and we had the horses to do that." On climate, he bragged, "I had those meetings in my office. When those folks came in to talk about global warming, they came in with five or six of their people, and I guarantee you they went out shaking their heads."

Chief of Staff John Sununu (*right*) with President Bush at the 1989 annual meeting of the National Governors Association, an organization Sununu headed the previous year while governor of New Hampshire. Courtesy AP Photo/Ralf-Finn Hestoft.

However, Sununu's arrogance combined with substantial misunderstandings of climate science. In Sununu's words, one peeve was that leading experts were using "these two-dimensional weather models that were designed primarily for short-term predictions that they were using for long-range predictions." He informed his visitors (he believed to their surprise) that they were ignoring the thermal capacity of the oceans. As a result, the models were "garbage in, garbage out." He declared that, on climate, "you don't make trillion-dollar decisions using hoaxed-up models." Despite Sununu's protestations, by 1989 the most influential climate models had long been three-dimensional and much more elaborate than two-dimensional versions. Leading climate scientists were well aware of the thermal capacity of oceans. They publicly acknowledged that it would take a long time for supercomputers (which wouldn't fit in Sununu's office) to simulate the complex dynamics of oceans, even before being chastised in the West Wing. In addition, Sununu somehow came to believe that James Hansen "changed his mind on global warming, and that was the original NASA scientist that pushed it." In fact, there is zero evidence that Hansen ever came to doubt the reality of global warming.

Sununu still had an EPA problem. In December, the agency published a lengthy report, *The Potential Effects of Global Climate Change on the United States,* containing analysis at odds with powerful figures at the White House.[36] The EPA based its findings on workshops with atmospheric scientists, ecologists, hydrologists, geographers, and forestry and agricultural specialists. It also ran scenarios on several leading general circulation models. The agency identified the "limitations" of such models but was unwilling to dismiss them because of remaining uncertainties. The report was cautious in its language and careful to point out the benefits as well as the dangers of global warming.

However, numerous factors led to the conclusion that net impacts would harm nature and human societies. One concern was that climate change from greenhouse gases would have substantial effects within a century— more quickly than natural climate changes of the past. As a result, natural ecosystems would have a very limited ability to adapt. According to the report, "The ultimate effects could last for centuries and would be virtually irreversible." These effects would be observed in the declining health of forests and the loss of biological diversity. The pace of climate change would also make human adaptation to rising sea levels more difficult. The EPA also projected that increases in mortality due to extreme summer heat extremes (particularly among the elderly) would outweigh decreases due to milder winters. However, the report added, "If people acclimatize by using air-conditioning, changing their workplace habits, and altering the construction of their homes and cities, the impact on summer mortality rates may be substantially reduced."

Like Hansen's congressional testimony, the EPA report was subject to interagency review, giving the White House potential influence over its findings. But this project responded to a specific request from the Senate, dating back to 1986, when the Republicans held a majority. As a result, aggressive editing by the OMB might draw additional attention. It was better to allow some leeway and hope the report wouldn't attract much notice.

In at least one important regard, the EPA report, private discussion at the White House, academic deliberations, and congressional hearings took a common approach to climate change assessment—downplay the impacts of U.S. emissions on other countries. Greenhouse gases were a rare environmental threat with global consequences, and the United States was the major emitter. The costs and benefits of U.S. emissions of greenhouse gases would look very different if the impacts outside its borders were not so conveniently laid aside. How would the citizens of the lowest-income countries cope with higher mortality due to extreme summer heating when they could not afford air-conditioning?

The arena for public discussion about climate change was expanding beyond the usual voices during 1989. In a September issue of *The New Yorker,* for instance, journalist (and later college professor) Bill McKibben published a lengthy piece called "The End of Nature." Like his book of the same name published by Random House twelve days later, the article was a pioneering effort to bring the threat of global climate change to the attention of a broader audience. Using skills honed by time as an editor of the *Harvard Crimson* and a staff writer at *The New Yorker,* the twenty-eight-year-old—who one colleague observed had "the laconic bearing of an Episcopal novitiate but worked with the metabolism of a hummingbird"—wrote with a clarity that had been missing from the ongoing skirmishes over climate science.[37]

McKibben provided a tutorial, for those who needed one, on the extant science: Revelle's warning about a "large-scale geophysical experiment," the data from Mauna Loa, Revelle's lesser-known work on the potential release of embedded methane and on the dehydration of the Colorado River basin, and Hansen's interpretation of the dramatic increase in global temperatures observed during the 1980s. Most scientists were reluctant to accept Hansen's assertion of "proof," McKibben noted. The Goddard researcher was "out on a limb, if a fairly stout one."

Much of McKibben's analysis could be found in reports of the National Academy of Sciences, though he cited some predictions of temperature rise from other sources that were less plausible. His major contribution was to frame climate change in terms nonscientists could understand, thus giving the risks greater urgency. He suggested, "Imagine every car on a busy freeway pumping a ton of carbon into the atmosphere [every year], and the sky seems less infinitely blue." He complained that scientific reports using Celsius instead of Fahrenheit and centimeters instead of feet made the threat appear less ominous for Americans unfamiliar with the metric system. Likewise, he resisted the view of the year 2000 as "the symbol of the bright and distant future . . . The turn of the century is no farther in front of us than Ronald Reagan's election to the presidency is behind."

According to McKibben, the human race faced tough choices. "Adjustments to the greenhouse world will not be easy; our addiction to oil is deep. Our every comfort—especially the freedom from hard labor, for those of us who enjoy such freedom—depends on fossil fuels. They allowed us to dominate the earth, instead of letting the earth dominate us . . . Our impulse will be to defy the doomsayers and press ahead into a new world."

Revelle had once said it might take some kind of religious revival for humans to reach a prudent accord with nature. McKibben said the solution would have to rely on human wisdom. "As birds have flight, [humans'] special gift is reason. Part of that reason drives the intelligence that allows us to master DNA or build big power plants." But reason could lead us on a different path by keeping us from "following blindly the biological imperatives toward endless growth and territory . . . Should we so choose, we could exercise our reason to do what no other animal can do: we could limit ourselves voluntarily . . . What an achievement that would be."

The creation of the Intergovernmental Panel on Climate Change, the proliferation of climate bills in Congress, and the expansion of climate advocacy to include celebrities like Robert Redford and Meryl Streep provoked a substantial reaction from the industries most threatened by new policies to slow down global warming. In 1989, the National Association of Manufacturers sponsored a coalition of fifty-seven business corporations and associations to lobby public officials. Eventually known as the Global Climate Coalition, the group acknowledged that many people regarded climate change as "the supertanker of environmental issues." Its views on what to do about it were largely compatible with those of the Bush White House. It emphasized model uncertainties and actions justified even if global warming turned out not to be a problem. The group took umbrage at occasional press reports suggesting more global warming than the scientific consensus and at proposed legislation that was "draconian interventionist," though it cited no specifics.[38]

In December, member companies helped GCC staff set up individual meetings with their representatives in Congress. Given the companies' stature, the potential influence on congressional deliberations was immense.

Not surprisingly, the largest bloc of members included companies that mined oil and coal, and their trade associations. Reorganizations at Exxon had already altered its approach to climate change. When Edward David headed its massive research program—somewhat akin to Bell Labs, where he had once worked—the corporate giant had developed peer-to-peer relationships with leading climate scientists. David talked directly with leading climate scientists like Revelle, Broecker, and Hansen and occasionally provided funding for their universities. Hansen, in particular, was favorably impressed with David's comments that moral considerations and time frames of at least fifty years should affect decisions about climate change and his recognition of the need to eventually transition from fossil fuels to renewable energy.[39] The oil giant couldn't tell university researchers what to say. But Exxon was in a position to provide an occasional nudge.

In the mid-1980s, however, Exxon chief operating officer Lee Raymond pushed David to focus on applied research with more practical value for the company. David resisted, not wanting to limit the creativity of the world-class people he had hired. Raymond stuck to his guns, declaring, "We're not going to be the next Bell Labs. We are going to quit building all these damn buildings for [the researchers], and they are going to have to start producing something that makes sense." In October 1985, David departed "to pursue other business and research interests." One effect of the slashing of the research program was diminishment of the company's reputation among academic scholars. In addition, the company began to rely more on its public and governmental affairs departments to deal with climate science.[40]

Industries that were intensive users of fossil fuels also joined GCC. These included electric utilities and producers of aluminum, iron and steel, petrochemicals, automobiles, and cement. Generators of electricity—usually regulated monopolies—had long sought influence with state legislatures and public service commissions. Reliant mainly on coal as a fuel, they already had to comply with the mandates of the Clean Air Act. New legislative action or international agreements on climate change could disrupt their existing fuel mix.

The interests of other GCC members were less apparent. One, the Coalition Opposed to Energy Taxes, signaled that business interests might see carbon taxes as the worst way to curb greenhouse gases, in contrast to academic economists viewing them as optimal policy. Other notable members were the American Nuclear Energy Council and the American Gas Association. For decades, some scientists had argued that the potential dangers of carbon emissions justified greater use of nuclear power and natural gas. In 1989, the nuclear and gas industries chose to align themselves with coal, arguing that there was no urgency to reduce carbon emissions.

The new collaboration of companies and organizations in the Global Climate Coalition introduced a powerful new force into the political calculus of climate change.

14

The World Speaks

Bush, January to June 1990

The calendar for 1990 included additional milestones in climate change history—the International Panel on Climate Change's initial science and policy response assessments, expected during the summer. Bush's advisors provided conflicting advice on what the president should tell a preparatory IPCC meeting scheduled for February 5 at Georgetown University.

In mid-January, Energy secretary Admiral James Watkins and EPA administrator Reilly sent Bromley their joint recommendations for the president's remarks.[1] The agency heads favored a commitment to "greenhouse gas stabilization as soon as possible, consistent with the requirement for global economic growth." They also urged assistance to developing countries, including the transfer of advanced technologies. Furthermore, they called for expanding support for noncarbon energy sources: renewables (hydro, solar, biomass, and geothermal) and nuclear power (with new reactor designs).

The Office of Management and Budget disagreed strongly with two ideas in the memo. The DOE and the EPA suggested saying, "The issue of global climate change is an important public policy issue . . . as the earth experienced some of the hottest years in the last century (five of the ten hottest years in the last 100 have occurred in the 1980s) and as evidence of a significant build-up in the atmosphere of certain 'greenhouse gases' became more widely known." A budget official wrote in the margin a large "NO." Another idea was to endorse recent statements by Prime Minister Thatcher, which drew an even more emphatic "NO. NO. NO."[2]

The Commerce Department also contradicted Energy and the EPA. It argued that the "relative uncertainty of long-run climate change" should be more clearly "juxtaposed with the relative certainty of the high costs of actions to substantially reduce the output of greenhouse gases." It argued that adopting climate policies right away removed the option of "doing nothing" or "postponing decisions until the results of research are known."[3]

The presidential speechwriting team circulated a draft for comment on February 1. Its combative language suggested that recent observations of polar ice had reduced the threat of climate change, criticized proposed climate responses based on "media-driven emotion or the politics of the apocalypse," and mocked the idea that climate models could predict the future. It said that policies based on models would rest "on the shifting sands of hypothesis and a chaos of conjecture." Bromley advised the creative wordsmiths to tone it down. The audience would be IPCC members, many of whom presumably thought climate change was a genuine threat.[4]

Three days later, the *New York Times* quoted "administration officials," who asserted that State and the EPA wanted the president to "call for strong measures to curb what environmentalists believe is an ominous warming of the world's atmosphere" in his talk to the IPCC. These anonymous sources said Sununu had "played a major role in writing an initial draft of Mr. Bush's speech" but the president had not yet approved that draft.[5]

As delivered, Bush's statement exuded positivity about the IPCC. He told the group, which included panel chair Bert Bolin and UN Environmental Programme executive director Mostafa Tolba, "By being here today, I hope to underscore my country's and my own personal concern about your work, about environmental stewardship, and to reaffirm our commitment to finding responsible solutions." Preaching to the choir, he acknowledged, "We all know that human activities are changing the atmosphere in unexpected and unprecedented ways."[6]

Bush's main thrust was the need to improve climate science. He proposed new U.S. investments in climate research, including support for the Mission to Planet Earth that would "advance the state of knowledge about the planet we share." A primary goal was to "reduce the uncertainty of predictive models." The president also gave quick nods to "a major reforestation initiative to plant a billion trees a year" across America and "initiatives to increase energy efficiency and the use of renewable resources."

But other ideas—one articulated and another left unsaid—raised doubts about whether Bush's statement provided the basis for a vigorous climate response. The president's approach assumed that climate change closely resembled other environmental issues. But the problems of rising CO_2 in the atmosphere were very different than those of dirty air and water. Bush also claimed that "strong economies allow nations to fulfill the obligations of environmental stewardship," including protecting the atmosphere. But wealthy countries had larger industrial sectors, drove bigger cars, and installed more air-conditioning. As a result, rising per capita incomes (and the ability, for instance, to purchase luxuries like electric or gas clothes dryers) were generally associated with increased CO_2 pollution.

Another feature of Bush's IPCC presentation was its failure to mention the setting of goals and targets for reducing the emissions of greenhouse gases. His stance continued to separate him from other industrialized nations pushing such measures.

The next day, the *New York Times* reported on its front page that the president had called for a "cautious approach" that "struck a middle ground between conflicting positions among Mr. Bush's aides." According to still anonymous sources, the speech didn't represent a victory for chief of staff Sununu, though he had exerted substantial influence.[7]

Sununu's go-slow approach to global warming drew praise from one secret admirer. On February 7, former president Richard Nixon sent him a handwritten note saying he was "right in holding the line against extreme environmentalists." He added, "As you know, I started EPA. I, like the president, am an environmentalist. But like him, I am not a nut. At least wait until the facts are in before going overboard on global warming."[8]

Ten days after the IPCC meeting in Washington, the American Association for the Advancement of Science gathered at the Renaissance Hotel in New Orleans for its annual meeting. After his presidency of the venerable organization in 1974, Roger Revelle had worked to keep climate change at the top of its agenda. Not surprisingly, six panels at the 1990 meeting addressed some aspect of the climate problem.

With the stoop common to an eighty-one-year-old body, Revelle no longer towered over his colleagues and had given up his role as the primary organizer of the climate sessions. But his presence loomed large. He presented a paper, "Greenhouse Warming: How Much Abatement? How Much Adaptation," and also presided over panels on "Better Models, Better Policies?" and "Cultural Change and Population Growth." Moreover, the meeting reflected Revelle's longtime view that scientists (including social scientists) should examine climate from many perspectives. Thus, the panels included prominent model builders—including pioneers Joseph Smagorinsky and Syukuro Manabe, along with younger scientists trying to resolve the persistent mysteries of climate. A half dozen economists also participated, including Bill Nordhaus.

The AAAS also explored potential solutions to climate change. Art Rosenfeld, director of the Center for Building Science at the DOE's Lawrence Berkeley Laboratory, discussed "Energy Conservation in a Warming World." Following Rosenfeld, H. M. (Hub) Hubbard, director of the Solar Energy Research Institute, talked about "Renewable Energy: Progress and Promise." On another panel, Amory Lovins, director of research for the Rocky Mountain Institute, explained "Abating Global Warming at Negative Cost." Reflecting an old Revelle concern, another session examined "the ethical responsibilities of the scientific community as it works toward greater scientific and technical understanding and, perhaps, resolutions to the global environmental crises."

Many conference attendees were active in the ongoing preparation of the IPCC reports. But the person most responsible for bringing climate change to the world's attention would not be one of them. In late March, Revelle's assistant, Christina Beran, wrote John Houghton, chair of the IPCC working group on climate science, that Revelle would not be able to comment on the draft circulating due to recent open heart surgery.[9] Indeed, when Revelle got back from New Orleans he had been taken straight from the airport to the hospital, where his doctors performed a triple bypass operation. And his prognosis was about to get worse. In early April, Revelle had an emergency hernia operation. The six weeks of strong antibiotics for a resulting staph infection almost killed him.

During the spring, the Intergovernmental Panel on Climate Change was trying to forge a consensus on what it would say about the world's climate. It was a messy process. At one meeting in Berkshire, England, a hundred scientists gathered for three days to edit the report on climate science. In the room, observers (who were sometimes allowed to make comments) represented diverse interests—Exxon, other mining and chemical companies, and the environmental organizations Greenpeace and the Natural Resources Defense Council. Outside, television crews and other journalists waited for the latest announcements.[10]

Governments had designed the IPCC to give themselves considerable influence over its reports. The dominant U.S. concern was the executive summary for policymakers—the most read part of the publication. American representatives feared that the summary would fail to adequately "reflect the current state of uncertainty regarding the ability of general circulating models to predict the magnitude and timing of climate change, either at the global, let alone regional scale."[11]

The United States attempted to showcase its leadership in April by hosting a two-day White House conference on science and economics research related to global climate change, featuring two appearances by Bush. In opening remarks, the president said that the gathering of top IPCC and U.S. officials "offered hope for a new era of environmental cooperation around the world." Despite the gracious rhetoric, his remarks carefully avoided any commitments to limit greenhouse gas emissions.[12]

In addition, Bush misrepresented the role of climate models. On April 17, he insisted, "What we need are facts, the stuff that science is made of—a better understanding of the basic processes at work in our whole world, better earth system models that enable us to calculate the complex interaction between man and our environment." This requirement was why, he said, he had recommended a 60 percent budget increase for climate science research. But in the realm of science, projections from climate models did not produce "facts," only estimates of future events based on probabilities. Several scientific reports over the years left the impression that certainty was around the corner. But when introduced into the policy arena, the idea that models could produce "facts" led to a quest for certainty that delayed action in perpetuity.

Once again, the president conflated greenhouse gas emissions with other forms of pollution. When talking about solutions to climate change, for instance, he spoke a lot more about reducing lead in gasoline, urban smog, and sulfur dioxide emissions than cutting greenhouse gases. Heat-trapping gases were different. They remained in the atmosphere for a much greater length of time and spread pollution around the world—even to areas with minimal emissions. They required distinctive, though sometimes complementary, solutions. Bush's talks made only vague references to how the United States and other nations could cut their output of greenhouse gases.

Finally, the president treated his call for incorporating economic analysis into decisions about climate policy as a new idea. In fact, Revelle had brought economists into the climate discussions at the AAAS in the 1970s, and their inclusion in National Academy and other studies had become common.

The lack of U.S. commitment to reducing greenhouse gas emissions rankled some European officials. West Germany and France joined the Netherlands in declaring they would take unilateral action to reduce carbon emissions. In closing remarks, Bush pushed back on the criticism by moderating his comments. He insisted, "We've never considered research a substitute for action."

In May, Thatcher said that if other countries did their part, the United Kingdom would also reduce its projected growth of carbon emissions enough to stabilize them at 1990 levels by 2005. The *New York Times* called the announcement "a blow to the Bush administration." Friends of the Earth called it too little, too late.[13]

Bush's reluctance to embrace climate targets drew support from organizations combating climate activism. One such think tank, the George C. Marshall Institute, published *Scientific Perspectives on the Greenhouse Problem* in April 1990. The book's editors were three prominent members of the institute's board—Robert Jastrow (founder of the Goddard Institution for Space Studies and once James Hansen's boss), William Nierenberg (director emeritus of the Scripps Institution of Oceanography), and Fred Seitz (past president of the National Academy of Sciences and an avid defender of the tobacco industry). Like the Bush White House, the authors tried to undermine the credibility of what they called "highly uncertain climate models," even the ones used in the 1983 Nierenberg study. However, the Marshall Institute book didn't reject all efforts to reduce carbon emissions in advance of scientific certainty. For instance, the book's economic analysis suggested that sustained research and development by the public and private sectors on solar energy, nuclear power, and energy efficiency "could reduce the costs of carbon constraints—perhaps by several trillion dollars."[14] By contrast, Bush called only for "exploring" nonfossil sources of energy, not sustained R&D. The Marshall Institute was more bullish on renewable energy than the White House science advisor.

Internally, the administration faced some dissent over its conservative stance on climate. In late June, Mike Deland wrote senior White House staff about how to deal with diverse views on climate at an upcoming Houston economic summit of the G7. The Council on Environmental Quality chief urged the president to announce that the United States would "essentially cap its emissions of 'greenhouse gases' until the year 2000." By that time, he said, "We will have better scientific understanding of the certainty, timing, and degree of climate change, and the relative effectiveness and efficiency of various response strategies." Some analysts at the EPA, also underplaying the time needed to change the existing energy infrastructure, agreed that such caps were feasible if they accounted for "greenhouse gas equivalents," such as tree planting and implementation of the Montreal protocol.[15]

In Houston, Bush reaffirmed U.S. support for an international framework climate change convention but stood fast against setting caps on greenhouse gas emissions or specifying what significant steps the United States might take on its own. Responding to a reporter about charges that his forestation program was a "fig leaf" and that Reilly should have been part of the U.S. delegation, the president expressed his frustration with criticisms from environmentalists. He replied feistily, "You cannot appeal—and I have to be careful because there were some reasonable people involved—but on the environmental extreme, they don't want the economy to grow." He added, "I did not rely heavily on them for support in getting elected President of the United States."[16]

The IPCC released its initial and much-awaited climate science and response strategies assessments in June. The science working group struck a confident tone. It boasted about the thorough preparation of its report: "One hundred and seventy scientists from 25 countries have contributed to it, either through participation in twelve international workshops . . . or through written contributions. A further 200 scientists have been involved in the peer review." It acknowledged its inability to accommodate all minority opinions. Still, it said, the extensive review helped "ensure a high degree of consensus" among the scientists involved. As a result, "The assessment is an authoritative statement of the views of the international scientific community at this time."[17]

United Kingdom scientists led the report preparation. But American experts filled many key positions. Several served as "lead authors"; well over one hundred others were official contributors or peer reviewers. Overall, the assessment sounded very different than the 1983 Nierenberg report or the Bush science office. Trying to overcome the language barriers between scientists and nonscientists, the IPCC said it was "certain" about a few aspects of climate science. The summary for policymakers asserted that it was certain that "emissions resulting from human activities are substantially increasing atmospheric concentrations of the greenhouse gases." It said, "These increases will enhance the (natural) greenhouse effect, resulting on average in an additional warming of the earth's surface."

The assessment identified three additional findings that it could "calculate with confidence." First, some greenhouse gases were more effective than others at changing climate. Of these, carbon dioxide had been responsible for more than half of the impact and would likely be so in the future. Second, continued emissions of long-lived gases (carbon dioxide, nitrous oxide, and CFCs) at current rates would commit the world to increased atmospheric concentrations for centuries ahead. "The longer emissions continue to increase at present-day rates, the greater reductions would have to be for concentrations to stabilize at a given level." Third, "the long-lived gases would require immediate reductions in emissions from human activities of over 60% to stabilize their concentrations at today's level. Methane would require a 15–20% reduction."

The assessment presented a lengthier list of predictions "based on current model results."

In a "business-as-usual" emissions case, global mean temperature would likely increase about 1 degree Celsius above current levels by 2025 and 3 degrees before the end of the next century. However, the rise would not be steady because of the influence of other factors. It noted that increasing controls on emissions could substantially slow the pace of global warming.

The assessment also predicted that land surfaces would warm more rapidly than the ocean, with the high northern latitudes heating up more than the global mean in winter. Global mean sea level would rise about twenty centimeters by 2030 and sixty-five centimeters by the end of the century, with significant regional variations.

The summary explained that uncertainties about patterns at the regional level were due to an incomplete understanding of the usual suspects—sinks, clouds, oceans, and polar ice sheets.

In the IPCC's judgment, rising surface temperatures and sea levels over the previous hundred years were "broadly consistent" with climate models. But they also could be due to natural variation. As a result, the unequivocal detection of the enhanced greenhouse effects "from observations" (rather than by models) was not likely for a decade or more.

The executive summary closed with a plea for more research and improved models.

The authors of the scientific assessment believed they had provided the world with a "firm basis" for developing strategies that would respond to climate change. But a different working group, headed by the U.S. State Department's Fred Bernthal, was working concurrently on the response strategies portion of the IPCC analysis. The response document appeared less polished than the work of the scientists. For instance, it provided no list of contributors or indications of peer review. It acknowledged being a work still in progress. IPCC chief Bolin later recalled that the response group posed more questions than it answered, to avoid "politicization."[18]

The working group said its assignment was to "lay out as fully and fairly as possible a set of response policy options and the factual basis for those options." Its purpose was not to "select or recommend political actions." To allow flexibility, it declared, "The great diversity of different countries' situations" would require "a wide variety of responses." The report noted that some countries had already decided to stabilize their emissions.[19]

One obvious way to respond to climate change was to limit the emissions of greenhouse gases. The first short-term measure on its list was "improved energy efficiency," which reduced carbon emissions while providing other benefits. High on this list was the "improved efficiency of road vehicles." It cited several ways to reduce vehicle fuel consumption without sacrificing safety or performance. These included electronic engine management and transmission control, advanced designs that reduced size and weight with composite materials and structured ceramics, and improved aerodynamics. On the supply side, lower emissions could also result from constructing gas-fired power plants or standardizing nuclear power plant design to improve economics and safety. Longer-term options included the development of not yet available technologies.

The report was bullish on the "technical potential" to achieve energy efficiency. It noted, for example, the possibility of 50 percent improvements over the average vehicle on the road. However, it tempered its enthusiasm for vehicle and other efficiency measures by concluding, "There is very little information available on the actual economic and social feasibility . . . of such options."

The report danced around the sensitive question of whether an international agreement should include targets for emissions reduction. But its stance that experts did not know enough about economic feasibility appeared to support the U.S. position that it was too early to do so.

The other response to climate change was adaptation. These options included enhancing emergency and disaster preparedness, reducing the vulnerability of coastal populations, and adapting crops to saline regimes.

The science and response assessments provided a foundation for future international meetings during the year. The Bush White House continued to carefully monitor the situation, with Sununu's office reviewing some drafts. The feedback from the White House had one recurring theme. It consistently instructed U.S. negotiators to get more emphasis on scientific uncertainty. In contrast, the European Community, late in the year, agreed it would adopt a proactive approach to reducing emissions.

Participants in the international talks realized that any agreements or differences among nations were a prelude to a framework convention scheduled for 1992. That would be the forum in which governments would decide whether to commit to emissions reduction targets and whether such targets would be binding.

15

Filibusters and PR Firms

Bush, August 1990 to July 1991

In the summer of 1990, Congress took up omnibus energy legislation that would affect greenhouse gas emissions, forcing Bush to take a firmer stand on what policies he favored or opposed. The White House was pleased that the energy committee (rather than the environment committee) would take the lead in the Senate. But it was prepared to buck Congress on measures it felt went too far.

The most visible item on the agenda was a bill from Senator Richard Bryan, Democrat of Nevada, scheduled for key votes for September. The legislation required dramatic increases in automobile mileage efficiency, upping standards 40 percent by 2001. Its primary goal was reducing dependence on oil imports, which in 1990 stood at a net 42 percent of consumption.[1] A secondary objective was to reduce the massive carbon dioxide emissions from burning gasoline and diesel fuel.

Many senior members had voted for the original auto efficiency legislation in 1975. Those standards, along with higher fuel prices, had helped slash oil imports from 1978 to 1985. But dependence on foreign suppliers had resumed a steady rise as the effects of those requirements began to wane.[2] Moreover, on August 2, Iraq launched an invasion of oil-rich Kuwait. Saddam Hussein's aggression led to a steep drop in global oil supplies. It also created the specter of an Iraqi advance into Saudi Arabia, the world's leading oil exporter, which would likely result in the deployment of U.S. military forces to the region. Events in the Persian Gulf provided a vivid example of how breaking news might bolster the rationale for legislation nearing critical votes. It appeared that efficiency standards might meet the Bush criteria for measures that could help slow climate change but could also be justified on other grounds. On August 23, White House staffers told John Sununu that the Senate would likely pass the bill by a wide margin.[3]

Despite the bill's momentum, the White House had reasons to resist. As top lawyer Boyden Gray frequently reminded White House personnel, Vice President Bush had headed the Reagan effort to reduce government regulation, which proposed axing the existing mileage efficiency standards. Behind the scenes, the Bush White House continued to study how to repeal those standards.

The administration was also reacting to the concerns of auto manufacturers, more active in opposing higher standards in their private communications than in public statements. Ford, for instance, told the White House it was already making substantial contributions to global warming concerns by eliminating chlorofluorocarbons in its air-conditioning, as required by the Montreal Protocol. It also pointed to studies concluding that the smaller vehicles required by previous standards had increased highway traffic fatalities.

The auto industry found a sympathetic ear from economists at the White House. Four years earlier, while still in academia, one senior staffer at the Council of Economic Advisers had joined three other economists in casting a skeptical eye on auto mileage standards. They concluded, "We see no need to compel further improvements in fuel efficiency. Car buyers can be left to choose their own fuel efficiency. If future improvements in fuel efficiency were desired, the most direct way to obtain them would be to raise gasoline prices by increasing the federal tax."[4]

The Senate began debating the auto efficiency bill on September 13.[5] With U.S. troops gathering in Saudi Arabia, Senator Bryan dramatized the need to reduce U.S. reliance on foreign fuels. He declared, "The cost of dependence on imported oil, once measured in billions of dollars sent abroad to foreign bank accounts, can now also be measured in the thousands of American lives being placed at risk in the Persian Gulf." He spent almost as much time talking about the environmental benefits of the bill, saying it provided an opportunity to deal "with another urgent policy question—increased emissions of carbon dioxide, a primary greenhouse gas, a contributor, many believe, to the onset of global warming." He reported that transportation alone contributed almost one-third of the country's CO_2 emissions, one reason the EPA had recommended vehicle efficiency as a near-term strategy for dealing with global climate change.

The chief Republican sponsor, Slade Gorton of Washington, also enthusiastically praised the multiple benefits of reducing fuel consumption. He proclaimed: "It is truly difficult to imagine that we in the Congress could be presented with a proposal which would decrease our dependence on foreign oil, reduce the trade deficit, save money for consumers all across the country, cut the emissions of greenhouse gases, and reduce smog, all in a single bill, and have anything other than an overwhelming show of support." He called the environmental impacts of the bill "equally, if not more important" than those on oil imports. "Of all the steps Congress can take to curb CO_2 emissions," he declared, "this bill is likely to be the most important."

Other senators saw standards in a different light. Don Riegle of Michigan led the opposition to the bill. The Democrat from the nation's chief auto-producing state said that demands for greater efficiency faced many unfortunate trade-offs, including pressure to produce smaller vehicles that would be less safe. He complained that if you wanted to withstand a collision, "the bigger the car is and the heavier it is, the more protection it gives you." He quoted a letter from the Department of Energy, saying, "Consumers would be unable to purchase the vehicles that meet their requirements and could face increased risks of injury."

Tim Wirth countered that the Senate shouldn't accept industry projections that they couldn't meet the stiffer efficiency targets without draconian cuts in other features. He said automakers had damaged their credibility in earlier debates by insisting that requiring air bags would be "an absolute disaster," but "we saved them from themselves." Now, at automobile showrooms, "a salesman comes up and markets air bags." He said that if the bill's critics followed the weight issue to its logical conclusion, the Senate should mandate that everyone drive an M1 tank.

Don Nickles, a Republican from the oil-producing state of Oklahoma, said that the best way to reduce oil imports was to open up drilling in the currently off-limits Alaska National Wildlife Refuge. He charged that the bill would be "a big move toward central planning by the government." Passing it might prove futile, in any case, he warned, because the president's advisors had told senators they would urge him to veto it.

Senate rules required two procedural votes with super majorities to move to final consideration. The next day, the Senate voted on the first (to proceed) with plenty of votes to spare. Despite White House opposition, twenty-three of the forty-two Republicans recorded agreed to the measure. The total 68 "yea" votes suggested that a milestone might be at hand. Combined with Reagan-era requirements for appliance efficiency and elimination of CFCs, mileage standards would send a strong signal to the world that the United States was serious about reducing greenhouse gas emissions.

Despite the evidence of overwhelming support for the bill, the White House was not about to fold. On September 19, its legislative office briefed Sununu about the potential to switch senators on the second procedural vote to block any filibusters. The list focused on the Republicans who had "voted against the administration." The memo listed seven who might have already decided to do so and asked the chief of staff to call six others who were wavering.[6]

On the morning of September 25, eleven senators (nine Republicans and two Democrats) who had voted to proceed switched positions and voted against cloture.[7] As a result, the number agreeing to move forward dropped to fifty-seven, three short of the sixty needed. The action—an impressive display of political muscle by auto companies, their unions, and the White House—killed the Bryan bill and any chance to add substance to the U.S. response to the IPCC reports during the 101st Congress. It also demonstrated that the Senate's requirement for supermajorities to end filibusters created high hurdles for any future climate legislation.

In August of 1990—just before the Senate took up the automobile mileage bill—the National Center for Atmospheric Research in Boulder, Colorado, had SWAT teams positioned on its roof and around its grounds. For the first time, the site was hosting the head of a national government. Margaret Thatcher had come to town.

The United Kingdom had just established a similar institution, which became the Hadley Centre for Climate Prediction and Research. In addition, the prime minister was prepping for her upcoming address at the second World Climate Conference in Geneva.

At NCAR, Thatcher displayed her customary attention to the details of climate. The task of presenting the briefing, scheduled for two hours, fell to meteorologist and senior modeler Warren Washington. Washington had in the 1960s become the first prominent African American climate scientist and later served with Roger Revelle on the panel responsible for the 1977 *Energy and Climate* report. When his two hours expired, everyone in the room stood up to leave—except for Thatcher. She had noticed a stack of unshown viewgraphs. She said in a commanding voice, "I will not leave here until I see all the viewgraphs."[8]

At the November climate conference in Geneva, the Tory leader warned, "The danger of global warming is as yet unseen, but real enough for us to make changes and sacrifices, so that we do not live at the expense of future generations." She observed, "We have treated the air and the oceans like a dustbin."[9]

Thatcher praised the American scientists she'd recently visited for their work to better understand the effects of the oceans on weather and improve the capability of climate models. But the need for more research should not, she said, "be an excuse for delaying much needed action now." The uncertainties about climate change were not all in one direction. "Climate change may be less than predicted. But equally it may occur more quickly than the present computer models suggest."

Thatcher went on to strongly endorse binding targets for reducing greenhouse gas emissions, still a sore point with the Bush administration. She declared that the United Kingdom was "prepared, as part of an international effort including other leading countries, to set itself the demanding target of bringing carbon dioxide emissions back to this year's level by the year 2005. That will mean reversing a rising trend before that date."

The prime minister's attention to climate and other policy issues in November reflected a certain obliviousness to the political troubles brewing in her party's caucus, in the midst of war in the Persian Gulf, no less. Faced with the possibility that Conservative backbenchers might vote her out, she threw her support to John Major, who she believed didn't have many enemies and would be most likely to continue her policies. On November 28—just twenty-two days after Thatcher's speech in Geneva—Major became Britain's new prime minister, replacing one of the world's strongest advocates for taking action to slow global warming.

In mid-November, Bush put on a full court press to repair relations with environmentalists upset by his tepid approach to climate change. On Tuesday, November 13, the president hosted what he called "a gathering of scientific and technological genius" in the East Room of the White House.[10] The guests of honor were ten scholars receiving the National Medal of Science. Awardees included Roger Revelle, who had recovered enough from his medical emergencies to attend. Bush used the occasion to highlight his significant budget increases to "investigate global climate change."

Revelle's recognition was, in many ways, long overdue. Numerous scientific organizations had already presented him with their top honors. During the Johnson administration, he had served on the advisory committee recommending names for the award he was now receiving. Still, the ceremony highlighted his contributions over more than three decades to advance the understanding of climate change.

Barely able to walk, the Californian needed special assistance to travel from his hotel to the luncheon. Revelle told the *Los Angeles Times* that he was scheduled to sit next to John Sununu, who was (in the reporter's words) "militantly reluctant to commit the United States to measures that would reduce global warming." Revelle arrived "prepared for verbal battle" with the chief of staff but complained "with evident disappointment" that the powerful insider failed to show. He said, "The president has got the message, but I wanted to work on Sununu."[11]

The next morning, flanked by the heads of the EPA and the Council on Environmental Quality, Bush presented the President's Environmental Youth Awards for the winning projects. In a comic reference to generational equity, he shared with his young audience a quote that he said he'd received from Bart Simpson: "When I mess up my bedroom, my mom comes in and yells, but eventually she cleans it up, and everything is cool. But when we mess up the environment, we're the ones who are going to be yelling, and it definitely won't be cool."[12]

On Thursday, Bush signed historic environmental legislation, the Clean

Air Act Amendments of 1990, cheered on by a bipartisan audience packed into the East Room of the White House. The administration, Senate majority leader George Mitchell, and the Environmental Defense Fund had pushed for the bill, which passed Congress overwhelmingly in October. Bush boasted at the signing ceremony that the new law "strengthened our clean air statutes, already the world's toughest." The president declared that the bill's severe penalties supported the principle "Polluters must pay." He handed the first pen to Bill Reilly.[13]

Although the legislation had a few side effects on climate change, both positive and negative, its primary targets were not greenhouse gases. However, the law did further limit the emissions of chlorofluorocarbons and tasked the EPA with identifying options for reducing global warming by reducing methane emissions.[14] One overlooked provision required power plants to have the capability to monitor CO_2 emissions in case the data were needed down the road.

On Friday, the president wrapped up the week's environmental messaging by approving eight less consequential bills to, in his words, "preserve America's natural beauty." One was the Global Change Research Act of 1990, which required additional studies on global warming and a report to Congress every four years on the environmental, economic, health, and safety consequences of climate change.

The Senate had passed the climate bill on a recorded vote of 100–0. But its popularity relied on the narrowness of its scope. It hewed to the Bush position supporting additional research on climate science but not much else. It displayed little awareness of extensive federal support for such research going back to Keeling's Mauna Loa project in the 1950s. Several federal agencies had greatly expanded their support for climate science over the years. The new bill's reporting requirements were a small step forward. However, they could create the impression that climate science was just getting started, thereby justifying further delays in acting to slow global warming.

In February 1991, climate scientists Patrick Michaels of the University of Virginia and Robert Balling Jr. of Arizona State University wrote President Bush about a recent conference with the working hypothesis that "the impact of future climate change is likely to be much less detrimental than generally expected, perhaps even neutral or beneficial." They asked him to bring the conference papers to the attention of federal agencies and private companies so these organizations would be "more disposed to support research that could lead to a balanced assessment of the issue." They asserted, "A premature attempt to deal with, what in fact may be, a misinterpreted problem could lead to precipitous economic dislocations."[15] The two professors likely did not anticipate how soon they would be inside the West Wing.

Michaels, the more visible of the two, was a professor of environmental sciences at the University of Virginia and the state climatologist. His résumé included many traditional academic publications and activities. But in 1990 he also was a guest speaker at three coal industry conferences, where people were increasingly apprehensive about the potential effects of new climate policy. Michaels had already accused people like Al Gore and James Hansen of establishing a "climate of fear" about global warming. He had even implied that the earth was actually cooling, in *Washington Post* opinion pieces titled "Greenhouse Effect? Then Why Is It Colder" and "Greenhouse Climate of Fear."[16]

The letter to the president was well timed. The White House had a new problem in its climate messaging, this time from the National Academy of Sciences. The NAS was responding to a congressional request for a study on the policy implications of greenhouse warming. Dan Evans—former Republican U.S. senator and a current board member of the Nature Conservancy—chaired the most substantive Academy study of climate change since 1983.

Due in part to the new study, the administration was trying to thread a needle on its approach to climate science. Two weeks after the Michaels/Balling complaints about mainstream scientists, Bush addressed the very mainstream American Association for the Advancement of Science. He made scant reference to climate science but did spy Roger Revelle in the audience and went out to shake his hand. Warren Washington said, "It appeared that Revelle knew the president personally because they engaged in conversation. Revelle appeared to be in poor health."[17]

The National Academy project involved a massive and diverse list of participants. The fifteen-member synthesis committee included two panelists from the Nierenberg report—Bill Nordhaus and Paul Waggoner. Also listed were Jessica Matthews of the World Resource Institute, Stephen Schneider from the National Center for Atmospheric Research, and Sir Crispin Tickell, former prime minister Thatcher's lead advisor on climate change.

On April 8, Allan Bromley shared with Sununu a warning that the Academy would soon recommend a more activist approach to policy than the White House preferred. It concluded, for instance, "Despite the great uncertainties, greenhouse warming is a potential threat sufficient to justify action now." However, Bromley—a highly regarded leader at the Academy for many years—recommended against taking a combative approach to its report. He told his boss that there were numerous commonalities between the views expressed in the report and those of the White House. Thus, he believed, "It will be important for us to be supportive overall when the Evans report is released, emphasizing the many points of overlap and agreement."[18]

The Committee on Science, Engineering, and Public Policy of the National Academy released the synopsis of *Policy Implications of Greenhouse Warming: Mitigation, Adaptation, and the Science Base* two days later. The eighty-two-page summary of the full Evans report (to be released later) contained findings and recommendations that would please almost every audience. Statements were heavily caveated. Its preemptive defenses against potential objections sometimes made it sound like the report was arguing with itself. The synopsis was also very thorough. For instance, it evaluated fifty-six mitigation options to slow global warming. The list even included placing 50,000 giant mirrors in the earth's orbit to reflect incoming sunlight or launching billions of aluminized, hydrogen-filled balloons into the stratosphere to provide a reflective screen. (Neither idea was recommended.)

The report was consistent with White House thinking on many points. The damages from global warming of less than 2 degrees Celsius were minimized, as were the impacts on other nations. The cost–benefit analysis used economists' discount rates, with a default rate slightly higher than the one adopted by the Council of Economic Advisers. The Academy also pointed to the slower growth in U.S. greenhouse gases during the 1980s and the progress in making energy use more efficient and limiting the use of CFCs.

Policy Implications also somewhat lessened the urgency to adopt mitigation measures, with its optimistic view of adapting to climate change. Two panel members publicized their dissenting opinions on adaptation, which was unusual for an Academy report. Matthews said the report's "flawed analysis" assumed that "human economic activities are largely divorced from nature and that modern technology effectively buffers us from climate." Jane Lubchenco, a professor of zoology at Oregon State University, complained later that the adaptation section's "complacent tone" was "unwarranted."[19]

Despite all its qualifications and caveats, the Academy's synopsis endorsed with considerable clarity "prompt responses" to address the potential threat of greenhouse warming. The first paragraph of its chapter "Findings and Conclusions" declared, "Investment in mitigation measures acts as insurance protection against the great uncertainties and the possibility of dramatic surprises. In addition, the panel believes that substantial mitigation can be accomplished at modest cost. In other words, insurance is cheap."

Because discount rates favored measures with shorter-term payoffs, the report focused mainly on end-use efficiencies. Its top two recommendations were to adopt nationwide energy-efficient building codes (including appliances) and to improve the efficiency of the U.S. automobile fleet by using "an appropriate combination of regulation and tax incentives." Both, it said, could lead to massive reductions in carbon emissions. Because lower operating costs could offset higher purchase prices, they were better than cheap; they could provide "net savings."

The next Congress was expected to revive the debate over corporate average fuel economy standards, so the NAS analysis of auto efficiency measures was particularly relevant to the ongoing energy policy debate. *Policy Implications* calculated that the standard could be raised from the existing 27.5 miles per gallon to 32.5 miles with current technology and would not require changes in size or other vehicle attributes. The standard could increase to 46.8 miles per gallon with changes such as downsizing.

The report also undercut several administration assumptions that justified its reluctance to adopt new climate policies. Sununu had called models that estimated future impacts of global warming "hoaxed-up." But the Academy said that despite their imperfections, general circulation models were "generally considered the best available tools for anticipating climatic changes." The president had argued that global economic growth helped reduce greenhouse gases. But the Evans report stated that economic growth produced additional emissions because of the increased demand for goods.

The administration avoided the idea that the United States had special responsibilities in leading the world's effort to slow climate change. But the report argued, "The position of the United States as the current largest emitter of greenhouse gases means that action in the rest of the world will be effective only if the United States does its share."

The *New York Times* headlined its coverage of the Evans report "Quick Steps Urged on Warming Threat." Environmentalists praised the study. Michael Oppenheimer said its recommendations went "a long way toward meeting the goal of keeping the climate from going haywire." Jessica Matthews described it as a "nimble policy." The White House also found reasons to like the result of the massive Academy study. Bromley said he was "delighted with it" and claimed that actions already taken by the Bush administration would result in emission reductions. A senior administration official, "who spoke on the condition of anonymity," said he was pleased the report hadn't recommended "draconian changes" or proposed target dates and quotas for carbon emissions reductions.[20]

Despite the Evans report's caution, it reflected a significant departure from the wait-and-see emphasis of the Academy's 1983 study.

During the spring of 1991, two eminent American scholars traveled abroad to offer their views on the findings of the Intergovernmental Panel on Climate Change.

In April, Bill Nordhaus addressed a meeting on economics and the environment in Linz, Austria. His paper (published the following year) retained many features from his earlier work. He cited the data on rising levels of atmospheric carbon and used discount rates to analyze optimal paths that would protect the environment and economic growth. He continued to favor carbon taxes over other approaches, such as government regulations or investments in advanced technologies, to reduce carbon emissions if they ever needed to be controlled.[21]

There were, however, significant differences. The Yale economist's pioneering papers on climate had stressed that his critical assumptions were extremely arbitrary and dismissed geoengineering options as unrealistic. By 1991, his assumptions had become standard economic orthodoxy, no longer requiring caveats. In addition, geoengineering had moved up to become an available alternative. His new paper, brimming with much greater certainty, was more explicit about various options and where they might lead.

Nordhaus began by asserting that the 1991 IPCC assessments created alarmism by calling for "drastic curbs on the admissions of greenhouse gases . . . without any serious attempt to weigh the costs and benefits of climate change or alternative control strategies." He assumed "that the goal of economic activity is *consumption* [consisting] of a 'bundle' of goods and services." He said, "The purpose of economic—and greenhouse gas control— is to enhance to the greatest extent the total consumption 'bundle'" across current and future generations. The high value that Nordhaus placed on consumption made the gross domestic product—the total monetary or market value of all the finished goods and services produced within a country's borders—the anointed arbiter of what was good and bad. Unquantified quality-of-life factors did not rate consideration.

The Yale professor also provided a rationale for placing most responsibility for climate change on future generations with his assumption that they would have higher per capita incomes. Later, wealthier humans would be better able to bear the costs of dealing with the changes in their environment. As a result, his "optimal control path" suggested that "the current generation (including notably those in developing countries) should not make large sacrifices to slow climate change today if future generations are richer and can slow climate change themselves." He didn't mention when the passing of responsibility for climate change to future generations might stop. Succeeding generation might assume that those who came after would be even more prosperous, alleviating themselves from responsibility for the carbon accumulating in the atmosphere.

His model results constituted a striking counterpoint to the IPCC. He concluded that the optimal policy, from an economist's viewpoint, called for minimal government intervention and had "only a minor effect on slowing climate change." This "optimal" approach resulted in temperatures rising 3.5 degrees Celsius by 2100, not much lower than doing nothing.

Nordhaus had once talked of a possible "danger zone" if global temperatures rose more than 2 degrees from preindustrial levels. In 1991, he argued that future generations might be worse off due to climate change but they would still likely be better off overall than current generations, even if temperatures rose more than 3 degrees and didn't stop there. Nordhaus had moved on from the wait-and-see position of the Nierenberg report and the "no regrets" approach of the Evans report—despite serving on both panels. In Linz, in effect, he made a case for wait-and-wait.

European economists at Linz adopted a different worldview. Two from Kiel, Germany, believed, "Environmental quality is a typical public good that cannot be so easily incorporated into our economic models."[22] Reliance on national economic data in these models amplified the problem of not giving the environment its due. In their view, "Environmental problems demand collective decisions on environmental values." The Germans also argued that "Nature is not unlimited." Therefore, proper maintenance required that "the services that nature provides be paid for by those who use them once these services become scarce." They concluded that their academic discipline needed "a model of sustainable development."

Far away, in another part of the globe, Bert Bolin accepted an invitation from energy-producing corporations to explain the IPCC's findings and future plans. When he addressed the World Petroleum Council in Buenos Aires in October, he was surprised to run into Bill Nierenberg, a known critic of the IPCC's conclusions, representing the Marshall Institute. Nierenberg accepted the idea that a doubling of atmospheric carbon dioxide might be a grave matter. However, he calculated that the global mean temperature would rise no more than 0.5 degrees Celsius. A shocked Bolin countered that the IPCC had considered the notion, which originated with Richard Lindzen, and found that it "systematically underestimated the observed change so far." The IPCC concluded that Lindzen's "back of the envelope" computations were "simply wrong." As Bolin remembered it, the confrontation with Nierenberg did not end cordially.[23]

On April 18, Bromley responded to Sununu's interest in meeting with Michaels's recommended scientists by providing the names of the "best known and most outspoken" of what the science office called the "scientific skeptics on global warming." Lindzen, Fred Seitz, Nierenberg, and Michaels led the list. The chief science advisor said these scholars had produced "some good stuff" but was nervous that such a meeting might become public knowledge. He warned of potential headlines that the White House was reaching out to "right-wing elements" to combat the Evans report.[24]

Several scientists Sununu wanted to see were coming to Washington in early June for a Cato Institute conference at the Capital Hilton, facilitating an hour-long meeting with five of them—including Lindzen, Michaels, and Balling—at the White House. The scientists expressed frustration because their unconventional views on climate led to discrimination in funding, scientific paper reviews, and hearings on the Hill. They wanted improved "marketing of the U.S. approach to the global warming issue" and challenged whether the National Academy truly represented the scientific community. Sununu offered suggestions on how they could gain greater visibility for their views. Science office staff feared "the probability of this meeting remaining confidential is close to zero."[25]

While the White House was setting up the meeting with global warming skeptics, Michaels and Balling were working with the Edison Electric Institute, the Southern Company (parent of Georgia Power, Alabama Power, and Mississippi Power), and a group called Informed Citizens for the Environment to discredit prevailing climate science. In May, ICE (an acronym intended to mock the evidence of global warming) had tested whether it could sway public opinion in several small towns with print and radio advertising based on Michaels's op-eds in the *Washington Post*. Newspaper ads in Flagstaff, Fargo, and Bowling Green carried headlines like "Who Told You the Earth Was Warming . . . Chicken Little?" and "If the Earth Is Getting Warmer, Why Is Kentucky Getting Colder?" Readers were urged to write or call Michaels in hopes that a larger ICE effort could build a database of potential climate skeptics. The creative design team for the project suggested targeting older, less educated males, younger low-income women, and the radio audience of Rush Limbaugh—whose ridiculing attacks on "feminazis" and "environmental wackos" were setting new standards for AM programming.[26]

The ICE campaign blew up when the *New York Times* disclosed in July that the organization was considering taking the experiment national. Its story, "Pro-Coal Ad Campaign Disputes Warming Idea," identified sponsors and the campaign's goal, according to a planning document, to "reposition global warming as theory," not fact. After the story broke, Michaels and Balling resigned from the three-person "science advisory panel." They said that cooling in some areas did not contradict the theory of global warming, but the public relations firm believed that an ad that discussed the contradictory state of the evidence would not have enough punch. Since the ads resembled Michaels's previous pieces in the *Washington Post,* his explanation raised the question of whether the PR pros had required that nuance be removed from those, too.

Public revelation of the plan led to several disavowals. The Arizona Public Service Company—an Edison Electric Institute member utility—told a local paper, "The subject matter is far too complex and could be far more severe than the ads make of it for the subject to be dealt with in a slick ad campaign."[27] At the White House, the science office undoubtedly breathed a sigh of relief when the story did not mention its recent hosting of Michaels at the White House.

Media disclosure dealt a death blow to the ICE campaign but did not eliminate some coal companies' and coal-based utilities' terror about potential clamps on their businesses. Nor did it discourage them from finding other ways to influence public opinion more discreetly. Deputy energy secretary Henson Moore reported to Sununu in late June about spending some recent "quality time" with the top coal industry CEOs. He found them most concerned about global climate change and "almost paranoid about it as they consider it clearly the most serious threat to their continued existence."[28] Fearful industry groups funneled funds to friendly scientists and intensified their efforts to influence public opinion.[29]

For debates over climate science, the rules of combat would never be the same.

16

Road to Rio

Bush, July 1991 to July 1992

On July 7, 1991, Roger Revelle suffered another heart attack. He died eight days later at the age of eighty-two.

The *Los Angeles Times* published numerous accolades for a Southern California icon. The state's Republican governor (also former U.S. senator and mayor of San Diego), Pete Wilson, called the professor a person who could look back on his life and say, "I made one *hell* of a difference." Allan Bromley declared, "Roger was as close to a Renaissance man as we have had in modern science." Bill Nierenberg observed, "Without his work, I don't think we would have oceanography as we know it today." The paper wrote extensively about Revelle's work establishing the University of California, San Diego but strangely had only one sentence (quoting Walter Munk) on "the greenhouse effect, where his observations on carbon dioxide became the cornerstone of further research." The *New York Times* obituary called Revelle "an early predictor of global warming" but devoted little space to the subject.[1]

No single article, of course, could capture Revelle's impact on the scientific understanding of climate change. He had numerous accomplishments in other areas. His pioneering and highly influential work on climate extended over three and a half decades, making it difficult to measure its cumulative effect. Moreover, many of his contributions were too technical to explain to the public. His successes in inspiring other scholars and getting his ideas heard in places like the White House were not easily documented. As a result, memorialists grossly underestimated how much one person had done to promote the understanding of an extraordinarily complicated puzzle.

In the 1950s, it was Revelle who figured out why, at least in theory, the current estimates of carbon dioxide in the atmosphere might be wildly inaccurate and supported the work of Dave Keeling to collect the empirical evidence. Through the 1970s, if leading scientists, presidential speechwriters, or other influential people knew about climate change, it was often because they learned about it from Revelle. In the 1980s, he investigated likely climate change impacts that others were ignoring, such as anticipated water shortages due to the drying out of major rivers like the Colorado.

Furthermore, Revelle led the efforts to make climate studies more interdisciplinary. His extensive foreign travels promoted a multinational understanding of the most global of environmental problems. When the Intergovernmental Panel on Climate Change assembled experts from around the world to assess climate science, it often took advantage of seeds Revelle had planted.

The prolific professor stayed in his lane by working vigorously to advance scientific knowledge while accepting the restrictions of not being "political." But when it came to elucidating the ramifications of discharging greenhouse gases into the atmosphere, no single person had moved the ball further down the field.

Scripps held a good-bye memorial service for Revelle on July 19. At the end of the ceremony, his ashes were scattered at sea from a research vessel.

The pace of international climate negotiations quickened in the year of Revelle's passing. Great anticipation surrounded the UN Framework Convention on Climate Change and Development, scheduled for the following year in Rio de Janeiro. This historic "earth summit" would move the proceedings from the discussion to the decision stage. Because the issues at stake were too complex and contentious to work out in Brazil, the United Nations set up a series of preparatory meetings, hoping that the road to Rio would lay a foundation for eventual consensus.

Nairobi hosted one preliminary session in September 1991. Again, the United States and other industrialized nations clashed on the fundamental question facing the negotiators—what to do about the rising emissions of greenhouse gases. The European Community and Japan advocated specific targets and timetables to stabilize carbon dioxide emissions at 1990 levels by the year 2000. A Dutch diplomat explained, "You have to start somewhere," because without any action, carbon dioxide would rise 15 to 20 percent above 1990 levels by the end of the century. The chief U.S. negotiator, the State

Department's Robert Reinstein, acknowledged that the world faced a long-term problem. But he countered, "You don't know what benefits you can expect from these short-term actions." The *New York Times* called the United States "the odd man out" in Nairobi. Senator Gore accused President Bush of "stonewalling the world."[2]

The year 1992 would reveal whether the U.S. Congress could pass a comprehensive energy bill, the world could agree on a climate strategy, and George H. W. Bush could win reelection. The three outcomes were intertwined.

Al Gore had announced in late 1991 that he would not seek the presidency this time. He made it clear, however, that his White House ambitions had not died. "I would like to be president," he conceded.[3] The former journalist (before going into politics) stayed in the national conversation with the January 1992 publication of a book about the global degradation of nature.[4] The *New York Times* top ten bestseller, *Earth in the Balance,* traced Gore's early awareness of the threats to the environment. At the family dinner table, his mother had emphasized the importance of Rachael Carson's pathbreaking *Silent Spring.* At Harvard a few years later, he learned from Professor Revelle about the first measurements of atmospheric carbon dioxide. In that course, Gore became convinced, "If this trend continued, human civilization would be forcing a profound and disruptive change in the entire global climate." He began a lifelong habit of tracking the annual reports from Mauna Loa.

The challenge, according to the senator turned book author, was recognizing that "the startling images of environmental destruction now occurring all over the world have much more in common than their ability to shock." Global warming, ozone depletion, the loss of living species, and deforestation all had a common cause: "the new relationship between human civilization and the earth's natural balance."

Gore delved deeply into the history of natural climate variability. Over the past 20,000 years, dramatic climate changes from natural causes had led to famines, political unrest, and massive migrations of humans from one geographic area to another. If unchecked, human-induced climate change would have similar effects, forcing enormous adaptations.

He directly disputed those denying the urgency of climate change, including the small band of scientists disputing the IPCC consensus, fossil fuel interests trying to convince the public the earth was cooling, and Bush's economic advisors.

Gore didn't see an easy path ahead for policymakers. Congressional hearings demonstrated that some barriers to addressing the climate challenge were almost insurmountable. For example, "The insistence on complete certainty about the full details of global warming" was, in effect, "an effort to avoid facing the awful, uncomfortable truth." The senator also noted polls showing that more than 90 percent of Americans endorsed U.S. leadership to get other countries to join together to take action against world environmental damage. However, their general approval didn't mean they agreed to the specific steps needed to accomplish the goal. "In fact, almost every poll shows Americans decisively rejecting higher taxes on fossil fuels, even though that proposal is one of the first logical steps."

Gore rued his failure to highlight the climate challenge during his 1988 presidential campaign, having instead heeded the advice of political advisors to emphasize other topics. Syndicated columnist George Will had ridiculed Gore's brief focus on climate change and the ozone layer as addressing issues that were "in the eyes of the electorate not even peripheral." Gore conceded that, in political terms, Will was correct. Despite the potential political fallout, Gore decried the tendency to focus exclusively on current and short-term needs. In his sixty-six-page penultimate chapter, he called for "A Global Marshall Plan" to accelerate the "transition to an environmentally responsible pattern of life."

In early February, the Senate was trying to finish a comprehensive energy bill—the first since the Carter presidency. The previous year, the Bryan bill on auto mileage standards seemed to be a potential cornerstone of the overall legislation. Since then, an IPCC assessment and the Evans report had both placed reduced fuel use for transportation near the top of their recommended measures to mitigate climate change. Moreover, a new National Academy of Sciences study on automotive fuel economy undercut one argument against stricter standards when it found that fatalities per miles driven had fallen steadily since 1930, including when the government required greater fuel efficiency. Also encouraging, the NAS report identified technical fixes to some risks from smaller vehicles.[5] Senators who led the successful fight against the Bryan bill as a stand-alone measure in 1991 claimed they had told the auto industry they would have to agree to some concessions later when the comprehensive bill took center stage.

Nonetheless, the political prospects for efficiency standards had plummeted due to stiff resistance from the White House and intense lobbying by car companies. Using their contacts in the Global Climate Coalition and the grassroots lobbying of the industry-backed Coalition for Vehicle Choice, automakers successfully broadened the opposition to the Bryan bill. They targeted constituencies such as cattlemen and campground owners who favored the comfort and hauling capacity of large cars and trucks.[6] The United Auto Workers had also worked to turn undecided lawmakers against the bill.

Energy committee chair Bennett Johnston wanted consequential legislation, and at one point floated the idea of pairing a version of the Bryan bill with oil drilling in the off-limits Alaska National Wildlife Refuge as part of a grand compromise. But the White House, which strongly advocated opening ANWR, dug in its heels on auto standards. The Sierra Club, the Union of Concerned Scientists, and the National Wildlife Federation, while pushing hard for fuel efficiency, considered the ban on drilling nonnegotiable. Consequently, auto efficiency standards were no longer under consideration when energy legislation reached the floor on February 6.

A long Thursday of deliberations produced two notable turning points in the Senate. Imam Wallace Mohammed of Calumet City, Illinois, offered the morning prayer, after which members of both parties praised the historic first invocation by a member of the Muslim faith. Second, the body came closer to a final version of its energy legislation, constrained by a tight schedule that required broad-based support for the numerous amendments trying to fit into a narrow window on the calendar. The debates with long-term ramifications for U.S. climate policy became the most contentious.[7]

Gore offered two amendments—one requiring a more rapid phase-out of an expanded list of ozone-depleting gases, the other stabilizing U.S. carbon dioxide emissions at 1990 levels. Due to previous discussions, the ozone measure had the full support of both parties. Despite the president's generally strong record on these gases, the senator launched numerous personal attacks on Bush's handling of ozone policy. He repeatedly alleged that the president didn't provide leadership on the ozone hole until predictions that it would affect the Bush family vacation home in Kennebunkport, Maine.

The attacks on the leader of the Republican Party didn't change any votes on the ozone amendment, which passed 96–0. But they incensed its GOP supporters. The energy committee's ranking Republican, Malcolm Wallop of Wyoming, called it "an all-out assault on the President of the United States when the President of the United States has embraced as policy the effects of the amendment" and accused Gore of delivering a "presidential campaign speech." Of more significant consequence, John Chafee, a lead GOP cosponsor of both Gore amendments, said he was willing to disagree with Bush on climate change but complained that Gore's attacks on Bush's ozone position were "highly inaccurate."

When Gore's climate change amendment came up, he continued his digs about Kennebunkport. More relevant to the topic, he observed that tree ring data indicated that the past fifty years were the warmest fifty-year period in the past 160,000 years. Moreover, government analysts were looking at what it would take to relocate more than half of the population of Florida. He argued that recent studies from the Office of Technological Assessment and the EPA showed that the United States could improve its economy by using efficiency measures to reduce CO_2 emissions. He urged the United States to follow the example of the European Community by agreeing to stabilize emissions.

Wallop, representing a leading coal-producing state, led the effort to block the amendment. He claimed the climate change "Chicken Little scenario" was designed to startle and terrify the American public. As Wallop remarks dragged on, Johnston interjected, "I sense a filibuster and kind of deal breaker coming on with this amendment."

After further debate, Gore agreed to pull down his proposal, deferring to the committee's need to avoid lengthy delays that might threaten the overall bill. Johnston praised the courtesy of the Tennessee senator, who "performed with presidential timbre this evening."

A bipartisan pro-environment bloc, which had once pushed the Reagan administration to agree to appliance efficiency standards and ozone layer protection, remained in the Senate. But the coalition was no longer able to block filibusters when administration and industry lobbyists fought to avoid both auto efficiency standards and commitments to stabilize the emissions of greenhouse gases.

The same month, John Sununu headed up to MIT in his new capacity as lame-duck chief of staff. He had announced his resignation as the president's top advisor in December, saying he didn't want to be a drag on Bush's re-election prospects. He was now awaiting his new post as cohost of CNN's combative prime-time show *Crossfire.* To avoid conflicts of interest, he had recused himself from decision-making at the White House but continued to pay a surprising amount of attention to the global warming debate, the subject that lured him back to the campus where he had earned his doctorate.

While working for the president, Sununu had taken pride in his background as a scientist, ability to go toe to toe with climate experts, and courage to reach his own conclusions. In 1990, for instance, the ex-governor had peppered Warren Washington at the National Center for Atmospheric Research (the lab Margaret Thatcher visited that same year) with technical questions about climate models. After Washington, who focused on the impacts of oceans on climate, briefed senior White House officials on climate, he reported that they had demonstrated an intense interest in the subject but were sometimes frustrated with his answers.

Sununu wanted to do his own calculations and asked the Colorado lab to prepare a one-dimensional climate model that could run on the Compaq 386 personal computer in his office. Editing down supercomputer-size three-dimensional versions for such use was far from an easy task. Moreover, NCAR feared that a simple model with limited capabilities could be subject to misuse. But Bromley told them, "Give Sununu whatever he wants." Washington brought the resulting floppy disk in his pocket to a DC conference and gave it to Bromley. He never heard back whether the chief ever ran it.[8]

Before moving out of the White House, Sununu doubled down on his probing of climate science. He and John Deutch—a professor of physical chemistry at MIT, undersecretary in Jimmy Carter's Energy Department, and later President Bill Clinton's CIA director—agreed to host a workshop with the nation's leading climate experts to explore whether it was possible to find some commonality of views about the issues of climate science, economics, and policy that had become so divisive.

The group, twenty in all, sat at tables arranged in a rectangle for three two-hour sessions interrupted only by lunch at the faculty club. There were two ex-provosts from Yale (including Bill Nordhaus) and one from MIT (Deutch). Almost every major National Academy climate study, from the weather modification studies of the 1960s to the recent Evans report, had at least one panelist at the MIT workshop. Skeptics about the work of the International Panel on Climate Change, such as Bill Nierenberg and Richard Lindzen, faced off against longtime advocates for climate mitigation Gordon MacDonald and Michael Oppenheimer. Exxon and the Electric Power Research Institute each sent a top scientist, as did the National Resources Defense Council. All participants had reached the pinnacle of training in their respective disciplines and learned the scholarly canons of scientific truth-finding. There was even an overhead projector in the middle to facilitate the display of dozens of viewgraphs that might resolve disagreements about the data. What could go wrong?[9]

Sununu provided an early indication of where he stood. He believed there was a "rush to judgment" on "trillion-dollar decisions" to deal with a likely overstated risk. Leading the rush were "some constituencies out there who constantly seem to be looking for calamitous surrogates to creating anti-growth, anti-dynamic investment policies." These groups included the press, which was trying to sell papers, and the Green parties in Europe, whose influence in coalition governments was disproportionate to their electoral strength. He called on scientists to define what they did and didn't know "with a little more precision than they perhaps are usually accustomed to." He predicted climate policy would be "framed in very specific terms in the next five years."

The deliberations turned out to be cordial, bordering on chummy. The professors knew each other on a first-name basis and often began their rebuttals with some compliment of their protagonist. The talks were also messy. Midsentence interruptions were frequent and sometimes simultaneous. People who'd worked together on National Academy studies disagreed on the methodologies used. The scholars often couldn't respond to questions off the top of their heads. When doubting some assertions, Sununu threw up his hands or said, "Picky, picky, picky," usually with an impish grin.

Deutch, who presided, tried to rein in his unruly academics and find some consensus. But many speakers were determined to introduce a long list of new or "emerging" issues, even if there was no time to discuss them. What about the hydroxyl free radical, thermohaline circulation in the world's oceans, or the nitrogen fixation of plants? These might be important topics, but it would likely take decades to get clear answers. The devotion to the weeds made it difficult to see the trees, let alone the forest.

During the Cambridge cacophony, no one denied that there were many unknowns about the earth's future climate. The critical questions were whether enough was known to support well-designed policies, and could humans adapt to their new circumstances if the projections of the leading models turned out to be correct? The answers to these questions could help determine the urgency, if any, of reducing the emissions of greenhouse gases.

Not surprisingly, Lindzen launched a frontal assault on the National Academy's and IPCC's basic findings. Several at the workshop cited Hansen's data on rising global temperatures. But the MIT prof didn't buy it. He confidently asserted, "We haven't seen warming"—a minority view even among his fellow skeptics.

MIT atmospheric chemist Ron Prinn said the uncertainties arose primarily because of the vast number of questions scientists wanted to investigate and the IPCC had failed to answer. As the study of climate science intensified, he foresaw investigators discovering more that they didn't know. Thus, the coming years would produce less, not more, certainty.

Lindzen had a cynical explanation for why uncertainties would likely grow. He doubted that scientists would ever provide clear answers to pending questions because it didn't benefit them professionally to do so. He predicted, "The more important the policy community says the problem is, the less likely the science community is to give an answer. Because [policymakers are] telling you it's very important, hang on to it, it will sponsor research forever. There is a prejudice against giving answers."

Several participants, generally more involved in current modeling and international negotiations, pushed back against the critics of the IPCC. Washington suggested that scientists already knew more about oceans than Sununu thought and would learn a lot more in the next few years. Any gaps in knowledge, he stated, should not be barriers to the adoption of policies to reduce emissions. The Environmental Defense Fund's Oppenheimer countered that the discrepancies between temperature data and projections that Lindzen had complained about could be effectively explained by factoring in the effects of cooling aerosols.

Nierenberg—seventy-three and still at Scripps, the cradle of modern climate change science—continued to favor a climate strategy based on human adaptation. He rekindled an argument from *Changing Climate:* American farmers had enhanced their yields at a pace that suggested they could handle changes in climate. He implied that such adaptation would prove successful in other endeavors, even those outside the United States. Nierenberg continued to strongly favor a wait-and-see strategy.

Jesse Ausubel, having left the National Academy staff for Rockefeller University, agreed with Nierenberg and made an even more robust case against government policies to reduce greenhouse gas emissions. He argued that energy systems were already "evolving in the right direction." Kilograms of carbon emissions per unit of economic output were going down in prosperous economies. Economic growth might, for instance, increase by 2 percent a year while emissions grew by 1 percent. Thus, human societies would, in his view, be "decarbonizing" while the carbon content of the atmosphere was actually increasing. Ausubel didn't acknowledge that even this modest achievement benefited from government investments in nuclear energy, auto and appliance efficiency regulations, hefty gasoline taxes in Europe, and offshoring industrial activities to other countries.

Ausubel also made the novel argument that climate risk was declining even if the climate models were correct about a warming planet. He claimed there was too much emphasis on climate mitigation and not enough on adaptation. He argued, "Societal vulnerability is lessening [because] the importance of climate has been diminishing." He reached this conclusion because the climate was mainly important "for what goes on outdoors." Since the share of outdoor activities in the gross national product was declining, "susceptibility to climate change will be much smaller than it is today." His logic resembled an earlier assertion that if humans stayed indoors or used better sunscreens, they wouldn't need to worry about the destruction of the ozone layer.

Two Ivy League scholars cautioned against Ausubel's rosy outlook. Yale's Richard Cooper, a veteran U.S. diplomat, said that the high tolerance for risk on the part of Ausubel and others led Cooper to "believe that, in fact, we are going to perform the experiment on the global basis of having continuing high greenhouse gas emissions"—a haunting reference to Revelle's 1957 article. Cooper predicted it would probably take a series of disasters to spur international action. He questioned any such choice to live with potential catastrophe and "do nothing." He pled, "At least let us have the option of deciding not to do it at that time. If we don't prepare for it, we don't have the option." Bill Hogan—who helped create the Energy Information Administration during the Carter years—shared an idea from a student at Harvard's Kennedy School of Government. Their paper suggested that the longer the time needed to resolve the scientific uncertainty, the more we should be doing in the near term. "If we knew that was going to be resolved next year, we should just wait. If we don't know what's going to happen for twenty years until we figure it out, we should do something now in anticipation of it." Deutch: "That certainly makes sense to me."

Exerting great influence, as often the case, was uber-economist Bill Nordhaus. If mitigating climate change was inexpensive, then nitpicking over the science became less relevant. But Sununu had been talking about "trillion-dollar programs," which raised the stakes for new policies.

Nordhaus, as usual when with his academic peers, acknowledged the fragility of his assumptions. When displaying an energy price outlook, for instance, he cautioned, "Don't take it too seriously." But in some halls of power, his projections were taken more seriously than those of the physical scientists, even though the latter rested on more solid epistemological foundations and more rigorous peer review. Rafael Bras, a civil engineer at MIT who later became provost at Georgia Tech, complained about the deference to economic models. Economists, he said, "make policy decisions on uncertainties that, to me as a non-economist, look higher and bigger than many of the scientific uncertainties I deal with."

Nonetheless, the Nordhaus model was persuasive for many when it showed that reducing emissions 20 percent below 1990 levels would be "very, very costly." Nordhaus favored a small carbon tax, which he said wouldn't do much good but wouldn't do much harm. He appeared to abandon his previous position that progressively higher taxes could eventually control threats to the atmosphere.

The Yale professor also agreed with Ausubel that the adverse effects of climate change were often overestimated. "I actually thought the impacts were going to be much larger than they've turned out to be," he said, "but the careful analysis that is taking place in this area has led me to think that people are actually much more adaptive than my mindset in thinking about this led me to believe." He conceded his damage assessment included only the United States. This omission, in effect made the harms everywhere elsewhere in the world what economists might call a "free good."

Representatives from environmental organizations questioned the results of the Nordhaus model. Dan Lashof—the thirty-two-year-old senior scientist at the NRDC—suggested that energy taxes had less negative impacts on the economy if they displaced other, less efficient taxes. Under questioning from Lashof and Oppenheimer, Nordhaus replied that his model did not allow for dynamic advances in carbon-reducing technologies; he was uncomfortable with "extrapolating way beyond the range of current experience" with such advances. Thus, the economist was willing to rely on future human inventiveness to adapt to climate change in unspecified ways but unwilling to consider similar ingenuity in making breakthroughs in solar or nuclear energy.

Near the end of six hours of discussion and with the Rio summit just four months away, a relieved Sununu said he had found the discussion "more than I ever expected." He said people were "beginning to move towards admission of uncertainties." He did not mention that every climate report from the National Academy since the 1960s, and even the punching bag IPCC, had consistently devoted extensive attention to uncertainties.

After final comments from Deutch, the attendees gave themselves a round of applause for disagreeing without being disagreeable.

Sununu's departure from the West Wing added a new twist to the Bush administration's plans for Rio. Some pro-environment GOP senators believed the administration might soften its opposition to climate targets and timetables at an IPCC negotiating session scheduled for New York during the last two weeks of February. They considered the meeting at UN headquarters possibly the last chance to hammer out an agreement. During the debate on the Gore climate amendment, Senator Chafee expressed his hope that the president would use the UN event to moderate the U.S. position. In addition, the usual anonymous administration sources told the *New York Times* that the situation was now more "fluid."[10]

Behind the scenes at the White House, however, at least one die was being cast. The week before negotiators gathered in New York, new chief of staff Samuel Skinner sent a memo to eleven senior officials on "Coordination of Administration Policy on Environmental Issues."[11] He announced that Clayton Yeutter was leaving his posts as secretary of agriculture and chair of the Republican National Committee to become counselor to the president. Skinner directed the recipients involved in the meeting at the UN "not to take any position in the coming weeks that might serve to limit the President's flexibility in this important area without first speaking to the Counselor to the President." The clear intent was to rein in people like Reilly, whose views often resembled those of the Europeans.

Yeutter's attitudes toward climate change were similar to those of the departing enforcer of White House discipline. He believed the United States shouldn't have agreed to participate in the Rio summit because the commitments proposed injected "an inordinate level of pain on the American economy, and that is just not tolerable." The former U.S. trade representative also feared that a multinational agreement would create a double standard. Treaties signed and ratified by the United States had the force of domestic law. In many other countries, unattained goals had little legal consequence.[12]

Yeutter wrestled with his recommendation on Bush's attendance at the summit. A trip to Rio would take a big chunk of presidential time and raise the stakes of the outcome. In mid-March, Yeutter wrote Bush that participation might be "a big loser. If so, we'll want to keep you out of there." On the other hand, "It sure won't be easy to defend your absence." Next to the latter comment, Bush penned in the margin, "So true!"[13] Although critical issues remained in play, there was little evidence so far that the shake-up of the White House staff would significantly alter the U.S. position going into Rio.

Nonetheless, presidential statements sometimes gave optimists like Chafee a glimmer of hope. A late March environmental message to Congress, for instance, proclaimed, "The nations of the world must ensure that economic development does not place untenable burdens on the earth's environment," sounding more like European economists than his economic advisors. Bush's priorities for Rio emphasized actions other than reducing carbon dioxide emissions. But he promised the United States would reduce its net emissions "by improving energy efficiency, developing cleaner energy sources, and planting billions of trees in this decade." He hoped the nations at the Rio summit would commit to action plans that addressed "sources and reservoirs of all greenhouse gases as well as adaptation measures."[14] His statement fell short of endorsing the targets and timetables the Europeans wanted, though it did suggest that the United States would act to reduce greenhouse gases, perhaps in the still pending energy bill.

On April 20, Yeutter informed Skinner there would be one more pre-Rio session in early May, again at the UN headquarters. He said the United States would seek language that was "deliberately vague and could not be construed as a commitment by us or anybody else." Eleven days later, Yeutter informed the president that U.S. negotiators were pushing wording that called on nations to "adopt policies and actions" to limit emissions of greenhouse gases. But this language made "no commitments as to what those policies are, or what actions will be taken." He assured Bush that the United States was not agreeing "to do *anything* we are not already doing or have proposed to do."[15] On May 3, Bush responded that neither environmentalists nor *Wall Street Journal* editorial types would like the proposed language. He said, "It doesn't hurt to have a sensible middle-road position on something this touchy. We never can keep the extremes happy."

Organizations on both sides of the climate debate continued to forcefully espouse their views in New York. Greenpeace and the Natural Resources Defense Council aggressively pushed for strong policy responses to climate change. On the other side, the Global Climate Coalition had a new executive director, John Schlaes, who'd left the Edison Electric Institute to give the organization a more strident edge. A GCC press conference at the UN featured University of Virginia professor Fred Singer, known for his skepticism about the risks of tobacco smoke, acid rain, and global warming. Singer was a prominent combatant in the campaign to undermine the widespread belief among scientists that the earth was warming and would continue to do so. In a 1991 letter to the editor of the *Washington Post,* the Virginia prof declared, "No one doubts that we will be seeing the onset of the next ice age soon . . . I'd be more worried about a future global cooling than about greenhouse warming." As the Schlaes effect took hold, the Global Climate Coalition transitioned from questioning the certainty of global warming to asserting an undoubted return to an ice age.[16]

The president also reported on his calls to leaders who would be influential in Rio—Helmut Kohl of Germany, Brian Mulroney of Canada, and John Major of the United Kingdom. The UK leader was less vocal on climate than his predecessor. Bush conveyed that he had a good chat with Major, "who wants to help."[17] Nonetheless, Yeutter told the president that U.S. negotiators remained under considerable pressure. "They had," he reported, "the business community telling them we were selling out to the environmentalists, the environmentalists telling them we were ruining the world, the EC [European Community] telling them we were totally isolated in resisting a stabilization commitment, and Senators Gore and Wirth breathing down their throats with legislative threats."

The wordsmithing concluded at four a.m. on Friday, March 8, with compromises that allowed for differing interpretations. The key paragraph read, "Participating countries shall adopt policies and measures to modify emissions of greenhouse gases." That was not in dispute. The final draft went on to say that returning to earlier levels of emissions by the end of the decade "would contribute to such modification." This amendment replaced earlier (stronger, in the U.S. view) language that such action "would be an appropriate signal" by the developed countries.[18]

The UN's chair of the talks, Jean Ripert, conceded that the acceptance of the draft treaty relied on its "constructive ambiguities."[19] The UN quickly translated the document into six languages and sent it to the printer.

Few people lobbying in New York were totally pleased with the results. But audience reaction tilted to the idea that Chafee may have been partially correct. A bit of moderation after the departure of Sununu and the need for all sides to compromise may have given the opponents of climate action less than they expected. Greenpeace's Jeremy Leggett decried the "sickening fudge on commitments" by the United States but still found some comfort in the agreement. He reluctantly concluded that the convention "established a process which required governments to regularly review the science of global warming, the impacts of climate change, and the implementation efforts to curb greenhouse gas emissions. This left the door open, in principle, for rapid action to manage the challenge of achieving deep cuts in emissions . . . should sufficient political will to do so emerge." He observed that the coal, oil, and automobile industries backing the GCC realized the danger to their interests of a world agreement that acknowledged the threat of climate change.[20]

Both houses of Congress had already passed resolutions urging Bush to go to Rio. The organizers of the earth summit feared that Bush's absence would undercut the stature of a massive international event, which gave U.S. negotiators greater leverage on substantive issues. With the decision on presidential participation pending, Reilly was in Rio for preparatory talks when Prince Charles (later King Charles III) invited him for a sail up the Amazon on Her Majesty's royal yacht *Britannia*. With great sincerity, Charles declared, "Mr. Reilly, it is vital to U.S.–Brazilian relations that the president of the United States come. There are at least 40 heads of state waiting to find out if he's going to come before they decide whether to come. I will give you my word of honor. I understand Mr. Bush will be in the middle of the campaign. I will do everything in my power to ensure that he is not embarrassed."[21] With understandings acceptable to the Americans in hand, the president announced on May 12 that he would join the world leaders in Rio.

Concurrent with the negotiations, dueling organizations painted different pictures about the urgency and the difficulty of slowing climate change. In April, the Alliance to Save Energy, the American Gas Association, and the Solar Energy Industries Association published a joint study, *An Alternative Energy Future*. The groups, not surprisingly, advocated reallocating research and development funding to natural gas, renewables, and efficiency technologies. They argued that these investments, combined with the implementation of existing appliance efficiency legislation and market forces, would significantly reduce greenhouse gases as well as other pollutants. Their alternative scenario projected that by 2010 primary U.S. energy consumption would decline slightly (1.7 percent) from 1990 levels, in contrast to the growth projected by other models. During this same period, there would be a 50 percent increase in economic activity. These optimistic results suggested it might not be very costly to make at least some progress in curtailing greenhouse gas emissions. Their analysis sharply contrasted with the economists who saw an inevitable conflict between economic growth and protection of the atmosphere.

The same month, MIT's Richard Lindzen addressed the members of the Organization of the Petroleum Exporting Countries at its Vienna headquarters. In Vienna, Lindzen clung to the idea that doubling atmospheric carbon dioxide would increase average global warming by only 0.5 to 1.2 degrees Celsius, well below the range of every peer-reviewed model. He blasted the IPCC's scientific assessment for involving too few U.S. scientists, even though more than one hundred Americans had served as lead authors, official contributors, or peer reviewers. The professor also panned the National Academy's Evans report for not having the right experts, though he cited with approval a part of its findings with which he agreed, making it look like a separate study.

Lindzen's lecture was music to the ears of the world's prolific oil exporters. OPEC president Jibril Aminu warned that the oil industry would come under an "almighty onslaught" in Rio. Buttressed by comments of a prominent American scholar, the organization's secretary general, Subroto, said reputable scientists differed in their assessments of climate change. He called for studies to "reduce the scientific uncertainties." The Cato Institute published a revised version of the professor's presentation.[22]

The week of the OPEC meeting, Bromley displayed his unease with his role during the Sununu era of giving a White House audience to the scientists attacking the findings of the National Academy of Sciences. When Yeutter asked about groups like Cato and Marshall, he replied that they represented "the extreme conservative end of the scientific spectrum."[23]

As notables around the world began packing for the earth summit, the White House worried about the optics of the press coverage and drafted talking points to ensure consistent messaging. Yeutter told the president and others they needed them to "put some curbs on Reilly."[24]

The media presence was massive. Yeutter told Bush on June 9 that 9,000 journalists had already gathered in Rio. One environmental leader boasted that the coverage exceeded that for the Olympic Games. He observed that Al Gore seemed to have a television camera pointed at him on a permanent basis.[25]

The rhetoric on the world's environmental and economic challenges had a different dynamic in Rio than in Washington. Most participating countries lacked the financial resources of the United States, the European Community, or Japan. They were not significant carbon emitters but would still feel the brunt of global warming. In Brazil, they were on the big stage. Malaysia's ambassador to the UN rejoiced: "For the first time, people are taking us seriously." Czechoslovakia's president, Vaclav Havel, articulated an interpretation of climate change not often heard at the U.S. Congress or the White House. He declared that the wealthier nations "have to accept substantial blame for environmental degradation in the poorer countries."[26]

In contrast, at a press conference in Rio, the Global Climate Coalition emphasized the likely growth of emissions in developing countries, implying that they should make parallel commitments in the climate agreement.[27] Despite the attempt to deflect attention away from the more affluent countries' impacts on the less prosperous ones, the summit adopted a "commitment" closer to Havel than to the GCC. In somewhat less accusatory language, it stated that the industrialized nations "shall also assist the developing country Parties that are particularly vulnerable to the adverse effects of climate change in meeting costs of adaptation to those adverse effects."[28]

In an impressive show of unity, 154 nations signed the Framework Convention on Climate Change. The twenty-two-page document had enough elements to avoid any country leaving without some victories. It began with the "concerns" of the parties "that human activities have been substantially increasing the atmospheric concentrations of greenhouse gases." It continued, "These increases enhance the natural greenhouse effect [and] will result on average in an additional warming of the Earth's surface and atmosphere and may adversely affect natural ecosystems and humankind." It believed "that the largest share of historical and current global emissions of greenhouse gases has originated in the industrialized countries, that per capita emissions in developing countries are still relatively low, and that the share of global emissions originating in the developing nations will grow to meet their social and development needs." This language reflected the ongoing work of the Intergovernmental Panel on Climate Change, though it might sound like "Chicken Little" to the Global Climate Coalition.

The framework's "Principles" section called on the industrialized nations to "take the lead in combating climate change and the adverse effects thereof." It also declared, "The parties should take precautionary measures to anticipate, prevent, or minimize the causes of climate change and mitigate its adverse effects." Furthermore, "Lack of full scientific certainty should not be used as a reason for postponing such measures." These steps should be "cost-effective," which could be best achieved if they were comprehensive and covered "all relevant sources, sinks, and reservoirs of greenhouse gases and adaptation." This section's delicate compromises met the U.S. goal of approving the trading of greenhouse gas reductions, allowing fossil fuel users to postpone their reductions by purchasing credits from those cutting their emissions of "other gases." European concessions to reduce the emphasis on CO_2 came at the cost of U.S. agreement that the lack of scientific certainty did not excuse a failure to act.

In the most fiercely contested part of the framework, the parties agreed to present their plans to mitigate climate change "with the aim of returning individually or jointly to their 1990 levels" of "anthropological emissions of carbon dioxide and other greenhouse gases." The Americans had removed the legal commitment to reach stabilization. Some observers, though, saw this language as establishing a moral obligation.

The framework also created a Conference of Parties to oversee its ongoing implementation. Bush promised that Reilly would be actively involved in fulfilling U.S. commitments.[29]

On June 12, Bush offered a positive take on the convention he had just signed. He praised the agreement as a "landmark" that required "countries to formulate, implement, and publish national programs for mitigating climate change by limiting net emissions of greenhouse gases." He also endorsed the convention's comprehensiveness in covering all sources and sinks of greenhouse gases, and its flexibility for national programs. He announced the intention of the United States to continue its leadership role and "to leave this earth in better condition than we found it." He predicted, "Our children . . . will be grateful that we met at Rio, and they will certainly be pleased with the intentions and the commitments made. But they will judge us by the actions we take from this day forward. Let us not disappoint them."

Despite Bush's lofty rhetoric, his remarks were not fully in sync with the thrust of the convention. Unlike other speakers, he did not talk about why limiting the global impacts of greenhouse gases was important. As on previous occasions, he devoted considerable attention to U.S. accomplishments under

President Bush (*left*) with EPA administrator William Reilly, addressing environmental organizations at the Rio earth summit, June 12, 1992. Courtesy AP Photo/Dennis Cook.

the Clean Air Act, which didn't have global effects. His vision also included a heavy reliance on volunteerism.

Another subtle difference involved the choice of baselines used to assess the success of limitations on greenhouse gas emissions. The framework convention required that nations track emissions from 1990 levels to assess their success in meeting stabilization and, eventually, reduction goals. Bush adopted a different method from the rest of the world—measuring progress against future expectations. Estimates of future emissions in a "business as usual" scenario were imprecise at best. Moreover, this metric allowed nations to claim that emissions were coming down when they were actually rising.

Bush made specific mention of only two new or enhanced U.S. measures that would affect atmospheric carbon dioxide—"forests for the future" and a "green lights" initiative at the EPA. Protecting forests would help the earth's carbon balance and protect biodiversity. Bush promised to double international forest assistance with a U.S. down payment of $150 million the following year and got specific pledges from Italy and Austria to support the effort. Some critics called the idea a "smokescreen" to distract from U.S. failures on biodiversity and climate change.[30] The lighting effort sought to create partnerships that would upgrade energy efficiency. Because the Bush lighting initiative was voluntary, it was unclear how much savings the EPA program would produce beyond what "partners" were already planning.

The battle over targets and deadlines had taken its toll, creating considerable public relations problems for the Bush team. Some European nations complained that U.S. political pressure during the deal making had been "unfriendly."[31] Despite the criticism, the convention had considerable content that contradicted White House thinking. Plus, the Americans had fully supported mechanisms that would institutionalize global cooperation on climate—a necessary though not a sufficient milestone for transferring the findings of science into public policy.

William Nitze looked back on the earth summit from dual perspectives. The son of Paul Nitze—famed defense policy advisor to several presidents— had served as deputy assistant secretary of state from 1987 to 1990. He left the department after Sununu accused him of overstating the Bush position on global warming during international negotiations. He then assumed the presidency of the Alliance to Save Energy, the organization arguing that nations could reduce carbon emissions and grow their economies with improved energy efficiency. Two years later, he said the framework convention had reflected "a failure of presidential leadership." Bush, he recalled, refused to get personally briefed on climate science, leaving Sununu "free to shape U.S. policy to his liking." Still, Nitze acknowledged that the administration had shown some flexibility in agreeing to a far-reaching long-term objective. It had approved language calling for the "stabilization of greenhouse gas concentrations in the atmosphere at a level that would prevent dangerous anthropogenic interference with the climate system . . . within a timeframe sufficient to allow ecosystems to adapt naturally." Rio had produced "greater opportunities" if not assured results.[32]

17

In the Gutter

Bush, July 1992 to January 1993

Political insiders had expected the announcement for several weeks.

The presumptive Democratic nominee for president, Arkansas governor Bill Clinton, revealed his choice of a running mate on July 9, just days before the party faithful gathered at Madison Square Garden for their quadrennial national convention. It would be Al Gore. A glowing story in the *New York Times* described the forty-four-year-old Tennessee senator as "tall and broadshouldered, somewhat resembling Christopher Reeve, the actor who played Superman."[1]

Gore's environmental credentials added to his appeal, since they complemented Clinton's focus on other priorities. The veep candidate even earned praise from a senator across the aisle for his diligence on climate protection. John Danforth of Missouri called Gore's knowledge of the environment "almost scholarly." "When he engages witnesses," observed Danforth, "it's not just a few superficial questions that have been prepared by a staffer." Some colleagues, however, complained about Gore's verbosity on the subject. They sometimes wished he didn't feel compelled to "share all that he knows."[2]

On the convention's opening day, delegates approved a party platform with dozens of planks, including one on "Preserving the Global Environment." It closely tracked Gore's comments in the Senate. The document charged, "As governments around the world have sought the path to concerted action, the Bush administration—despite its alleged foreign policy expertise—has been more of an obstacle to progress than a leader for change, practicing isolation on an issue that affects us all." The Democrats promised to "act now to save the health of the earth, and the health of our children, for generations to come." Resurrecting Gore's failed amendment to the energy bill, they urged the country to "join our European allies in agreeing to limit carbon emissions to 1990 levels by the year 2000."[3]

In his Thursday evening acceptance speech, Gore's remarks on climate change were brief but even more hard-hitting.[4] He told a packed arena and national television audience that the Bush administration had "mortgaged our children's future to avoid the decisions they lack the courage to make." After leading the delegates in a chant, "It is time for them to go," he continued, "They embarrassed our nation when the whole world was asking for American leadership in confronting the environmental crisis." After a pause, he again encouraged the refrain, "It is time for them to go."

Minutes later, Gore reflected, "For generations, we have believed that we could abuse the earth because we were, somehow, not really connected to it. But now we must face the truth. The task of saving the earth's environment must, and will, become the central organizing principle of the post–Cold War world." The Garden rocked with boisterous cheers and applause.

Five weeks later, the Republicans returned fire at their convention, with a full-throated attack on big government. They pictured the end of the Cold War as a lesson for the United States. It showed that "collectivism" didn't work. The GOP platform stated, "Here at home, we warned against Big Government because we knew concentrated decisionmaking, no matter how well-intentioned, was a danger to liberty and prosperity."[5] The document was filled with praise for individualism and charged Democrats with supporting a paternalistic state, government bureaucrats, deniers of personal responsibility, party bosses, red tape, and an imperial Congress. It complained, "The Democrats have controlled the House of Representatives for 38 years—five years longer than Castro has held Cuba."

The Republicans' antigovernment message set the template for their approach to greenhouse gases. The environmental plank said climate change should be "the common concern of mankind." But it praised President Bush "for personally confronting the international bureaucrats at the Rio Conference," where he refused "their anti-American demands for income redistribution." It credited him for winning a treaty that relied "on real action plans rather than arbitrary targets hostile to U.S. growth and workers."

The party endorsed spending more on climate science research, including

Bill Reilly's idea for a Mission to Planet Earth. It also supported research and development for renewable energy, hastening the move to "the next generation of nuclear power plants," and massive reforestation

However, the platform spent more time talking about what it was against. It even declared that a Republican Senate would not ratify "any treaty that moves environmental decisions beyond our democratic process or transfers beyond our shores authority over U.S. property." Experts of many persuasions had generally agreed that a rare global pollutant like CO_2 would require a binding international agreement to limit global emissions. Now, with only thirty-four votes needed in the Senate to block a treaty, the GOP appeared to be driving a stake into the idea.

The GOP also opposed "any attempt to impose a carbon tax as proposed by liberal Democrats." They also blistered "the Democrats' draconian plan for higher Corporate Average Fuel Economy (CAFE) standards," which they argued would lead to "unsafe vehicles, reduced consumer choice, higher car costs, and a loss of 300,000 jobs in the auto industry here at home."

After the national party conventions, Republicans launched repeated attacks on Gore's *Earth in the Balance.* In late August, Vice President Dan Quayle portrayed the Democratic presidential ticket as "environment elites" opposed to job creation. He asserted, "We envision man as the harvester, working with the earth to raise up homes and cities, respecting nature but shaping it to legitimate ends." Based on his interpretation of the Gore book, Quayle charged, "They view man as an uninvited guest here on earth, setting down and consuming resources to which he has no right." Campaigning in Michigan, the veep targeted Democratic support for automobile efficiency standards. He charged that the plan would lead to the loss of 40,000 jobs in the state and that the National Academy of Sciences had found that "this level is not even technically achievable."[6] The chair of the report cited by the vice president countered that the NAS found such standards feasible but did not discuss whether American buyers would be willing to pay higher prices for cars that used less gasoline.

The partisan divide on climate change resurfaced during a September 18 Senate Foreign Relations Committee hearing. Gore—chair of the Senate delegation monitoring the Rio summit and now a candidate for vice president—testified in favor of treaty ratification. He said the agreement could help reduce "risking disruptions in the carbon balance more severe than any in the past 10,000 years."[7] With the election less than two months away, he also skewered Bush as "the single largest obstacle to progress." He charged that U.S. "intransigence" has left the final agreement "completely devoid of any legally binding commitments to action."

After welcoming "Al" back from the campaign trail, ranking Republican Mitch McConnell punched back, using *Earth in the Balance* as his cudgel. He told the committee his state of Kentucky had more than 100,000 coal-related jobs and claimed, "At least half the scientific community thinks the concern about global warming is greatly exaggerated." He accused Gore of ignoring "scientific doubt" and endorsing "absolute solutions." Later in the proceedings, McConnell cross-examined Gore on whether his book advocated a carbon tax. Gore replied that early actions on climate change could rely on nontax measures. However, for the long run, it would be rational to have a revenue-neutral swap of taxes. This approach would reduce payroll taxes to increase incentives for work and shift the burden to pollution as an incentive to emit less. McConnell pounced on the chance to tie the Democratic ticket to energy taxes, asking, "Did you or did you not advocate a carbon tax in your book?"

During October, members of Congress, like the president, had to balance their reelection campaigns with the job of governing.

On October 5, the House of Representatives adopted the long-delayed conference report on the Energy Policy Act of 1992. The overwhelming 363–60 vote, with healthy majorities in both parties, reflected the multiple compromises in the wide-ranging final version.

Chief sponsor Phil Sharp of Indiana said the bill went beyond its principal goal of increasing the reliability of fuel supplies. He asserted, "Congress has acted affirmatively to address the issue of global climate change. The global warming title, with its studies and voluntary reduction programs, is only part of the story. Concerns about global warming have been woven into the fabric of this bill through the efficiency and renewable provisions, the alternative auto fuels programs, the clean coal technology export provisions that can increase the efficiency of coal combustion in developing countries, and more." The bill still needed to pass the Senate, where the threat of a filibuster remained.[8]

Two days later, the Senate conducted an anticlimactic vote to ratify the UN Framework Convention on Climate Change. With no demand for a recorded tally, a voice vote confirmed U.S. approval of the Rio agreement, suggesting the messy negotiations had hit a political sweet spot. Senators convinced there was little urgency to address the risks of climate change had won the battle to avoid legal mandates, so they accepted language saying that greenhouse gases were a problem, that developed countries should try to stabilize emissions by the end of the century, and that less affluent nations deserved special assistance. Those more concerned about the risks of climate change rued the lack of firm targets and timetables but found plenty in the pact they favored.[9]

The next day, with a third of the body eager to return home for their final campaign push, the Senate took up its final consideration of the omnibus energy bill. An 84–8 cloture vote snuffed out a filibuster threat from Nevada's senators, who opposed the targeting of their state for a high-level nuclear waste depository. With that settled, the product of two years of bargaining passed on a voice vote. Interest groups of various persuasion had pledged their support for the bill. The administration was on board, almost every senator had put something they liked in it, and controversial measures had been abandoned long ago.

Tim Wirth, a Senate leader on climate change who was not seeking re-election, called the bill "an extraordinary change of policy" and a "wonderful shift from where we have been going ever since the dawn of the fossil fuel revolution." He praised the inclusion of "stringent conservation standards," steps toward alternative fuels with "plentiful natural gas" as a bridge fuel. He said the legislation has "for the first time a beginning of a balance between the environment and energy policy."[10]

The bill's length and complexity made it difficult to discern whether its provisions would constitute an effective response to the call in Rio for nations to change the trajectories of their greenhouse gas emissions.[11] The section on global warming did not provide much assurance of progress. It called for studies that would assess various feasibilities. It required the secretary of energy to produce more strategies and plans. It did not prevent the repetition of previous experiences when studies produced calls for more studies, plans suggested more planning, and new strategies recommended developing additional strategies. As Sharp and Wirth advised, the substance of a climate policy was more likely to be found in the sections on energy efficiency and alternative fuels.

The bill's primary target for energy efficiency was buildings. It required the energy secretary to establish minimum efficiency standards for federal government buildings and the states to adopt commercial building standards that met or exceeded model industry standards. Congress was reluctant to mandate more efficient homes but encouraged states to update their residential building codes to improve energy efficiency. The bill also increased the coverage of efficiency standards for certain types of lighting. The initiatives on buildings fell well short of the appliance efficiency standards legislation of 1987 in anticipated impact. But they did move the ball forward a bit by reducing carbon emissions somewhat below what they would have been otherwise.

Though Congress had squashed dramatic increases in automobile efficiency, it did not ignore transportation. The new act authorized funds for cost-sharing programs with industry to provide subsidies to demonstrate the viability of electric vehicles. Partnerships with states would start developing the infrastructure needed for conversion to these new vehicles. It also called on the secretary of energy to create a five-year program for research and development on technologies such as advanced batteries, high-efficiency electric power trains, and hybrid power trains. If the market was not yet ready to move away from traditional internal combustion engines, these initiatives might accelerate the learning curve for a transition further down the road.

The energy act also promoted both fossil fuels and the fuels that might replace them. It authorized demonstration programs at the Energy Department to stimulate the development of geothermal, solar, and wind energy. It also allowed an incentive of 1.5-cents per kilowatt-hour for electricity produced from renewables. For nuclear power, the legislation streamlined the licensing process to reduce delays in construction and authorized research and development on advanced reactors that were safer and more efficient than existing designs.

As an added feature, the legislation opened a new door for renewables to enter energy markets. It provided small to medium-size wholesale electric plants special access to transmission systems controlled by investor-owned monopolies. But these independent producers had to supply electricity from renewables or natural gas to qualify. This provision created the potential to shake up the fuel mix for electric generation and move the country toward less dependence on coal.[12]

The Energy Policy Act provided numerous tools to deal with greenhouse gas emissions from energy. But there was little assurance they would be used. In most respects, the "new" measures were not all that new and were far less bold than energy laws adopted from 1978 to 1987. However, it was still possible to detect incremental progress, particularly if the executive branch and congressional appropriators aggressively implemented the provisions in the 1992 act.

Third-party candidate Ross Perot was a wild card in the race for the White House. The Texan made a strong early impression but dropped out in July. Then, in early October, he jumped back into the race. As a result, the sole vice presidential debate scheduled for October 13 on the Georgia Tech campus in Atlanta included three combatants—Vice President Quayle, Senator Gore, and Perot's running mate, retired vice admiral James Stockdale.[13]

Stockdale—famed for his military record and brutal imprisonment in North Vietnam—was flanked by two more aggressive candidates, neither of whom was reluctant to interrupt. At one point, Stockdale said he felt like an observer at a Ping-Pong game. Nonetheless, he sometimes tilted the balance in the clash of ideas, sometimes toward one side, sometimes toward the other.

Thirty-eight minutes into the debate, moderator Hal Bruno of ABC news turned to the subject of the environment. The two major party candidates, unsurprisingly, stuck to their recurring messages. Quayle blasted *Earth in the Balance* and Democratic plans he said would raise taxes and cost jobs. Gore argued that Bush had done a poor job with both the economy and the environment.

The debate's back-and-forth was hard to follow for the estimated 51 million television viewers. Though the verbal sparring during this part of the program was primarily about climate change, no one used the term. For tactical reasons, both sides believed it wise to avoid a concept the public might not understand, so limited their comments to "the environment." The "invisible blanket" was the political wedge issue whose name would not be spoken.

Stockdale created even greater confusion with an abrupt attempt to undercut Gore's credibility. The admiral said, "I read somewhere Senator Gore's mentor had disagreed with some of the scientific data that is in his book." Gore's response evoked the evening's only jeers from the audience. The brief exchange was not a trivial matter. It was part of an attack on Gore based on the charge that Roger Revelle had changed his views on climate change just before his death.

The background for Stockdale's allegation began with Revelle's participation in the 1990 American Association for the Advancement of Science conference. After returning home, he received a handwritten note from the University of Virginia's Fred Singer.[14] It wished him a quick recovery from his recent operation and asked for an okay on a joint essay to be signed by Singer, Revelle, and Chauncey Starr and published in the "Outlook" section of the *Washington Post*—a frequent publisher of climate denier op-eds. Singer said he wrote the draft to include comments Revelle had made during a breakfast conversation in New Orleans. The path to publication and its aftermath became a drawn-out affair marked by bitter controversy, brutal litigation, and massive distortions of climate change history.

Fred Singer was born in Vienna in 1924. His family fled the Nazi occupation there, living for a time in England, where he acquired a British accent. Eventually settling in the United States, he earned a doctorate in physics from Princeton in 1948. After various academic and governmental positions, he became one of the earliest employees of the Environmental Protection Agency. During this time, he expressed pro-environmental views, advocating "lifestyles which permit 'growth' of a type that makes a minimum impact on the ecology of the earth's biosphere." In 1971, he began his longtime residence in Charlottesville.[15]

By the mid-1970s, Singer was focusing on climate change as a critical environmental challenge. He edited *The Changing Global Environment,* which included contributions from notable climate change scientists, including Syukuro Manabe, Murray Mitchell, and George Woodwell. In his commentary, Singer stressed that many questions about natural and human-induced climate change remained to be answered. While the jury was out, he was "persuaded to think that any climate change is bad because of the investments and adaptations that have been made by human beings and all of the things that support human existence on this globe." He hoped that nuclear reactors would eventually supplant fossil fuels.[16]

During the 1980s, Singer's pronouncements took on a sharper contrarian and antiregulatory edge. He frequently attacked the scientific studies supporting government controls on tobacco smoke, gases destroying the earth's ozone layer, and carbon dioxide emissions. His defense of polluting industries earned him a leading role in Naomi Oreskes and Erik Conway's later book *Merchants of Doubt.*

In 1989, Singer produced another edited work, *Global Climate Change: Human and Natural Influences.* His publisher was Paragon House, owned, like the *Washington Times,* by Sun Myung Moon's Unification Church. This book reflected an evolution in Singer's views. Establishing himself as a witty iconoclast, he opined, "An agreement among the majority of scientists does not mean that the majority is correct. It only means that the majority of scientists agree on something." Despite differences among scientists on the impacts of global warming, Singer was confident in his view: "The results are not catastrophic."[17]

The following year, Singer expanded his ties to the Unification Church. Moon had established the nonprofit Washington Institute for Values in Public Policy to support his generally libertarian views of government and establish relations with American academic institutions, sponsoring seminars at places like Stanford and the University of Chicago. In 1991, the institute began funding a new think tank at the University of Virginia, Singer's Science and Environmental Policy Project. SEPP's mission was to examine "questionable government policies, no matter how popular, to determine if [they] are based on the rigorous application of the scientific method and not just a passing fad." When Singer wrote Revelle proposing a joint article, he did so on the stationery of the Washington Institute.[18] He wrote the proposed essay on an institute computer.

Singer also had close ties to the Global Climate Coalition. The GCC and several of its members compensated Singer for "consulting" as part of their opposition to regulating coal emissions.[19]

The draft Singer sent to Revelle for their proposed joint article opened with "a simple message: The scientific base for a greenhouse warming is too uncertain to justify drastic action at this time." As a result, it said, policymakers should wait for improved understanding before taking "panicky actions." The essay chastised the "just plain ignorance" of current climate science and claimed that scientists could not be sure whether the next century would bring warming that was negligible or significant, good or bad. Such sentiments resembled those of Allan Bromley and John Sununu.

The Singer piece contained explanations of basic science not in dispute. And it was no secret that Revelle had been expressing his reservations about climate models since the 1970s, though not with Singer's provocative disdain.

But several other sections ran directly counter to the views expressed by Revelle. For one, Singer said the most likely global warming in the next century would be between 1 and 2 degrees Celsius. None of the leading climate models, National Academy studies, or statements by Revelle supported such a low number. Singer also claimed that replacing coal with natural gas would be a mistake, directly contradicting Revelle's frequent statements advocating the transition. Another example: Revelle never commented publicly on specific legislation in Congress, but Singer launched a sharp attack on the auto efficiency regulations contained in the Bryan bill.

According to Singer, there was little risk in delaying action to prevent the buildup of atmospheric carbon. He shared the good news that "If all else fails, there is always the possibility of putting 'Venetian blinds' satellites into earth orbit to modulate the amount of sunshine reaching the earth."

The draft showed scant evidence of incorporating Revelle's views. Instead, it looked like an attempt to give scientific credence to the arguments of automobile and coal industry lobbyists. Moreover, the *San Diego Union* had interpreted the AAAS conference differently than Singer. Its story of February 17 on Revelle's public comments appeared under the headline "Revelle Urges Nations to Cooperate in Fight on Global Warming." According to the newspaper, he told the scientists in New Orleans that immediate steps should be taken to decrease reliance on coal and oil, increase the use of natural gas, and turn to nuclear power. As usual, Revelle had argued against the wait-and-see approach of Singer and others because of the long residence of carbon dioxide in the atmosphere—at the very conference that Singer was using to argue that Revelle now agreed with him.

* * *

Before the end of March, after getting no response from a still ailing Revelle, Singer said to ignore the first draft and submitted two updates that he called "shorter and better organized." He said the revisions were "improved (I hope) by many comments from colleagues." His note raised the question about the identity of these unidentified collaborators. The archival record reveals that Singer discussed the proposed article with Bill Nierenberg—like Singer, on the warpath against the IPCC assessments—and Chauncey Starr, the third author.

Starr was a giant in the field of electrical engineering. After receiving his doctorate from Rensselaer Polytechnic Institute, he worked on the Manhattan Project at the Oak Ridge labs. After the war, he transitioned to the peaceful use of the atom, collaborating with Alvin Weinberg on developing reactors that could power a nuclear navy or civilian electric plants. The Newark-born scientist earned a reputation as an expert in risk analysis and became dean of engineering at UCLA in 1967. In 1973, Starr founded and served as the first president of the Electric Power Research Institute in Palo Alto, California. He received numerous prestigious awards during his career, including President Bush's National Medal of Science—in the 1990 cohort that included Revelle. Starr's stature made him an impressive addition to Singer's project.

There was, however, a significant obstacle facing the proposed trio of authors. Unlike other climate change skeptics, Starr believed that the apparent relationship between the use of fossil fuels and atmospheric concentrations of carbon constituted a "circumstantial correlation" based on "happenstance."[20] He invoked the principle "Correlation in science is not proof, presumptive as it may appear." According to the Electric Power Research Institute's president emeritus, the residence of carbon in the atmosphere was only about five years, not the hundred years or more estimated by most scientists. Thus, his calculations suggested, "Anthropogenic emissions contribute only a fraction of the observed atmospheric rise." Starr's conclusions ran counter to those of Revelle, Dave Keeling, Bill Nordhaus, National Academy studies, the IPCC, and even most scientists in the skeptics' network. Starr's extreme outlier view was not a problem for Singer, however. The skeptics didn't need to agree on their theory of the case. They only needed to demonstrate that some prominent scientists disagreed with some aspects of the consensus emerging from extensive peer-reviewed studies.

Singer later recalled sending the drafts to other scientists for comments and suggestions but could not remember who had participated in this process. Under oath, he later testified it was "possible" the reviewers included Richard Lindzen, Patrick Michaels, and Robert Balling.

Version 3, the draft Singer sent to Revelle on March 20, 1990, removed the section attacking conversions to natural gas and new auto efficiency standards.[21] Otherwise, it was similar to the original. The message was still "Look before you leap."

In August, still not hearing back from Revelle, Singer appeared to end his pursuit of a triauthored article with his solo publication of "What to Do About Greenhouse Warming" in the American Chemical Society's journal, *Environmental Science and Technology*.[22] This work was essentially the draft 3 sent to Revelle in March, with some paragraphs deleted for length and a few minor changes in wording.

In November, Revelle recovered enough to chair a National Research Council panel meeting in Phoenix, just two days after receiving the National Medal of Science in Washington. Speaking in his own voice, the still frail scholar outlined once again the basics of climate science.[23] He said it was difficult to say how much warming would result from increased concentrations of atmospheric carbon dioxide. However, it was "probably somewhere between 2 degrees and 5 degrees Celsius at the latitudes of the United States—probably a greater change in average temperature at higher latitudes and a lesser change at lower latitudes." In particularly confident language, he asserted, "One thing we can be certain of is that whatever climate change there is will have a profound effect on some aspects of water resources." Revelle said he was also worried about the increasing emission of other greenhouse gases, which were harder to handle than CO_2. Revelle's principal suggestion for reducing carbon emissions was to "use nuclear power as a substitute for fossil fuels and use hydrogen as a fuel for transportation produced by electrolysis of water, with the primary energy coming from nuclear reactors."

Once again, Revelle's views differed sharply in tone and substance from Singer's.

On February 6, 1991, to the surprise of Revelle's assistants, Singer appeared at Revelle's La Jolla office. The editor of the new journal of the Cosmos Club—

a DC institution created in the nineteenth century, whose members included distinguished academics and government elites—had approached Singer about writing an article on climate change. The Virginia professor was asking Revelle yet again, and this time in person, to be one of three authors on a paper Singer had written.

The draft—already typeset by the journal and listing Singer as the author—incorporated the bulk of Singer's shorter article published six months earlier by the American Chemical Society. He filled it out with material—again, mostly verbatim—from earlier drafts he'd sent to Revelle. The two haggled over the wording for several hours, after which Singer departed with a marked-up copy. He said he would negotiate Revelle's proposed changes with Starr and his editor.

The Cosmos Club published the article "What To Do About Greenhouse Warming: Look Before You Leap" in April, listing Singer, Revelle, and Starr as its "authors." The infant journal was circulated principally to members, limiting its initial impact. The piece didn't make much of a ripple. When Revelle died in July, his assistant, Christina Beran, remained unaware of its publication.[24]

A knowledgeable reader, however, might view the article as a potential bombshell. It came at a particularly critical time. In the spring of 1991, the Global Climate Coalition, electric utilities, coal companies, the Heartland Institute, and the Cato Institute were intensifying their effort to undermine the IPCC assessment and a major National Academy study on climate science. The Cosmos article's citations of Michaels, Balling, and Sherwood Idso—considered by the Bush science office as "on the fringe" of climate scientists—as authoritative sources gave them a boost in stature from their recognition in an article "written by" Revelle. The month after the Cosmos article, the Edison Electric Institute launched its test radio campaign in small towns, using Michaels, Balling, and Idso to convince people the earth was cooling.

Moreover, the insertion of large-print pull quote boxes, not included in Revelle's review, highlighted the text's bent toward climate change skepticism. One blared, "The scientific base for a greenhouse warming is too uncertain to justify drastic action at this time." In certain contexts, Revelle agreed with this statement. Al Gore and the *New York Times* editorial board were on record opposing "drastic" actions such as quickly cutting back on the use of coal. But in the world of the skeptics, "drastic" meant any activity that did not immediately pay for itself—or even those that did. The word also implied that the world was on the verge of acting too quickly to slow climate changes, a view Revelle never espoused. A second box (longer but in regular-size type) was headlined "Another Ice Age Coming?," resurrecting the debate of the 1970s. The final insert opined that global warming's net impact "may well be beneficial." According to the marked-up draft retained by Singer, Revelle did not challenge the highlighted assertions in the draft, although they didn't sound like anything he had ever written or said.

The collaboration faced its highest hurdle when Singer and Revelle tried to reconcile their differences on how much global temperatures were likely to rise with a doubling of carbon dioxide. Over the years, Revelle had agreed with MacDonald in estimating a range of 2 to 3 degrees Celsius, sometimes adding that the geographic variance in the increase was more important than the average. At other times, Revelle deferred to the National Academy of Sciences' and the IPCC's wider range of 1.5 to 4.5 degrees Celsius (sometimes rounded off to 2 to 5 degrees), as he had done at the recent Academy event in Phoenix. By contrast, like others in the contrarian camp, Singer had moved to Lindzen's view that the increase would be below 1 degree Celsius. The distinction was of vital importance. If the change was under 1 degree, it might be due to natural variation, and humans might not, after all, be responsible for global warming. But all the models accepted as credible by the Academy and the IPCC contradicted the natural variability thesis, suggesting it was the emissions of greenhouse gases that drove climate change.

Singer's draft for Cosmos assumed "a modest average warming in the next century of less than one degree Celsius—well below normal year-to-year variation." It then suggested that the adjustments to this warming would be "easy and relatively cost-free." Revelle drew a line through "of less than one degree Celsius" and wrote in the margin "one to three." Singer crossed out "one to three" and later claimed that, after a verbal back-and-forth, the two verbally "compromised by leaving out any reference to any number." This alleged "compromise" in the Cosmos article was a total concession to Singer's estimate of temperature change of less than 1 degree, because of its continued reference to "natural variation."

With deceptions small and large, the Virginia professor had won his battle. If Singer's recounting of the discussion was correct, a titan of the scientific hierarchy had agreed that the peer-reviewed assessments of the National Academy of Sciences and the IPCC no longer held sway. Instead, Michaels, Balling, and Lindzen had become the arbiters of likely warming. Revelle's graduate assistant, Justin Lancaster, and his secretary, Christina Beran, later made a convincing case that Singer took advantage of the geniality and ill health of a man no longer able to work a full day in the office. They believed that Singer was determined to stay in Revelle's office for as many hours as it took to get his way. Singer agreed that the conversation that day was "spirited," although he had refused to accept any substantive suggestions from Revelle.

Revelle had been somewhat of a celebrity scientist earlier in his career. But in 1991, his name was less known to most Americans. Hence, the "Look Before You Leap" article in an obscure journal had little news value. The situation changed dramatically in 1992, when *Earth in the Balance* credited Revelle for educating Gore on climate change and Gore emerged as a likely pick for a vice presidential nod. All of a sudden, Revelle became known primarily as "Al Gore's mentor."

Gore became aware that his critics were starting to hold up Revelle's coauthorship of the Singer article to embarrass him. To preempt any political damage, Katie McGinty, Gore's assistant, contacted Lancaster about the situation. Lancaster told her that he knew from his conversations with Revelle "that he was proud of the early influence he had on Gore when Roger was a professor at Harvard and that Roger approved of Gore's environmental position. I was, and am, certain that if Roger had been alive, he would have been very distressed at this turn of events, and he would have taken steps to stop it."[25]

Gregg Easterbrook at the *New Republic* was the first journalist to recognize the increased news value of "Look Before You Leap." *TNR* identified its contributing editor as a "liberal skeptic." The Northwestern University–educated journalist was indeed a contrarian, but from a different branch than Singer. In an April article, Easterbrook faulted environmentalists for their "Chicken Little approach" to threats to nature and their failure to acknowledge impressive achievements cleaning up U.S. air and water resources in recent decades. But he also attacked those who projected overblown economic losses from environmental protection. They often complained on Capitol Hill "about being 'steamrollered' by environmentalists, a group with perhaps a billionth of their money's clout." Easterbrook concluded, "Environmental legislation often causes business disruptions. But to my knowledge, no economist has ever demonstrated net job loss." He agreed with Bush's budget office that environmentalists did not pay enough attention to economic analysis but countered, "The trouble is that freedom from pollution is a social good, deserving of the special protection granted other freedoms."[26]

During the first week of July, Easterbrook launched additional missiles at environmentalists in another *New Republic* article, "Green Cassandras."[27] Again, his primary thrust was the failure of environmentalists to celebrate past victories and their exaggeration of the remaining challenges. "As the movement has advanced from a low-budget operation," Easterbrook said, "the need to acquire ever larger sums has driven many green groups to rely on direct mail. The direct-mail business is based on scare tactics, conspiracy theories, bogeymen, and preposterous levels of exaggeration."

Easterbrook specifically targeted Gore, then in the final stages of vetting for the VP nomination. He credited the Tennessean for an environmental concern that was "genuine" and becoming "involved long before the issue was trendy." But, he complained, "Increasingly, his environmental oratory is out of control."

Easterbrook criticized Gore for not mentioning in *Earth in the Balance* that his mentor Roger Revelle had published a paper saying that the "scientific base for greenhouse warming is too uncertain to justify drastic action at this time." In his own voice, the journalist asserted, "A lifetime of study persuaded Revelle that carbon emissions should be restricted but are far less hazardous than initially feared."

The *New Republic* piece was misleading in several respects. Saying that Revelle "published" the article without mentioning the other authors, greatly exaggerated his role. A thorough review of available evidence leads to the conclusion that Revelle did not write a single sentence, that he tried for almost a year to avoid agreeing to the article, and that Singer had overruled the essential correction that Revelle requested. Revelle, a man known for his reluctance to tell people no, agreed to be an author, but only under duress. Moreover, was anybody in the Senate really calling for "drastic" action?

The idea that the risk of climate change hazards was abating was accurate in only a limited sense. It had taken analysts a while to understand that per capita energy consumption was stabilizing in industrialized nations, mainly because of responses to the 1970s oil disruptions. Lower-than-expected energy use led to somewhat lower carbon emissions than in early projections, providing a little more time to identify climate responses. Revelle and other experts had agreed this was the case. However, the midcase estimates of global warming from the doubling of carbon concentrations hadn't varied much over the years, whether the source was the JASONs, the National Academy, the IPCC, or Revelle. Indeed, during the 1980s, Revelle had been a leader in highlighting, in his own very professorial style, the hazards of climate change from the loss of strategic water flows and the leaking of methane trapped in permafrost or oceans.

A journalist might have called Revelle for clarification—had he not died almost a year earlier. Another avenue for fact checking was reading what Revelle had written in his own voice during the previous two years, which would have revealed that Singer's article did not reflect Revelle's views. Additional research would have shown that Singer had led a press conference for the Global Climate Coalition attacking the IPCC, a fact that might have given readers of "Green Cassandras" needed context.

A couple of days after the release of the Easterbrook article, Gore called Lancaster, who reiterated his belief that the Cosmos article did not reflect the opinions of Professor Revelle.[28]

On September 3, *Washington Post* columnist George Will launched an attack on Gore based partly on Singer's manipulation of Revelle.[29] Educated at Oxford and Princeton, Will often employed his literary skills to expose the alleged hypocrisies of environmentalists. Starting his barb barrage filled with accusatory alliterations, he worked his way from a Kipling poem to the claim that Gore's book was "wastebasket-worthy." Gore's environmentalism obviously reflected his "green guilt." Just as bad, Gore had hung around Washington so long that he overflowed "with the certitude characteristic of that circle."

Will used the Singer article to undercut the vice presidential candidate's climate change credentials. He wrote, "Gore knows that his former mentor at Harvard, Roger Revelle, who died last year, concluded, 'The scientific base for greenhouse warming is too uncertain to justify drastic action at this time. There is little risk in delaying policy responses.'" Echoing Easterbrook, he declared, "The planet is more resilient, the evidence about its stresses more mixed, and the facts of environmental progress more heartening than [Gore] admits."

Revelle's family reacted quickly and forcefully to Will's suggestion that he had renounced his beliefs on global warming. In a response carried by the *Post,* Revelle's daughter, Carol Revelle Hufbauer, declared, "Nothing could be further from the truth."[30] Hufbauer challenged Will's characterization of her father: "When Revelle inveighed against 'drastic' action, he was using that adjective in its literal sense—measures that would cost trillions of dollars. Up until his death, he thought that extreme measures were premature. But he continued to recommend immediate prudent steps to mitigate and delay climatic warming. Some of those steps go well beyond anything Gore or other national politicians have yet to disclose."

She cited recent examples of his recommendations for action that were eminently reasonable. These included using more natural gas in place of coal and oil, increasing the tax on gasoline by a dollar a gallon, and using nonfossil energy sources to cut back on coal, "a nasty, dangerous substance." Revelle had praised Bush's proposal to plant a billion trees a year for the next ten years but was disappointed that he never got his chance to tell Bush's chief of staff Sununu that the administration needed to do more.

Will's readability and syndication to papers across the country amplified the core message of Singer and the Global Climate Coalition. But his column and Stockdale's confusing reference to it created only a ripple in an ocean of campaign rhetoric. Nonetheless, Will exposed many Americans to the idea that Revelle believed Gore was moving too fast. The thought long lingered, even when people couldn't remember where they'd heard it.

The Cosmos incident showed that some people attacking the climate science espoused by the National Academy and the IPCC were willing to inflict substantial collateral damage. It was a Shakespearean tale; it was Samson and Delilah. The conniving Singer had besmirched the legacy of a brilliant scientist (too weak at the age of eighty-two to say no) and an American hero of the twentieth century.

* * *

In the closing weeks of the presidential campaign, Republican paid advertising continued to use *Earth in the Balance* to attack the Democratic ticket.
A Bush national radio blitz alerted voters to how many jobs they would lose
in their states if Gore's ideas came to fruition. The ads claimed, for example,
"That would threaten 24,000 jobs right here in Georgia," "That would
threaten 13,000 jobs right here in Maine," or "This proposal would threaten
200,000 Louisiana jobs."[31]

GOP alarms achieved some success in states dependent on the mining
of fossil fuels. West Virginia's *Bluefield Daily Telegraph* headed one editorial
"Gore's Warnings Bode No Good Will for Coal." It urged miners to "think
twice about backing a ticket that might bring down the roof on their liveli-
hoods." It regretted that the United Mine Workers of America had "heartily
endorsed Clinton–Gore."[32]

Bush still had presidential duties to attend to, and on October 24 he
signed the Energy Policy Act of 1992. Earlier, some legislators had thought
the bill should provide a partial response to the Rio agreement urging nations
to limit their greenhouse gas emissions. But Bush's statement didn't mention
Rio, greenhouse gases, global warming, or climate change.

Three months later, the Energy Information Administration—the policy-
independent data and analysis arm of the Department of Energy—would release its carbon dioxide emissions projections. It incorporated the likely impacts of the new legislation but assumed no future policy initiatives. Its most
likely estimate showed U.S. emissions from 1990 to 2000 rising 12 percent,
well above the stabilization goal advocated by many European countries.[33]

On Tuesday, November 3, Clinton–Gore emerged victorious over Bush–
Quayle and Perot–Stockdale by an impressive margin. The Democrats garnered 370 of 538 electoral votes and outpaced the Republicans by 5.6 percent
in the popular vote. As it turned out, Gore's opponents' use of *Earth in the
Balance* as a political weapon failed to gain much traction. Clinton–Gore
won the popular vote in coal powerhouse West Virginia by a comfortable
13 points.

The "Bush 41" years contained several turning points in the history of climate
change science and politics. Academic experts exerted a considerable influence on two of the critical questions: First, did the science, as understood at

the time, provide a basis for acting to limit greenhouse gas emissions? And second, did cost–benefit analysis justify government actions to limit greenhouse gas emissions that provided net benefits for Americans?

During the Cold War, the nation had highly valued the work of its scientists. Thus, it was not surprising that the government turned to its national science labs, its great universities, the National Academy of Sciences, and later, the National Atmospheric and Oceanic Administration to better understand climate. It should not be surprising that a scientist with Revelle's accomplishments had met four American presidents face-to-face.

By the Bush years, however, the old order was crumbling. In the climate arena, individual scientists, often funded directly or indirectly by fossil fuel industries, worked to undermine the consensus of scientific elites. The president's chief of staff, situated just outside the Oval Office, turned for his climate advice to professors like Michaels and Balling, who were trying to fool the public into believing that the earth was cooling (even if they did not personally believe it was). Groups like the Global Climate Coalition and the Heartland Institute identified academics who would work with them to advocate against climate action with the argument that scientific knowledge was just too uncertain to justify government intervention or that the earth was actually cooling. Everyone deserved to be heard. But what was the public to believe?

The problem was more fundamental than the rise of a coterie of scientists skeptical of mainstream scholarship. Even before the founding of the GCC, most scientists sought an understanding of science that met their standards of "certainty." Within the ivory tower, this quest made sense. There was always more to be learned. Why not ask for more research funding to resolve remaining questions, even when a lot was already known? But outside academia, standards that guided the actions of individuals, organizations, and governments were usually based on probabilities. As Revelle and Gore sometimes lamented, the scientists' rhetoric of certainty, when transferred to the policy arena, justified postponing action on climate change, even before the rise of industry-funded skeptics.

The demands in the United States for scientific certainty were boosted by fears of massive costs of constraining greenhouse gas emissions that would jeopardize human well-being. To assess costs, they turned to economists.

After the 1992 elections, economists continued to debate the appropriateness of the Nordhaus modeling that exerted great influence in university departments of economics and the DC halls of power. A few economists continued to suggest that discount rates and other assumptions built into the models created a bias against climate action.[34] But the sharpest critique came from atmospheric scientist Stephen Schneider. On December 22, the now Stanford prof sent a letter to the editor at the AAAS journal *Science* raising issues about the Yale scholar's recent article. In his transmittal, Schneider said he didn't want his comments to be interpreted "as criticism of Nordhaus' major methodological advances" in integrating climate and economic models. Nonetheless, Schneider argued, "The policy debate which will be fueled by this paper is desperately in need of discussion." He sent a copy of his submission to Nordhaus, with whom he sometimes collaborated. He also blind-copied the Clinton–Gore transition team in Little Rock, which was closely tracking the issue.[35]

Schneider praised Nordhaus for acknowledging some inadequacies in his approach. These included underrepresenting developing countries, omitting other market failures (such as damage to the ozone layer) that could be leveraged, and overlooking opportunities to mitigate climate change more cheaply with technological advances. Schneider wanted to add to the list of caveats. He cited the absence of "surprise scenarios" of changes in climate and security issues such as the creation of environmental refugees and political instability.

Schneider made a further argument that, despite these deficiencies, Nordhaus had stuck with a viewpoint that his admittedly simple tools could "underpin any rational decision" on climate action. Schneider objected to Nordhaus's confident assertion that climate stabilization "would appear enormously expensive." Schneider recommended that policymakers not rely solely on Nordhaus's judgments about these costs. In Schneider's view, the Nordhaus model and all other quantitative attempts then available or likely to be available in the next decade or so were "woefully incapable of providing a sole basis for policy choice simply because what these tools omit or treat poorly could well be as or more important as what they include well."

Schneider's most telling evidence of bias (while heaping praise on Nordhaus) came with his explanation of the economist's portrayal of economic damages if governments attempted to limit greenhouse gas emissions. Schneider called for a closer examination of Nordhaus's claim of a 20 percent loss of economic benefits, which sounded draconian to the layperson, a significant threat to the quality of life. Nordhaus based his claim on a projection, less transparent to readers, that over 140 years, the economy would grow by only 450 percent if it constrained emissions, compared to 470 percent in an unconstrained (free market–only) case. If Nordhaus had described his estimate more accurately as a slight drop in the economy's annual growth rate, more people might have been willing to pay that price to slow down climate change. Nordhaus was following the protocols of his profession in his portrayal of the "20 percent loss." It was yet another example of how the prevailing assumptions and analyses of many academic economists tended to favor the virtues of unregulated markets and devalue government regulation of pollution. Their influence across administrations of both parties was often unseen but large and growing.

Studies of climate change during the Bush years often focus on the international negotiations, animated by the thousands of government officials, nongovernmental organizations, and news media in attendance. There was the breathless anticipation that if the proceeding could be extended for another day or even a few more hours, the participants could change the course of history. The effects of these meetings were indeed far-reaching, but secondary in impact to the decisions of national legislatures.[36] Unfortunately, the making of laws appeared baffling to many observers and often failed to get the coverage it deserved.

The 1992 earth summit merited the attention it received. But the collapse of the bipartisan, pro-environmental coalition in the Senate, when it couldn't prevent a filibuster of higher auto efficiency standards, revealed more about where the United States in the 1990s was headed on limiting greenhouse emissions than its stance at Rio. Congressional lethargy and the need for sixty positive votes in the Senate signaled that it might be a long time before the world's greatest emitter of greenhouse gases took significant steps to reduce them.

18

After

The Years 1993 to 2023

Although America's elder statesman of climate science, Roger Revelle, died in 1991, most of his fellow climate pioneers were considerably younger and lived into the twenty-first century. They were able to observe, as Revell could not, the obvious physical effects of climate change without a need for model projections. These survivors' reflections in later life provide valuable perspectives on what happened from Mauna Loa to Rio (the primary focus of this book). This later story also shows how the early roots of climate science and politics continue to impact decision-making on a challenge for the planet often overshadowed by other challenges that can appear more pressing.

Dave Keeling was sixty-three and still pursuing what he called "the carbon dioxide molecule in all of its ramifications" when the first President Bush left the White House. The master of atmospheric measurements remained active, but the torch was gradually passing to a new generation. The Intergovernmental Panel on Climate Change's peer reviewers for its 1994 report included longtime giants in the field—Keeling, Robert Watson, and James Hansen of the United States, John Houghton of the United Kingdom, and Bert Bolin of Sweden. The list now also included younger scholars like Ralph Keeling, who had earned a 1988 doctorate in applied physics from Harvard for his work on the accurate measurement of atmospheric oxygen and then joined his father at Scripps.

The senior Keeling had studiously shunned the politics of climate science over his long career. He once noted, "Even to publish scientific findings that suggest a peril in rising CO_2 or temperature can be construed as taking a prejudicial position." For those needing research funding, taking sides might lead to a loss of support. But in the 1990s, personal experiences spurred him to rethink his longtime policy neutrality.[1]

Keeling encountered a meteorologist working for a utility company and, on another occasion, a nonscientist in Montana on land once owned by his grandfather, both claiming to have found some flaw that undermined the generally understood science of climate change. The Montanan got the information from a small local newspaper publishing an article from a "national center" in Washington. Keeling believed these naysayers were getting their ideas from unreliable sources and ignoring the preponderance of evidence.

He reflected, "A safe approach is just to remain an interested observer of the unfolding scientific evidence of man-made global change and its possible significance to future human welfare." But he came to believe he could no longer maintain a dispassionate posture. He concluded, "A more prudent attitude would be to heed the rise of atmospheric CO_2 concentrations as serious unless proven to be benign." He urged his fellow scientists to "make clear to the public the wisdom of this cautious approach."

Keeling was confident the available data supported his view. He observed in 1998, "The consumption of fossil fuel has increased globally nearly three-fold since I began measuring CO_2 and almost six-fold over my lifetime." He dismissed the idea that the connection between rising carbon concentrations and temperature was speculative. Instead, "I am convinced that temperatures are rising significantly, not just from viewing the temperature data . . . but also because the atmospheric CO_2 record makes any other interpretation difficult."

Keeling's new willingness to engage in the broader dialogue about the risks of climate change led to his attendance at President Clinton's Round-table on Climate Change in 1997. His decision to go was based in part on his respect for the vice president. In his letter to Gore accepting the invitation, he recalled first hearing him speak in the 1980s and concluding that "No ghostwriter could have fashioned the concepts which you were expressing. Later, when I read your book, it became evident that you were formulating a vision of government, science, and society which goes beyond the thinking of most of your contemporaries and is probably unique for a high public official in our nation."

On July 23–24, the Californian spent a busy twenty-four hours in Washington. Wednesday night, he attended a small private dinner at the home of Tim Wirth, now under secretary of state for democracy and global affairs—in effect, the U.S. quarterback for international climate negotiations. The next day, the roundtable with Clinton, Gore, and prominent climate scientists (including Stephen Schneider) in the Old Executive Office Building lasted an hour. In his remarks, Gore reiterated the importance of the Mauna Loa data. Then Keeling joined Gore for a private meeting lasting twenty minutes.

Amid a heated national debate about pending climate negotiations in Kyoto, Japan, Keeling's participation in the DC events was a sharp reversal. He followed up by signing a Union of Concerned Scientists "call for action" stating, "Human beings and the natural world are on a collision course." Keeling had now joined a political battle he had long avoided.[2]

In May 2002, President George W. Bush ("Bush 43") announced the year's National Medal of Science laureates would include Keeling. The citation read, "For his pioneering and fundamental research on atmospheric and oceanic carbon dioxide, the basis for understanding the global carbon cycle and global warming." A few scientists grumbled that Keeling had just done one thing over and over, even if he had done that thing extraordinarily well. But the Californian had realized better than most of his peers the importance of unassailable, continuous time series data that were unassailable.

On June 20, 2005, while engaged in his beloved pastime of hiking, Keeling died of a heart attack. His obituary in the *New York Times* said, "The current debate about global warming centers on how much warming the increased carbon dioxide will generate, but few have disputed Dr. Keeling's underlying carbon dioxide data."[3]

Keeling's death didn't end the praise for what he had accomplished. In April 2015, the American Chemical Society designated the Keeling Curve records at Mauna Loa as a National Historic Chemical Landmark. The society called the curve "an icon of modern climate science." Zooming even further out, the National Oceanic and Atmospheric Administration deemed Keeling's measurements "the most widely recognized record of mankind's impact on the earth."[4]

*   *   *

Wally Broecker outlived the four other panelists who had produced the 1965 climate change report for LBJ. Like Revelle and Keeling, he received the Presidential Medal of Science—in Broecker's case, from Bill Clinton in 1996.

Broecker released a memoir in 2008 in which he foresaw that solar, wind, and nuclear fusion would ultimately make it possible to eliminate carbon dioxide emissions. But the world couldn't count on that happening in the twenty-first century. In the interim, there had to be another answer.[5] For Broecker, that solution was a "scrubber" that could remove CO_2 from the ambient air. Because carbon was well spread, such a machine could operate anywhere in the world to benefit the entire world. He said this device was not yet commercially viable, but the underlying science was not complicated. If successful, carbon scrubbers would allow humans to follow the moral imperative of "cleaning up after ourselves" and enable later generations to look back on the climate challenge as "the one that allowed us, as a species, to grow up."

Broecker's idea appeared to be a desperate Hail Mary pass when other solutions appeared elusive. But in August 2023, Secretary of Energy Jennifer Granholm announced the first projects approved for a new $3.5 billion fund to demonstrate direct air carbon capture at commercial scale. The technology is "essentially a giant vacuum that can suck decades of old carbon pollution straight out of the sky," she exclaimed.[6]

Another scientist who warned about the effects of atmospheric pollution, James Hansen, emerged as a celebrated climate scientist during the 1980s. What made the NASA researcher stand out was his commitment to making science understandable to the public, his willingness to risk criticism, and his visible anger when he believed people were ducking their responsibilities to protect the planet.

In his later years, Hansen gravitated to a new role as a political activist. He publicly confronted contrarians like Richard Lindzen, arguing that their claims were indefensible. His 2009 book, *Storms of My Grandchildren*, broke the mold of a cautious scientist, warning about "the coming climate catastrophe and our last chance to save humanity."[7] He urged young people to become involved in politics and later joined them in demonstrating against the Keystone Pipeline. Two protests outside the Barack Obama White House led to Hansen's arrest.

In 2010, Hansen's book inspired the creation of Our Children's Trust. The nonprofit organization filed lawsuits in all fifty states against the federal government on behalf of the children born onto an increasingly hotter planet. The thrust of the litigation was "they knew" and didn't do much to stop an impending crisis. Hansen's granddaughter Sophie became a plaintiff in the suit, and he served as a pro bono expert witness.[8]

In hindsight, the pioneers of climate science deserve high marks for their curiosity and prescience. Chemists understood the heat-trapping properties of carbon dioxide, and Revelle and Keeling provided the theoretical and empirical evidence of its significant accumulation in the atmosphere—all by 1960. From the expected growth in population and per capita wealth, it was easy to extrapolate, at least in rough terms, how severe global warming might become, while there was still time to deal with the threat in a gradual and measured way.

No one was satisfied with the precision of climate models, including the modelers themselves. But by the late 1970s, the most respected models provided guidance that stood the test of time. The projections didn't all agree on temperature sensitivity to carbon accumulations. However, recent data suggest that the ultimate numbers will likely land well within the broad range of projections from four decades ago, and not very far from the midpoint. A detailed study of the performance of models since the 1970s concluded in 2020, "We find no evidence that the climate models evaluated in this paper have systematically overestimated or underestimated warming over their projection period."[9]

Models also did an admirable job of projecting what parts of the globe would experience temperature increases above or below the mean. They also identified with considerable accuracy the regions most likely to suffer extreme droughts.

The overall performance of the models is more remarkable since they initially focused on carbon dioxide without fully incorporating the impact of other potent greenhouse gases like methane or the complex interactions with clouds and oceans. Though hitting a bull's-eye forty years in advance was virtually impossible, the models turned out to be pretty accurate, and the factors not accounted for seemed to balance themselves out.

* * *

During the Clinton years, the forces denying the consensus science remained strong, often financed by corporate interests through think tanks with civic-sounding titles. The most visible leader of this group was Fred Singer.

In September 1993, Singer and Justin Lancaster resumed their debate over the authorship of the Cosmos article, this time face-to-face. The venue was Middlesex County Superior Court, Commonwealth of Massachusetts. Singer had sued Revelle's former assistant for libel for saying that Revelle was not a genuine author of the still controversial piece. Under oath, Singer often cited a faulty memory and could not provide any evidence that Revelle had "written" any of the disputed article.[10]

Lancaster claimed he was 95 percent certain he would win the case. But the Harvard postdoctoral student was outgunned in court, where he represented himself as counsel. Moreover, the potential financial threat to his family from an adverse verdict put him in a precarious position. To avoid the risk of damages, Lancaster agreed to apologize for any damage he had caused to Singer's reputation and retracted his implications about Singer's conduct, character, and ethics.

Years later, when Lancaster's nondisclosure agreement lapsed, he declared, "Fred Singer is the most unethical scientist, in my opinion, that I have ever met . . . The worst decision I ever made in my life was to provide a retraction of my statements in the early 1990s about Singer's nastiness. The retraction was coerced."[11]

Despite criticism of Singer's claim that Revelle had "written" the 1991 Cosmos article, the University of Virginia scholar's reputation as the rock star of climate change contrarians soared. He established the Nongovernmental International Panel on Climate Change to attack the findings of the UN body that published scientific assessments based on the peer-reviewed consensus of hundreds of scientists worldwide. He was also a prolific producer of books and articles.

Singer's signature book became *Hot Talk, Cold Science: Global Warming's Unfinished Debate.* The libertarian-leaning Independent Institute in Oakland, California, published three editions (1997, 1999, and, posthumously, 2021)—all frequently cited in contrarian circles. Funding from the Electric Power Research Institute gave the book credibility, as did blurbs from Richard Lindzen; Thomas Schelling, the 2005 Nobel laureate in economics; and William Nierenberg. Senate energy committee chair Frank Murkowski called it "an important scientific contribution to the global warming debate."

Hot Talk attempted to undermine the findings of the Intergovernmental Panel on Climate Change and most American scientists. Singer argued, "Global temperature is not rising, even though greenhouse theory says it should." In his preface to the second edition, he declared, "The economic impact of a possible global warming has been re-evaluated and found to yield positive benefits rather than losses."[12]

Singer's articles frequently appeared in both academic and nonacademic outlets, none a more powerful platform than the opinion section of the *Wall Street Journal*. The paper's massive circulation made Singer one of America's most widely read scientists. His most provocative essay in the *Journal* was his 2018 "The Sea Is Rising, but Not Because of Climate Change."[13] In it, he concluded, "Contrary to the general wisdom . . . the temperature of sea water has no direct effect on sea-level rise. That means neither does the atmospheric content of carbon dioxide."

Revelle—the scientist Singer claimed shared his views on climate science—wrote the sea level chapter in the 1983 National Academy of Sciences report *Changing Climate* and on many other occasions explained why the growing concentration of atmospheric CO_2 led to rising sea levels. It fell to still-living scientists to refute Singer's dissent.

In a letter to the editor, professors Andrea Dutton of the University of Florida and Michael Mann of Penn State asked, "Would the *Wall Street Journal* run an op-ed 'Objects are Falling but Not Because of Gravity?'" They asserted, "Careful peer-reviewed research . . . shows that sea levels are rising and human-caused climate change is the cause. Don't take our word for it; help yourself to the mountain of scientific literature showing as much. When water warms, it expands. When ice warms, it melts. To deny these facts is not just to deny climate change. It is to deny basic physics."[14]

After Singer died in 2020 at ninety-five, his friend and fellow scientist Patrick Michaels compared him to Hubert Humphrey, the former senator from Minnesota and vice president, because he was another "happy warrior,"

THE PRESIDENTS AND THE PLANET

enthusiastic about his advocacy. Michaels urged people to give the latest edition of *Hot Talk, Cold Science* to their "curious or scientifically inclined children and friends, however woke they may be."[15]

The *Washington Post* obituary cited Singer's early scientific achievements. But it also observed that somewhere along the way he "found a new purpose as a scourge who sought to denigrate other scientists who warned the public about secondhand smoke, greenhouse gas emissions, acid rain, and the dangers of a steadily warming climate. 'It's all bunk,' he often said."[16]

Andrew Dessler—a professor of atmospheric sciences at Texas A&M, who has written a widely used textbook on climate change and occasionally invited Singer to speak to his classes—compared him to "an irascible uncle who would say the most preposterous things at Thanksgiving, and everyone would just roll their eyes." He added that while Singer's early career was respectable, his climate work was deeply flawed. "None of the scientific claims he made stood the test of time."[17]

Bill Nordhaus, the first economist to address climate change, continued to be a prolific and influential writer into the early decades of the twenty-first century. Every administration after the first President Bush continued to feel his influence. When experts debated the appropriateness of discount rates for guiding climate policy, their eyes turned to the erudite professor from Yale.

In 1999, Nordhaus joined another discussion, sponsored by Resources for the Future, on calculating the costs and benefits of climate action. Yet again he urged his fellow economists not to abandon discount rates that mimicked the rate of return for private investments when assessing potential government policies. However, Nordhaus did concede that economic logic, when justifying no action to prevent future damages, sometimes "violates our ethical intuition." In such cases, "a society may decide that such activities are intrinsically important in a way that cannot be captured by market valuations."[18]

Nonetheless, he remained adamant that it shouldn't be economists who decided when to override discount rates when establishing policies whose benefits were long term. Nordhaus placed the responsibility on politicians to make the tough calls. In the political arena, Nordhaus's "inefficiencies" translated into "damage to the economy." The conventional data portrayal of economists often exaggerated such damage, putting politicians in a tough spot. Plus, elected officials didn't have tenure.

Across the Atlantic, economists adopted a very un-Nordhausian view of the climate challenge. In 2005, the government of the United Kingdom authorized a massive study of the "economics of moving to a low-carbon global economy." In 2007, it published the results in *The Economics of Climate Change: The Stern Review,* named after Nicholas Stern, the head of the UK Government Economic Service and former chief economist of the World Bank.[19] Despite the book's length of close to seven hundred pages, its clear writing and color graphs made it accessible to noneconomists. It was meant to have an impact.

The *Stern Review* concluded, "Climate change is the greatest market failure the world has ever seen." Like many European economists, Stern believed the global climate should be considered a "public good." The current problem was "Those who create greenhouse gas emissions as they generate electricity, power their factories, flare off gases, cut down forests, fly in planes, heat their homes, or drive their cars do not have to pay for the costs of the climate change that results from their contribution to the accumulation of those gases in the atmosphere."

Using economic models from Stanford University, Resources for the Future, the IPCC, and elsewhere, the Stern economists painted a stunningly different picture of the costs and benefits of climate action and inaction than their U.S. counterparts. According to Stern, "If we don't act, the overall costs and risks of climate change will be equivalent to losing at least 5 percent of global GDP each year now and forever . . . In contrast, the costs of action— reducing greenhouse gas emission to avoid the worst impacts of climate change—can be limited to around 1 percent of global GDP each year." To help reduce mitigation costs, Stern recommended doubling global public funding for energy research and development. Accentuating the divide in economic thinking on the two sides of the pond, Stern declared, "Delay makes the problem much more difficult and action to deal with it much more costly."

Nordhaus, listed as an advisor to the *Stern Review,* unleashed a rapid response full of economist jargon and uncharacteristically bombastic attacks.[20] The Ivy League prof accused the report of being "a thicket of vaguely connected analyses" and the work of "the world social planner, perhaps stoking the dying embers of the British Empire, in determining the way the world should combat the dangers of global warming." He also asserted that it led to "bizarre results" and "deep mathematical problems such as nonconvergence of the objective function and incompleteness of the functional." Another stern rebuke: model results should have been "Pareto-improving" rather than "Pareto-deteriorating."

Nordhaus concentrated his review of the *Review* on what he regarded as its major flaw: a very low time discount rate of 0.5 percent a year. Nordhaus, who at the time favored a rate of 6 percent, said, "With near-zero discounting, the low damages in the next two centuries get overwhelmed by the long-term average over the many centuries that follow." Nordhaus provided little analysis or evidence backing up his assertion that unconstrained growth of carbon emissions over the next two centuries would lead to "low damages."

Climate change skeptics often cited Nordhaus to bolster their arguments, as Singer had done in the Cosmos article. But if they thought he'd joined their ranks, they got a rude awakening with his 2012 essay in the *New York Review of Books.* Using nontechnical language, he blasted a recent opinion piece in the *Wall Street Journal* by Richard Lindzen, Edward David, and fourteen other prominent scientists. Echoing the public relations output from coal-dependent industries, Lindzen and colleagues argued against global warming panic and the assumption that rising concentrations of atmospheric CO_2 were necessarily harmful. Climate change alarmism, they warned, "offers an excuse for governments to raise taxes."[21]

First off, the economist demolished the writers' claim that the data on rising temperatures were inconclusive. He presented a simple line graph with data from 1880 to 2011 on global mean temperature, averaged from three credible sources. In case skeptics wanted to argue that there was a massive conspiracy to falsify the data, Nordhaus and a colleague conducted an audit based on information from six weather stations around the world. As expected, the data were volatile, dropping during some years and rising during others. But the overall pattern was unambiguous. The world was substantially warming. The trend was so clear, he said, that no statistical analysis was needed to verify what the eye could see was obvious.

In similar fashion, Nordhaus objected to five other arguments of the skeptics. He took particular umbrage at the assertion in the *Journal* that "Nordhaus showed that nearly the highest benefit-to-cost ratio is achieved for a policy that allows 50 more years of economic growth unimpeded by greenhouse gas controls." Nordhaus accused the essay of "an elementary mistake in economic analysis." He retorted that his study was just one of many "showing that economic efficiency would point to the need to reduce CO_2 and other greenhouse gases right now and not to wait for a half-century."

Yale professor William Nordhaus arriving at a news conference to discuss his 2018 Nobel Prize in economics, awarded for his work on climate change modeling. Courtesy AP Photo/ Craig Ruttle.

As with Keeling in the 1990s, the extreme claims of the climate skeptics had provoked the Yale scholar to push more forcefully against the idea that policy responses to climate change could wait.

Six years later, Nordhaus flew to Stockholm to receive the Nobel Prize in Economic Sciences. The selection committee cited his success in the 1990s in merging his economic model with one (Stephen Schneider's) from the physical sciences. At seventy-seven, the new laureate reflected on what he had learned. Delivering his lecture in a quiet voice, he boldly called climate change a "menace to our planet." Championing his preferred method for finding solutions to this menace, he declared, "Mathematics is the future."[22]

In 2021, Nordhaus published *The Spirit of Green,* the latest in a long series of books about the economics of climate change. In it, he decried the ineffectiveness of international agreements to slow climate change and the subsidies for fossil fuels, due to the failure to impose carbon taxes based on their environmental damage. He also acknowledged the limitations of discount rates for analyzing benefits that stretch far into the future and moderated his recommended levels for those rates.[23]

In April 2023, the federal Office of Management and Budget revised its Circular A-4—an event unlikely to attract the attention of even devout policy nerds. The old version, last updated in 2003, had mandated discount rates of 3 to 7 percent when assessing the costs and benefits of federal regulations. The new guidance set a general rate of 1.7 percent—moving closer to the *Stern Report* and toppling the long reign of Nordhaus-like levels.[24]

The change affected a wide range of government activities, including those targeting climate change. For instance, the OMB said it had used a discount rate of 3 percent when it determined in 2021 that the social cost of carbon should be $51 per ton of carbon dioxide emissions. With a rate of 2 percent, it said, the price would have risen to $125 per ton. Over time, the lower discount rate will likely have an immense impact on federal policy—equivalent in some ways to a major breakthrough in technology. Like advances in technology, lower discount rates could have come earlier and helped cope more efficiently with the risks of climate change. But late is better than never.

Among elected officials, Al Gore stood out as a pioneer in the understanding of climate change, much as professors Roger Revelle and Bill Nordhaus had in their fields of expertise. As vice president from 1993 to 2001, Gore brought several champions of climate action—including Tim Wirth and Rafe Pomerance—into the government. But at the White House, the second-in-command encountered constraints on his ability to address climate. Behind the scenes, some economists and budget analysts advising the Bush administration carried over into the Clinton–Gore years.

One big blow to the prospects for climate action came in the 1994 elections, when Republicans won the House of Representatives. The new speaker, Newt Gingrich of Georgia, placed many members skeptical of government regulations and climate science in critical positions. The flip of the House resulted in an appropriations rider to strip the administration's authority to increase vehicle efficiency by rule. This anti–fuel economy measure remained in effect from 1995 to 2000.[25]

Growing GOP opposition wasn't the only political hurdle confronting climate action. Coal states like West Virginia remained critical components of the Democratic coalition. Moreover, the Mountain State's Robert Byrd now chaired the powerful Appropriations Committee. Not surprisingly, the administration avoided any show of hostility toward coal. Traditional oil use also had its Democratic-leaning protectors. The United Auto Workers union, for instance, opposed tightening auto efficiency standards, dampening Democratic motivation to fight for them. Like Republicans, Democratic members of Congress did not relish facing the electorate after imposing a carbon tax.

In the closing days of Clinton's second term, Secretary of Energy Bill Richardson and Assistant Secretary Dan Reicher finally got the White House to approve consequential updates to appliance efficiency rules.[26] On the whole, however, the Clinton–Gore agenda did not prioritize measures likely to reduce domestic greenhouse gas emissions.

The absence of new U.S. policies showed in decadal trends. Instead of stabilizing at 1990 levels by 2000, U.S. emissions rose an embarrassing 17 percent. Many signers of the Rio protocol tried to cap their emissions at 1990 levels but fell short. In contrast, the United Kingdom chopped theirs by 11 percent.

During its second term, the Clinton–Gore team focused on the negotiations leading up to a December 1997 environmental summit in Kyoto. Rio had called on the developed economies to stabilize their greenhouse gas emissions; the agreement in Japan would call for their reduction. American diplomats successfully lobbied for retention of an international trading scheme to encourage the early adoption of low-cost options.[27]

Senate action in July 1997 gave a strong indication of the treacherous path ahead for the next stage of international negotiations. A resolution offered by Byrd and Nebraska Republican Chuck Hagel expressed the sense of the Senate that the United States should not sign a treaty that "would result in serious harm" to its economy. In addition, the agreement should mandate specific reductions in greenhouse gas emissions for developing countries within the same compliance period as the developed nations. The economic issue was a matter of semantics, allowing the various sides to argue the meaning of "serious." In contrast, the requirement for developing nations—aimed at China, which would surpass the United States in annual CO_2 emissions in 2006—challenged a widely held assumption that poorer countries had so far not been significant contributors to the greenhouse problem on a per capita basis and needed to industrialize to elevate their standards of living. Since the 1987 formation of the IPCC, international talks had hardly discussed the idea that these emerging economies should shoulder a heavy burden during the initial stages of global controls. As a whole, the resolution was a dagger in the heart of the Kyoto negotiations.

On its surface, the Senate's 95–0 vote in favor of Byrd–Hagel was baffling. Longtime advocates of effective international agreements to curb climate change—such as Joe Biden, John Chafee, and John Kerry—sided with the opponents of such action in endorsing the resolution. Not wanting to fail an embarrassing test just months before departing for Japan, the administration recommended support for a measure opposing its negotiating stance.

In Kyoto, the United States agreed to cut its 2012 emissions by 7 percent. Offsets (some might call them "loopholes") would ease the burden. But on close inspection, the treaty's target was still an enormous lift. The base year for reductions was 1990. As a result of steep U.S. increases during the decade, Americans would have fifteen years to lower their emissions by about a quarter. The good intentions of international cooperation clashed with the realities of domestic politics, embedded infrastructure, and a long delay in adopting significant domestic climate policies.

The chair of the Senate energy committee, Frank Murkowski, declared the treaty "dead on arrival." But it never arrived. Knowing the votes for ratification would fall short, the Clinton administration never submitted it for Senate approval.[28]

When Gore ran for president in 2000, he won the popular tally but lost in the electoral college by a mere five votes. It was the first time since 1888 that the popular winner had failed to grab the big prize. Capsule summaries of the results focused on the controversial, razor-thin loss of Florida's large bloc of twenty-five electors. But in a contest so close, there were many decisive factors. Several pointed to the perils of climate change politics.

Green Party candidate Ralph Nader chipped away at environmentalist support for the author of *Earth in the Balance.* The damage was enough to have probably tipped the balance against Gore in Florida and New Hampshire, either of which could have given him the presidency. On the other side of the political spectrum, Gore lost the coal states of West Virginia, Kentucky, and his home state of Tennessee—all in the Democratic column for 1992 and 1996. A different result in any of these states would also have flipped the national outcome.

No longer shackled by elective office, Gore returned to his role as the most prominent American spokesman on climate change. In 2006, Paramount Classics released *An Inconvenient Truth*—a movie based on the climate change slide show Gore had been giving since the 1980s, now with the stunning high-resolution visuals made possible by advanced technology and more dramatic physical evidence of a changing climate. In his introduction, Gore said with a straight face, "I used to be the next president of the United States." On a more serious note, he rued his failure during his time in government to get the message across that it was imperative to make big changes to protect nature. With a haunting musical score in the background, he recounted learning about climate warming at Harvard from Professor Revelle—the man who, in the 1950s, "saw where the story was going" and hired Dave Keeling to collect the needed data. With considerable reverence, Gore said of Revelle, "I respected and learned from him so much."

The movie hit hard on the press coverage of climate science, warning that devotion to telling both sides of any argument often underplayed the understanding of what the most credible scientists had concluded. A large box office and two Oscars reflected the film's reach to a broad audience. The following year, a book based on the movie again used startling graphic imagery to communicate its points.

The release of *An Inconvenient Truth* led journalist Gregg Easterbrook—who once brought Fred Singer's Cosmos article to national attention—to ask, "Has anything happened in recent years that should cause a reasonable person to switch sides in the global-warming debate?" His answer: "Yes: the science has changed from ambiguous to near-unanimous. As an environmental commentator, I have a long record of opposing alarmism. But based on the data, I'm now switching sides regarding global warming, from skeptic to convert."[29]

The Nobel Peace Prize for 2007 was awarded jointly to Gore and the Intergovernmental Panel on Climate Change "for their efforts to build up and disseminate greater knowledge about man-made climate change, and to lay the foundations for the measures that are needed to counteract such change." The honor didn't slow Gore's ongoing organization of training workshops for people wanting to campaign for climate action or his public pleas for the world to act more quickly to protect the earth's atmosphere.

The establishment of the Intergovernmental Panel on Climate Change in the late 1980s was a historic step toward facilitating the global cooperation many scientists and environmentally inclined officials had sought. But the organization hobbled into the twenty-first century, lacking the support of the nations who mattered most—the United States, China, and Russia. Climate assessment tomes every five or six years provided increasingly detailed projections of the risks of a warming planet. But the annual meetings to figure out how the world could reduce emissions generally earned poor reviews from environmental advocates.

In 2015, the twenty-first Conference of the Parties (COP21) in Paris took a different tack from Rio and Kyoto. The changes both weakened and strengthened the prospects for coordinated international action. The Paris Agreement set a goal of "Holding the increase in the global average temperature to well below 2°C above pre-industrial levels and pursuing efforts to limit the temperature increase to 1.5°C above pre-industrial levels."[30] The language challenged a longtime, often implicit assumption that the world could tolerate a 2 degree increase and heightened the urgency of rapidly reducing greenhouse gas emissions. If the goal of 1.5 degrees, coming so late, appeared impossible due to the time needed to replace the existing energy infrastructure, striving to achieve it might help hold the increase to 2 degrees.

After Paris, nations no longer tried to hammer out binding joint commitments in international negotiations. Instead, they would submit verifiable and "increasingly ambitious" climate action plans every five years, containing their "nationally determined contributions."

The new system was, in effect, a Get Out of Jail Free card for the United States. Countries could now select their individual years from which to measure progress and ignore the Rio-established 1990. The United States set 2005, when its emissions neared their peak, as its base year, allowing it to claim credit for later reductions in emissions, even when they were still well above 1990 levels. Countries that had taken the Rio goals seriously chafed at the transition to the new, more flexible way of measuring progress. But in exchange, they had won the battle to obtain more robust responses from the Americans and Chinese.

President Obama had informed world leaders privately that winning Senate ratification for internationally negotiated limits on U.S. emissions would be impossible. Thus, the compromise in Paris was the best viable alternative.[31] The second-term president considered this an example of ensuring that the perfect did not become the enemy of the good.

In April 2022, a new IPCC report warned that the world was on track to warm an average of 3 degrees Celsius (well over 5 degrees Fahrenheit) from preindustrial temperatures, given current energy policies. The announcement, said *Bloomberg News,* indicated that the goal of limiting warming to 1.5 degrees Celsius was "officially on life support." The current trajectory would, it said, "painfully remake societies and life on the planet."[32] The forecast wasn't entirely bad news. It did reduce the likelihood of an even hotter planet, foreseen in some earlier projections. Moreover, many options remained available to slow global warming.

In late November 2023, 70,000 official and unofficial participants gathered in Dubai for the IPCC's twenty-eighth Conference of the Parties to work out the next steps to slow global climate change. Island nations, particularly vulnerable to rising sea levels, believed that the minimal progress in previous years did not bode well for COP28. Al Gore criticized the location of the conference in the United Arab Emirates, a leading oil exporter, and the selection of its chief oil official to chair the event. The former vice president charged that the conference, like its predecessors, was stymied by the need for consensus from all countries. As a result, "We have to beg for permission from the petrostates," he complained, to "protect the future of humanity."[33]

Climate activist Bill McKibben, who turned sixty-three during the conference, also condemned oil interests blocking tough language to phase out fossil fuels. After intense haggling about the future of coal, natural gas, and oil, however, the conference agreed to compromise language calling on countries to transition "away from fossil fuels in energy systems" and to accelerate action "in this critical decade, so as to achieve net zero by 2050"—language that did not totally please anyone but passed unanimously to widespread applause. McKibben believed the result was, in one sense, meaningless, because it contained no enforcement mechanism. But he argued that words still mattered. For him, COP28 had declared that the fossil fuel era was coming to an end.[34]

By the time of first President Bush's departure from the White House in 1993, numerous congressional witnesses—including Gordon MacDonald, Stephen Schneider, and Michael Oppenheimer—had testified in favor of energy efficiency as a low-cost response to climate change risk.

The dangers of atmospheric carbon dioxide had not been part of the rationale for adopting automobile mileage standards established in the 1970s—the nation's boldest early effort to increase energy efficiency. Only later did analysts begin linking such measures to climate change. By the early 1990s, the old mileage requirements were showing their age and having a reduced impact on vehicle choices. During the decade, sports utility vehicles wreaked havoc on the nation's fuel economy. These taller, roomier vehicles increased in popularity and were allowed by the regulations to burn more fuel. Manufacturers raised some models further off the ground to qualify for the SUV loophole. The policy response was weak, and the average mileage of vehicles on the road began to drop.

In 2007, renewed interest in U.S. energy policy led to the passage of the Energy Independence and Security Act.[35] Its primary provision was the first statutory increase in vehicle fuel economy standards since 1975. The new legislation mandated that by 2020 automobiles sold in the United States and weighing less than 8,500 pounds (including SUVs) had to meet a minimum fleetwide standard of thirty-five miles per gallon, a 40 percent increase over current averages. Climate change was still not a strong impetus for U.S. energy policy, but it was part of the mix. The overall bill won the support of forty-four Democratic senators, forty Republicans, and two independents. Its signing by President George W. Bush restored, at least for a while, the sometime bipartisanship of the 1970s and 1980s.

Presidents Barack Obama and Joe Biden built on the 2007 legislation and a Supreme Court decision the same year to establish tighter vehicle efficiency standards by rule. These measures accelerated the purchase of hybrid and eventually plug-in and all-electric engines. An option to reduce carbon dioxide emissions with larger batteries, discussed before a congressional committee in the 1950s, began entering the market in force. The penetration of all-electric vehicles was even more impressive in some other countries, representing a global response to a global problem.

Mandates for more fuel-efficient transportation over the years helped constrain CO_2 emissions below what they would have been otherwise. But their adoption faced many delays, including the big policy gap between 1975 and 2007 and the opposition of President Donald Trump. Moreover, the irrepressible bent of U.S. rulesmiths toward allowing laxer standards for larger vehicles and the political barriers to passing carbon taxes in America limited their value. U.S. emissions from transportation did not peak until 2007. By 2022, they had come down by about a glass-less-than-half-full 9 percent.[36]

The Inflation Reduction Act of 2022 provided hefty new incentives to accelerate the adoption of all-electric vehicles. These "carrots" lacked the efficiencies of the "sticks"—carbon taxes and regulations. But they will, nonetheless, reshape the car and truck market in the coming decades. Competition between Asian, European, and North American automakers to capture the electric car market will undoubtedly add to the global momentum for human mobility that pollutes less.

Efforts to improve the energy efficiency of "buildings" (defined to include the structures and the equipment in them) didn't achieve effective national standards for appliances such as air conditioners and refrigerators until Reagan's second term. Successive administrations tightened appliance standards by rule, sometimes under pressure from lawsuits launched by the National Resources Defense Council. The 2007 energy legislation phased out incandescent lights, which helped reduce electricity use and virtually eliminated the workhorse of indoor illumination since the days of Thomas Edison. It was not long before new light-emitting diode bulbs (LEDs), which could do the job much more efficiently, became readily available at Home Depot and Lowe's.

Later federal rulemaking brought further improvements in appliance standards, even though the motivations involved more than climate protection. Emissions from the residential sector peaked in 2005. Over the next seventeen years, they dropped by a quarter.[37] So far, the country has a better record of lowering carbon emissions from homes than from cars and trucks.

The Inflation Reduction Act will speed up the existing trend toward lower carbon emissions from the building sector. It attacks the challenge in multiple ways. For instance, it will stimulate stricter building codes. But it will additionally support training for contractors on the advanced technologies that will facilitate meeting goals in a more timely and cost-effective manner. This subtlety—one of hundreds in the bill—illustrates how policymakers have finally begun to deal with climate change more systematically.

Transitioning to carbon-free sources of energy was both potentially more significant and, in practice, more difficult than improving energy efficiency. For the troubled nuclear power industry, the lack of any orders for new reactors in the United States began in the 1970s. The pause continued well into the twenty-first century.

Despite the halt in orders, the nuclear share of U.S. electric generation held steady after 1992 at about 20 percent, with impressive improvements in capacity utilization offsetting occasional reactor retirements. However, high capital and operating costs, plus a culture that resisted advice to build smaller, modular reactors, impeded the nuclear industry's path to growth, or even maintenance of its share of U.S. power generation. When Georgia Power finally began commercial operation of two new large reactors in 2024, it made a big dent in the company's carbon emissions, and the experience might prove useful for future construction. However, the enormous delays and cost overruns hardly inspired others to pursue a similar course.

Nuclear reactors retain their appeal as options to reduce carbon emissions. They run twenty-four hours a day and don't require much land, compared to renewables. They have a constituency in Congress willing to provide subsidies for new construction. A brighter future might result from the learning gained from China's massive construction of reactors or from the investments of venture capitalists (including former Microsoft CEO Bill Gates) who with government support have made bets on smaller designs, using potential breakthrough technologies that the established industry had ignored.

At the turn of the century, the renewables share of U.S. energy consumption stood at 6 percent, with hydropower and biomass providing most of it. The number was even lower than when Carter set his bold goal of 20 percent. With the growth of renewable and nuclear energy stagnant and auto fuel efficiency in decline, the outlook for technology solutions to the climate problem appeared bleak. Many energy experts in the early 2000s avoided conversations about how to keep the earth's average temperature rise below 2 degrees. Such talks were too depressing.

The failure of wind and solar to expand their market penetration never indicated a total lack of progress. Although far below the levels envisioned by Carter, U.S. government support was helping chip away at lowering costs and improving efficiencies. At times, other countries implemented crash programs that helped advance renewable energy technologies. In the late twentieth century, American states began to create renewable portfolio standards for electric generation, hoping to take advantage of the "wind belt" in the central part of the country. The next needed step was to reach sufficient market penetration to stimulate mass production that would, in turn, bring further reductions in cost and increases in efficiency.

Considerably later—February 8, 2017, to be exact—a beaming Jimmy Carter helped dedicate 3,852 shiny new solar panels covering ten acres of his former farmland in Plains.[38] Modern technology allowed the panels to track the sun's light to maximize output. The ninety-two-year-old former president, who helped build the first nuclear submarines in the 1950s, made daily inspections of the build-out. Ironically, the one megawatt of solar power fed into the grid of Georgia Power, previously one of the most ferocious

Jimmy Carter with grandson Jason at the 2017 dedication of a photovoltaic solar farm on his property in Plains, Georgia. Courtesy AP Photo/David Goldman.

corporate opponents of government measures to slow global warming and establish renewable portfolio standards.

This small facility was part of something much bigger—a global solar and wind power boom.

A 2018 report from MIT explained the exploding if belated private sector enthusiasm for solar cells—the price of photovoltaics had fallen by a stunning 99 percent over the previous four decades.[39] The researchers attributed the dramatic drop to government policies around the world and progress in improving PV conversion efficiency, or the amount of power generated from a given amount of sunlight. Gregory Nemet—a professor at the University of Wisconsin, Madison and not involved in the MIT study—added further perspective. He observed, "Policies in Japan, Germany, Spain, California, and China drove the growth of the market and created opportunities for automation, scale, and learning by doing." A burst of research and development spending early in the Obama presidency also helped.

In April 2022, the U.S. Energy Information Administration spotlighted another milestone in renewable energy history. It announced that 2021 electricity data showed, "The growing number of large solar and wind energy projects resulted in renewable generation beating out nuclear generation." The tortoise had finally passed the hare.

The Inflation Reduction Act of 2022 provided new incentives for renewable and nuclear energy that were massive enough to further advance their prospects. The long-term impacts of the legislation will greatly affect future efforts to slow global warming. Unlike many earlier policy initiatives, it could build on existing commercial momentum toward renewable energy (combined with increasingly efficient storage, such as batteries, to compensate for its intermittent availability). U.S. action is already spurring other countries who want to remain competitive to up their subsidies for climate-friendly technologies.[40]

In August 2023, the International Energy Agency in Paris issued another upbeat global outlook for renewables. It projected that they would overtake coal as the planet's largest source of electricity by 2025. "It's astonishing what's happening," observed IEA executive director Faith Birol. "Clean energy is moving faster than many people think, and it's become turbocharged lately." Al Gore told the *New York Times*, "The nature of these exponential curves sometimes causes us to underestimate how quickly changes occur once they reach these inflection points and begin accelerating."[41] In other words, the market penetration of wind and solar was very slow until it happened very fast.

This book tells a four-decade story of the origins of contemporary debates about climate change in the United States. To be fair, we should evaluate the actions and individuals in this narrative based on what they knew then, not what we know today. However, since we now have the record of later events, it is tempting, with hindsight, to speculate briefly on how things might have turned out differently, for better or worse.

What if the ever-inquisitive Roger Revelle hadn't helped secure the funding for Dave Keeling's atmospheric carbon monitoring, or if he had lacked the connections to ensure that top government officials quickly became aware of the careful measurements from Mauna Loa? It is easy to imagine a scenario in which legitimate confusion over human impacts on the atmosphere could have continued for additional decades.

What if the Arab oil embargo in the 1970s hadn't strengthened the cartel of oil-exporting countries, pushed energy prices higher, and inspired new energy policies in the consuming nations, at least for a while. The energy crises of the 1970s led to additional emphasis on energy efficiency and investments in alternative fuels, including the creation of the nation's renewable energy research lab in Colorado. Similarly, what if bipartisan coalitions in Congress hadn't passed auto and appliance efficiency standards or ratified a treaty controlling ozone-depleting greenhouse gases? Climate change added little to the motivation for such actions. Nonetheless, government policies did slow somewhat the accumulation of atmospheric carbon from what it would have been otherwise.

What if Margaret Thatcher, a hero to conservatives worldwide, hadn't championed effective climate policies? Even after she was no longer prime minister, there was little pushback from conservatives against climate action in the United Kingdom or the European Union, at least at the scale seen in the United States. The narrowing of the partisan divide on climate made it easier for some countries to avoid controversy and make steady progress in reducing greenhouse gas emissions.

It is a misreading of history to assert that the United States did nothing to delay the arrival of average temperature increases of 2 degrees Celsius and above. However, it is also false to claim that the country has ever been in danger of acting too quickly, as the contrarians have continuously alleged. U.S. policies responding directly to the climate challenge were virtually nonexistent when Bush the father left the White House, and this continued to be the case well into the twenty-first century. This record was not inevitable.

What if scientists had explained climate change with greater clarity to enhance public understanding? When respected scientists dangled out the goal of scientific "certainty"—a worthy aim in the confines of academia—to policymakers, they encouraged perpetual delay. Some testifying before Congress interjected that absolute certainty might never be possible. But there was still a tendency to avoid rocking the boat and accept the assertions of elected officials that they would address the climate problem when the science was "settled." Similarly, natural scientists had only limited success in explaining to the public that the long residency of greenhouse gases in the atmosphere made it extremely risky to assume that the effects of global warming could be dealt with as they arrived.

What if American economists had not buried assumptions in their models, like discounting, that strongly favored adaptation over mitigation as a longterm strategy for dealing with a changing climate? What if these economists hadn't established the gross national product as virtually the sole measure of a society's success, in contrast to views earlier and elsewhere that regarded nature as a value transcending quantification? What if they hadn't focused their cost–benefit analyses on their own country, making it convenient to ignore the damage from U.S. pollution in the poorer nations?

What if companies relying on fossil fuels hadn't supported think tanks with noble-sounding titles that spewed crafty deceits the companies didn't want to be associated with publicly?

Delaying climate change required massive alterations of the existing energy infrastructure. Consequently, solutions were never going to be quick and easy for policymakers. Still, some viable and relatively painless options could have provided partial answers if they had been adopted earlier or more consistently. What if the country had persisted with the pedal-to-the-metal intensity of Jimmy Carter's funding for wind and solar technology, advancing their market readiness by a decade or two? What if auto efficiency standards had been regularly updated when those in force became less consequential? What if Congress had sharply limited methane leakage from energy production and transportation—hardly a draconian task? These steps would not have fully substituted for stiffer, more politically formidable actions down the road. But they would have reduced greenhouse gas emissions, come with a relatively low price tag, and made the United States a more reliable partner in the global effort organized by the International Panel on Climate Change.

If some choices demanded fundamental changes in infrastructure and the daily habits of the earth's people, it is essential to recognize there have been and will continue to be choices. It is never too early or too late to take action. As witnesses told Congress in 1980s, the optional time for action depends on how much anticipated damage you want to avoid. It also depends on how much insurance you want to buy to reduce the risk that changes could come more abruptly and severely than models forecast.

On this test of human wisdom, the report card is still "incomplete."

Acknowledgments

This book was not a solo effort. Hundreds of people assisted along the way, enough that thanking them all properly might necessitate another book.

My special appreciation to the scores of awesome archivists and analysts who helped me acquire new information and think more clearly about it. The family of presidential libraries—part of the National Archives—provided textual and audiovisual records that don't rely on distant memories (but do require sitting for long hours on hard wooden chairs). New explorations with the help of incredible custodians of the massive records at the University of California, San Diego's Special Collections and the American Association for the Advancement of Science demonstrated that not all critical conversations among scientists can be found online. My earlier years at the highly regarded Energy Information Administration deepened my understanding of energy and the environment as interrelated systems, and (perhaps even more challenging) of how to explain them to congressional committees and curious reporters.

Family and friends always had my back. Sometimes it was just an encouraging word or understanding of why I spent so much time in my office with the door closed. Several with relevant expertise provide insightful critiques of draft chapters or even the whole book.

The editors at the Louisiana State University Press have been a joy to work with as we moved from manuscript to book. I am also thankful to those who published my previous books and articles, which I hope were steps toward what I hope is my best effort.

Any errors that remain are of my own doing.

Over the years, I have met dozens of people mentioned in this book—sometimes for a brief social exchange, on other occasions for discussions of climate change. The hindsight of history is sometimes unkind to those in the arena, who are facing complexities not well understood by those in the stands. But honor is due to those who choose to serve, whether in ivory towers or the corridors of power. As this book illustrates, it is easier to opine about what the world should do about a daunting challenge to life as we've known it than to implement the concrete steps needed to sharply reverse the trajectory of rising greenhouse gas emissions. Still, we must also recognize the missteps if we want to create a better future.

There is one name that I must specify—that of my wife, Anita Zervigon-Hakes. She has to put up with me 24/7. She shares with me the hope that our children and grandchildren will understand better than previous generations how to make wise choices about protecting our planet.

Notes

Prologue

1. Steele, *Galileo,* 3, 28–29.

2. McCullough, *Wright Brothers,* 262.

3. Isaacson, *Einstein,* 1–2, 137–39.

4. Isaacson, *Leonardo Da Vinci,* 4, 407–9, 423–24.

5. Spencer Weart, "Money for Keeling: Monitoring CO_2 Levels," *Discovery of Global Warming,* May 2023, online; Charles Keeling, "The Concentration and Isotopic Abundances of Carbon Dioxide in the Atmosphere," *Tellus,* June 1960: 200–03.

6. Charles Keeling, "Is Carbon Dioxide from Fossil Fuel Changing Man's Environment?" *Proceedings of the American Philosophical Society* 114 (February 1970): 17.

1. Measuring the Invisible Blanket

1. For even earlier references in nonacademic publications with less readership, see Brad Johnson, "A Timeline of Climate Science and Policy," *Medium,* September 27, 2016.

2. Eunice Foote, "Circumstances Affecting the Heat of the Sun's Rays," *American Journal of Science and Arts* 22 (1856): 382–83; John Roberts, "Overlooked No More: Eunice Foote, Climate Scientist Lost to History," *New York Times,* April 27, 2020.

3. For a detailed bibliography of the early science, see Spencer Weart, "Bibliography by Year Through 2001," *Discovery of Global Warming,* May 2023, online; Weart, "The Carbon Dioxide Greenhouse Effect," *Discovery of Global Warming,* May 2023, online. For key documents, see Howe, *Making Climate Change History.* See also Bolin, *History of the Science and Politics of Climate Change.*

4. G. S. Callendar, "The Artificial Production of Carbon Dioxide and its Influence on Temperature," *Quarterly Journal of the Royal Meteorological Society* (April 1938): 223–40; Fleming, *Callendar Effect,* 65–87.

5. Sharp, *Roger Randall Dougan Revelle: Preparation,* 35.

6. Detailed biographies of Revelle include "Roger Randall Dougan Revelle (1909–1991)," San Diego History Center, online; Spencer Weart, "Roger Revelle's Discovery," *Discovery of Global Warming,* August 2021, online; Day, *Revelle.* See also Helvarg, *Blue Frontier,* 53–57.

7. Roger Revelle Papers, box 28, folder 66.

8. A summary of the conference can be found in Pfeffer, *Dynamics of Climate.*

9. Ibid., 3.

10. Ibid., 93–95, 133–34.

11. Ibid., 105–6, 132–33.

12. Day, *Revelle,* 16.

13. Sharp, *Roger Randall Dougan Revelle: Preparation,* 74–76.

14. Oreskes, *Science on a Mission,* 37–38, 45–49, 175–81, 191–92.

15. "6 Scientists Accept 'IGY' TV Invitation," *New York Times,* October 7, 1957.

16. Sharp, *Roger Randall Dougan Revelle: Preparation,* 11.

17. Weart, *Discovery of Global Warming,* 43. The book has a companion website with additional information.

18. Weart, "Roger Revelle's Discovery."

19. Charles Keeling, "A Chemist Thinks About the Future," *Archives of Environmental Health* 21 (1970): 765.

20. Benjamin Franta, "Early Oil Industry Knowledge of CO_2 and Global Warming," *Nature Climate Change* 8 (December 2018): 1024.

21. Charles Keeling, "Rewards and Penalties of Monitoring the Earth," *Annual Review of Energy and the Environment* 23 (1998): 32–34; Christianson, *Greenhouse,* 152–53.

22. Keeling, "Rewards and Penalties," 33.

23. Ibid., 35–36.

24. Ibid., 37.

25. Spencer Weart, interview with Harmon Craig, April 29, 1996, Niels Bohr Library and Archives.

26. Spencer Weart, "Global Warming, Cold War, and the Evolution of Research Plans," *Historical Studies in the Physical and Biological Sciences* 27, no. 2 (1997): 341.

27. Roger Revelle and Hans Suess, "Carbon Dioxide Exchange between Atmosphere and Oceans and the Question of an Increase of Atmospheric CO_2 During the Past Decades," *Tellus* 9, (1957): 18–27.

28. Ibid., 26.

29. Day, *Revelle.*

30. Hearing of the Subcommittee of the Committee on Appropriations on "Report on the Geophysical Year," House of Representatives, 86th Congress, 1st Session, February 18, 1959.

31. American Presidency Project, January 18, 1960; National Research Council, *Oceanography, 1960 to 1970,* 1, 4.

32. Broecker, *Fixing Climate,* 73–76; Christianson, *Greenhouse,* 153.

33. Weart, "Money for Keeling"; Keeling, "Concentration and Isotopic Abundances of Carbon Dioxide in the Atmosphere," 200–03.

34. Christianson, *Greenhouse,* 157.

35. Charles David Keeling Papers, box 9, folder 14.

36. Isaacson, *Einstein,* 472–76.

37. American Presidency Project, December 8, 1953.

38. Seaborg, *Atomic Energy Commission,* 156.

39. Hakes, *Energy Crises.*

40. Duncan, *Rickover,* 116–49.

41. Rhodes, *Energy,* 289; Hewlett, *Atoms for Peace and War,* 493.

42. American Presidency Project, January 18, 1960.

43. Benjamin Franta, "On its 100th Birthday in 1959, Edward Teller Warned the Oil Industry About Global Warming," *Guardian,* January 1, 2018.

44. Hearing on "Frontiers in Atomic Energy Research," Joint Committee on Atomic Energy, 86th Congress, 2nd Session, March 24, 1960.

45. The idea of nuclear-powered cars was not as speculative as it might appear. In 1958, Ford Motor Company unveiled a prototype of such a vehicle but never attempted to produce a commercial version. See James Gilboy, "Inside the Impossible Dream of the Nuclear-Powered 1958 Ford Nucleon," *The Drive,* July 5, 2021.

46. "Science: On the Way, Genuine Fusion," *Time,* April 4, 1960.

47. M. A. Matthews, "The Earth's Carbon Cycle," *New Scientist* 6 (October 8, 1959): 644–46. Document collection is available at climatefiles.com.

48. For a summary of Einstein's theory and the applicable science that preceded it, see Isaacson, *Einstein,* 94–101; Maycock, *Photovoltaics,* 169–70; Nemet, *How Solar Energy Became Cheap,* 56–57.

49. "Power from the Sun," *New York Times,* April 26, 1954.

50. Nemet, *How Solar Energy Became Cheap,* 57–61. Eight years later, Bell would introduce another innovation that would alter human life: the first commercial touch-tone phone.

51. Maycock, *Photovoltaics,* 63.

2. JFK's Early Mention of Climate Science

1. John F. Kennedy speeches, 1958, John F. Kennedy Presidential Library, online.

2. American Presidency Project, May 25, 1961.

3. Jerome Wiesner, "Kennedy," in Rosenblith, *Jerry Wiesner,* 267–68.

4. Ibid., 268–69; Kistiakowsky, *Scientist in the White House,* 392–93.

5. Theodore Sorensen, "A View from the White House," in Rosenblith, *Jerry Wiesner,* 39.

6. Wiesner, "Kennedy," in Rosenblith, *Jerry Wiesner,* 285.

7. Ibid., 281.

8. "Oceanography," President's Office Files, Kennedy Library; American Presidency Project, March 29, 1961.

9. Day, *Revelle,* 21; Spencer Weart, interview with Harmon Craig, April 29, 1966, Niels Bohr Library and Archives.

10. American Presidency Project, April 25, 1961.

11. Interview with Robert White, April 25, 1989, Clarence E. Larson Science and Technology Oral History Collection, box 4.

12. Wiesner, "Kennedy," in Rosenblith, *Jerry Wiesner,* 286–90; Wiesner, "How Roger Revelle Became Interested in Population and Development Problems," in Dorfman and Rogers, *Science with a Human Face,* 1–5.

13. Day, *Revelle,* 21–22.

14. "Roger Revelle Awards," American Association for the Advancement of Science, Climate Program Records, box 13.

15. The study was not completed until Revelle returned to California. See "Energy, Diplomacy, and Global Issues," document 132, *Foreign Relations of the United States, 1964–1968,* vol. 34.

16. Conservation Foundation, *Implications of Rising Carbon Dioxide Content of the Atmosphere,* 1.

17. Keeling Papers, box 6, folder 15.

18. Bernard Stengren, "Pollution of Air Viewed by Copter: Muskie Surveys Problem to Help Senate Study," *New York Times,* September 14, 1963.

19. American Presidency Project, February 23, 1961.

20. Ibid., February 7, 1963.

21. Congressional Quarterly, *Congress and the Nation, 1945–1964,* 1148. Information on congressional action on clean air issues going back to the 1950s can be found in Halvorson, *Valuing Clean Air,* 16–17.

22. A digital copy of the film (not released until 1964) is available through the Kennedy Library, Digital Collections, Moving Images, "Oceanography."

23. National Academy of Sciences, "John F. Kennedy Addresses the National Academy of Sciences Oct. 22, 1963," YouTube, August 3, 2016, http://youtu.be/BaYzqysyOcE; transcript at American Presidency Project, October 22, 1963.

24. Sharp, *Roger Randall Dougan Revelle: International Scientist,* 65.

25. Douglas Martin, "Donald Hornig, Last to See First A-Bomb, Dies at 92," *New York Times,* January 26, 2013.

26. Harold Schmeck Jr., "Kennedy Aide Asks Closer Ties Between Scientists and Public," *New York Times,* November 16, 1963.

27. Sharp, *Roger Randall Dougan Revelle: International Scientist,* 64–65.

3. LBJ's War on Pollution

1. Goldman, *Tragedy,* 152–55, 184–86; Roger Revelle to President Kerr, March 28, 1964, Revelle Papers, box 31, folder 11.

2. Joseph Fisher, Paul Freund, Margaret Mead, and Roger Revelle, "Notes Prepared by Working Group Five, White Group on Domestic Affairs," April 4, 1964, Revelle Papers, box 31, folder 11.

3. Revelle to Kerr, April 14, 1964, Revelle Papers, box 31, folder 11.

4. Goldman, *Tragedy,* 155.

5. Dallek, *Flawed Giant,* 80–83; Goldman, *Tragedy,* 194–97; filmed excerpts from the speech available in C-SPAN archives; David G. McComb, Donald Hornig Oral History, Interview 1, December 4, 1968, Lyndon Baines Johnson Presidential Library, online.

6. *San Diego,* January 11, 1965, in Revelle Papers, box 79, folder 4.

7. "Scientists Named to Advise Congress," *New York Times,* September 24, 1964.

8. American Presidency Project, November 1, 1964.

9. American Presidency Project, February 8, 1965; Marianne Lavelle, "A 50th Anniversary Few Remember: LBJ's Warning on Carbon Dioxide," *Daily Climate,* February 2, 2015, online.

10. President's Science Advisory Committee, Environmental Pollution Panel, *Restoring the Quality of Our Environment,* 1.

11. Ibid., 14–15.

12. Ibid., appendix Y4, 112–33.

13. Broecker, *Fixing Climate,* 22–24; Kevin Krajick, "Wallace Broecker, Prophet of Climate Change," Columbia Climate School, *State of the Planet,* press release, February 19, 2019, online.

14. President's Science Advisory Committee, Environmental Pollution Panel, *Restoring the Quality of Our Environment,* 112–13. This sentence was first brought to my attention by Roston, *Carbon Age.*

15. Evert Clark, "Johnson Panel Urges 'Polluters Tax,'" *New York Times,* November 7, 1965; Howard Simons, "Report to LBJ Urges Taxing All Polluters," *Washington Post,* November 7, 1965.

16. Franta, "Early Oil Industry Knowledge of CO_2 and Global Warming," 1024.

17. Interview with Robert White, April 25, 1989, Clarence E. Larson Science and Technology Oral History Collection, box 4, online.

18. National Research Council, *Weather and Climate Modification* (1966), 1:10, 2:88–89.

19. American Presidency Project, February 18, 1966.

20. Papers of LBJ, EX SP 2/3/66 to 3/9/66, box 77, Johnson Library.

21. Seaborg, *Science, Man and Change,* 54–55, 79–81.

22. James Gustave Speth, "Reference Page for *They Knew,*" Our Children's Trust, online.

23. Seaborg, *Atomic Energy Commission,* 154.

24. Speth, "Reference Page for *They Knew.*"

25. Congressional Quarterly, *Congress and the Nation, 1945–1964,* 1148.

26. American Presidency Project, October 20, 1965.

27. Ibid., November 21, 1967.

28. *New York Times,* January 27, December 16, 1966; September 24, December 31, 1967.

29. J. O. Fletcher, "Controlling the Planet's Climate," in Revelle, Khosla, and Vinovskis, *Survival Equation,* 446.

30. Donald Hornig, "Future Energy Needs vs. the Environment," *Edison Electric Institute Bulletin,* June–July 1968, 209–10. A scan of the report was put online in 2017 by the Energy and Policy Institute.

31. Robinson and Robbins, *Sources, Abundance, and Fate of Gaseous Atmospheric Pollutants,* 7, 9, 14, 25. Prepared for the American Petroleum Institute. The first part of the study was published in 1968.

32. Earl Droessler, interview with Roger Revelle, February 3, 1989, Niels Bohr Library and Archives, online.

33. Revelle Papers, box 56, folders 5–6.

34. Robert Calvert, interview with Roger Revelle, July 4, 1976, Revelle Papers, box 2, folder 2.

35. I was given the class syllabus and student roster by Professor Richard Rochberg, one of the three section leaders for the course. Additional information on the course was supplied by Professor Maris Vinovskis, leader of the section with Gore.

36. Gore, *Earth in the Balance,* 4–6, 91.

37. Congressional Quarterly, *Congress and the Nation,1965–1968;* Divine, *The Johnson Years,* vol. 2.

4. The Global Cooling Detour

1. Hakes, *Energy Crises,* 19, 28.

2. Hess, *Professor and the President.*

3. Moynihan Files, box 2, Richard Nixon Presidential Library.

4. Energy Policy Staff, Office of Science and Technology, October 24, 1969, Whitaker Files, box 110, Nixon Library.

5. Centigrade is the old name for Celsius. A change of one degree Celsius (or centigrade) is equal to a change of 1.8 degrees Fahrenheit.

6. Moynihan Files, box 3, Nixon Library.

7. Flippen, *Nixon and the Environment,* 50–52.

8. Oreskes, *Science on a Mission,* 402.

9. Walter Munk, Naomi Oreskes, and Richard Muller, "Gordon James Fraser MacDonald," in *Biographical Memoirs* 84: 225–50; James Fleming, interview with Gordon MacDonald, March 21, 1994, Niels Bohr Library and Archives, online.

10. American Presidency Project, January 22, 1970.

11. Halvorson, *Valuing Clean Air,* 45–48.

12. Council on Environmental Quality, *Environmental Quality,* 18, 93–94.

13. This legislation was titled the Clean Air Amendments of 1970, due to the previous legislation on the subject. Because of its impact, however, it is generally called the Clean Air Act, although this act itself was later amended.

14. *Congressional Record,* 91st Congress, 2nd Session, September 21, 1970.

15. Other technical reports used by the Congress in writing the Clean Air Act also referred to the effect of pollution on climate. See *Congress and the Nation's Environment,* Senate, Committee on Interior and Insular Affairs, 92nd Congress, 1st Session, February 10, 1971, 67–68.

16. Congressional Quarterly, *Congress and the Nation, 1969–1972,* 763.

17. Philip Shabecoff, "Nixon Offers Broad Plan for More 'Clean Energy,'" *New York Times,* June 5, 1971.

18. Seaborg to Nixon, July 23, 1970, David Files, box 8, Nixon Library.

19. Hakes, *Energy Crises,* 17–29.

20. Schneider, *Science as a Contact Sport,* 17–21.

21. S. Ichtiaque Rasool and Stephen H. Schneider, "Atmospheric Carbon Dioxide and Aerosols: Effects of Large Increases on Global Climate," *Science* 173 (July 1971): 138–41.

22. Victor Cohn, "U.S. Scientist Sees New Ice Age Coming," *Washington Post,* July 9, 1971.

23. Dana Nuccitelli, "A Remarkably Accurate Global Warming Prediction Made in 1972," *Guardian,* March 19, 2014; Alter, *His Very Best,* 178.

24. American Presidency Project, February 14, 1973.

25. Hakes, *Energy Crises,* 59–138.

26. Roger Revelle, "Population and Environment," Research paper no. 4, Harvard University, Center for Population Studies (May 1974): 22.

27. Hakes, *Energy Crises,* 150, 179.

28. Tolo, *Beyond Today's Energy Crisis,* 36–37.

29. National Research Council, *Weather & Climate Modification* (1973), 13.

30. For discussion of the road to the Environmental Modification Convention of 1977, see Hamblin, *Arming Mother Nature,* 204–20.

31. National Research Council, *Understanding Climatic Change,* vi.

32. Ibid., xi, 5, 13–16.

33. Ibid., 80.

34. Ibid., 2; Harold Schmeck Jr., "Understanding Climate Change," *New York Times,* January 19, 1975.

35. Walker, *Three Mile Island,* 13–14.

36. Wallace Broecker, "Climatic Change: Are We on the Brink of a Pronounced Global Warming?," *Science* 189 (August 1975): 460–63. This article is sometimes credited with the first use of the term "global warming." It is, more precisely, a declaration that the argument about global warming versus global cooling was over.

37. Letter with enclosed paper, Wallace Broecker, "The Implications to Climate of a Coal Economy," March 17, 1976, Revelle Papers, box 53, folder 2.

38. Hearings of the Subcommittee on the Environment and the Atmosphere on "The National Climate Program Act," House of Representatives, Committee on Science and Technology, 94th Congress, 2nd Session, May 18–20, 25–27, 1976.

39. National Climate Program Act, H.R. 3399, 95th Congress (February 9, 1977), 278.

40. Robert Calvert, interview with Roger Revelle, July 4, 1976, Revelle Papers, box 2, folder 2. Revelle made similar comments in "The Politician and the Scientist," *Science* 187 (March 21, 1975): 1100–05.

41. Thomas Peterson, William Connolley, and John Fleck, "The Myth of the 1970s Global Cooling Scientific Consensus," *Bulletin of the American Meteorological Society* 89, no. 9 (September 2008): 1325–37.

42. National Climate Program Act, 585. For a more detailed look at Mitchell's shift in views on global warming, see his "A Reassessment of Atmospheric Pollution as a Cause of Long-Term Changes of Global Temperatures," in Singer, *Changing Global Environment,* 149–73.

43. Schneider, *Science as a Contact Sport,* 19–21, 42–43.

5. Energy and Climate in the Carter Years

1. Carlton Neville Files, container 3, Jimmy Carter Presidential Library.

2. Revelle Papers, box 53, folder 6.

3. James Schlesinger Files, container 8, Carter Library.

4. Ibid., container 5.

5. Working paper, Frank Moore Files, container 28, Carter Library.

6. Council on Environmental Quality, *Solar Energy,* iii–iv, 6.

7. William Nordhaus, "Can We Control Carbon Dioxide?" International Institute for Applied Systems Analysis, working paper 75-63, Laxenburg, Austria (June 1975); Nordhaus, "Economic Growth and Climate: The Carbon Dioxide Problem," Cowles Foundation Discussion Paper no. 435, Yale University (October 1, 1976), also published in *American Economic Review* 67 (February 1977): 341–46.

8. Environmental leaders at the time often accepted the need for cost–benefit analysis in the assessment of environment measures, and even favored them at high levels when applied to Corps of Engineers projects that impacted nature. However, some expressed concerns about adopting such high discount rates for climate models. In November 1974, Maurice Strong, executive director of the United Nations Environmental Programme, stated, "Traditional cost–benefit analysis, with the present rates at 9 or 10 percent, gives no present value to benefits which accrue beyond thirteen or fourteen years. Thus, on a strictly economic basis it does not pay to save the planet . . . Cost–benefit analysis is a very good tool, but the religion of cost–benefit analysis is driving us to disaster." See Tolo, *Beyond Today's Energy Crisis,* 131.

9. Lind et al., *Discounting for Time and Risk,* 5.

10. Walter Sullivan, "Climate Peril May Force Limits on Coal and Oil," *New York Times,* June 3, 1977.

11. Hearings of the Subcommittee on the Environment and the Atmosphere on "Environmental Implications of the New Energy Plan," House of Representatives, Committee on Science and Technology, 95th Congress, 1st Session, June 8–9, 1977.

12. National Research Council, *Energy and Climate.*

13. Ibid., 3, 27.

14. Ibid., 58.

15. Ibid., 1, 3, 14, 23.

16. *Energy and Climate,* second draft, June 18, 1975, Revelle Papers, box 37, folders 10–11.

17. Walter Sullivan, "President's Science Advisor: Frank Press," *New York Times,* March 19, 1977.

18. Frank Press Files, container 6, Environment folder, Carter Library.

19. Walter Sullivan, "Scientists Fear Heavy Use of Coal May Bring Adverse Shift in Climate," *New York Times,* July 25, 1977.

20. "NREL at 40: It All Started with a Desire to Harness the Sun," *NREL News,* July 5, 2017.

21. American Presidency Project, May 3, 1978.

22. Yanek Mieczkowski, "'The Toughest Thing': Gerald Ford's Struggle with Congress over Energy Policy," in Lifset, *American Energy Policy in the 1970s,* 19–46; Arthur Rosenfeld, "The Art of Energy Efficiency: Protecting the Environment with Better Technology," *Annual Review of Energy and Environment* 24 (1999): 45–46.

23. Carter, *White House Diary,* 258–59.

24. Alter, *His Very Best,* 300.

25. Revelle Papers, box 45, folder 10.

26. Walter Sullivan, "Climatologists Are Warned North Pole Might Melt," *New York Times,* February 4, 1979.

27. American Presidency Project, March 27, 1979.

6. A Jolt from JASON

1. Rich, *Losing Earth,* 15–18; JASON, *Long Term Impact.*

2. Scientists in the 1970s, as in this report, often expressed temperatures in Kelvin, which is interchangeable with Celsius. An increase of 2 or 3 degrees, whether Kelvin or Celsius, would be the equivalent of 3.6 to 5.4 degrees Fahrenheit, the measure more commonly used by non-scientists in the United States.

3. JASON, *Long Term Impact,* 25–26. Concerns about Antarctic ice relied heavily on an article by Ohio State University glaciologist John Mercer, "West Antarctic Ice Sheet and CO_2 Greenhouse Effect: A Threat of Disaster," *Nature* (January 1978): 321–25.

4. JASON, *Long Term Impact,* 27–28.

5. Ibid., 5, 9–12.

6. Ibid., 28.

7. Staff Secretary's Files, March 29, 1979, container 111, Carter Library. Available in digital library.

8. Rich, *Losing Earth,* 20–24.

9. Ibid., 25; Statement of Philip Smith, White House Office of Science and Technology Policy, before the House Subcommittee on Natural Resources and Environment, July 10, 1979, in Revelle Papers, box 120, folder 2.

10. Speth, *Red Sky at Morning,* 2.

11. A detailed summary of the panels and copies of the contributed papers is available in U.S. Department of Energy, Carbon Dioxide Effects Research and Assessment Program, *Workshop on Environmental and Societal Consequences.* The DOE published a second volume on the conference, in December 1980, under the title *Environmental and Societal Consequences of a Possible Co2–Induced Climate Change: A Research Agenda.* These documents do not have continuous pagination. Additional organizational information can be found in "Climate Variability and CO_2, 1979–1986," American Association for the Advancement of Science, Climate Program Records, box 1.

12. Howe, *Behind the Curve,* 112.

13. Revelle Papers, box 53, folders 1–6.

14. Howe, *Behind the Curve,* 114.

15. Atomic Energy Commission oversight of civilian nuclear power was turned over to the new Nuclear Regulatory Commission in 1975. Research and development of atomic energy for military and civilian went first to the Energy and Research Development Administration and then, in 1977, to the new Department of Energy.

16. Alvin Weinberg and Gregg Marland, "Some Long-Range Speculations About Coal," Oak Ridge Associated Universities, Occasional Paper 77-22 (August 1977).

17. Keeling, "Rewards and Penalties," 56.

18. Ibid., xiii.

19. "Henry Shaw," *Inside Climate News,* September 15, 2015; Written Statement to the Congressional Oversight Committee by Dr. Edward A. Garvey, October 23, 2019, House documents. After priorities changed, Exxon ended the data collection in the mid-1980s, before its data could be published.

20. In post-workshop discussions with leading climate modeler Syukuro Manabe, a colleague of Joseph Smagorinsky at Princeton, Revelle complained that the big computer models on climate had already become so complex that it was hard to find anybody who could understand them. Manabe was exploring how to incorporate the kind of regional analysis that Lave said he needed, but he didn't display much optimism that this would prove successful in the foreseeable future. See William Nierenberg Papers, box 139, folder 1.

21. Nierenberg Papers, box 138, folder 9.

22. Edward David to William Carey, June 27, 1979, American Association for the Advancement of Science Archives, Climate Program Records, box 15.

23. R. J. Campion to W. W. Madden, July 9, 1979, Industry Documents Library, University of California, San Francisco Library, online.

24. American Presidency Project, April 23, 1979.

25. Hakes, *Energy Crises,* 281–82.

26. Markey was later elected to the U.S. Senate, where he also became a leader on climate.

27. In the 1970s, the term "solar energy" was often used interchangeably with "renewable energy," based on the understanding that hydropower, biomass, and wind power all depended in some way on the resources of the sun.

28. Nemet, *How Solar Energy Became Cheap,* 65–79.

29. Minutes of the Tokyo Economic Summit Meeting, *Foreign Relations of the United States, 1969–1976,* "Energy Crisis, 1974–1980," Document 221.

30. *Foreign Relations of the United States, 1977–1980,* "Foreign Economic Policy," Document 222.

31. Woodwell et al., *Carbon Dioxide Problem,* 7–8. The reprinted version includes a forward by Gus Speth.

32. Ibid., 11–13.

33. Stuart Eizenstat Files, container 286, Carter Library.

7. More Reports from the Academy

1. Several months after the completion of the report, Charney was diagnosed with cancer, from which he died in June 1981. See *New York Times,* June 18, 1981.

2. Rich, *Losing Earth,* 34.

3. National Research Council, *Carbon Dioxide and Climate,* 1–2, 19.

4. Ibid., 10–11, 16.

5. U.S. Senate, Committee on Government Affairs, *Carbon Dioxide Accumulation in the Atmosphere,* 2.

6. Ibid., 17, 20.

7. Ibid., 19. Revelle's interpretation had some similarities to Exxon's but did not go as far as the oil company, which had suggested that societies could "cope *readily* with *whatever* problems ensue." Emphasis added.

8. Ibid., 19.

9. Ibid., 8.

10. Hakes, *Energy Crises,* 295–97.

11. "Excerpts from Presidential Panel's Report on the Three Mile Island Accident," *New York Times,* October 31, 1979; Walker, *Three Mile Island,* 209–15; Glenn Zorpette, "From Three Mile Island to Fukushima Daiichi," *IEEE Spectrum,* February 28, 2014.

12. Joanne Omang, "Industry Pleased by What the Report Didn't Say," *Washington Post,* October 31, 1979.

13. American Presidency Project, December 7, 1979.

14. Mondale and Schlesinger to the President, "LMFBR Compromise," Staff Secretary's Files, container 89, Carter Library; American Presidency Project, December 2, 1977; Rod Adams, "Light Water Breeder Reactor: Adapting a Proven System," *Atomic Insights,* October 1, 1995.

15. Revelle Papers, box 154, folder 1.

16. Nicolas Nierenberg, Walter Tschinkel, and Victoria Tschinkel, "Early Climate Change Consensus at the National Academy: The Origins and Making of Changing Climate," *Historical Studies in the Natural Sciences* 40, no. 3 (Summer 2010): 327.

17. Memo from Press to Carter with attachments from Thomas Schelling and Philip Handler, May 5, 1979, Office of Science and Technology Files, container 2, Carter Library.

18. Energy Security Act, Public Law 96-294 (June 30, 1980), Title VII.

19. Revelle Papers, box 154, folder 4.

20. Council on Environmental Quality and Department of State, *Global 2000 Report,* 3.

21. Hakes, *Energy Crises,* 307; American Presidency Project, August 11, 1980.

22. Council on Environmental Quality, *Global Energy Futures,* v–vii, 52, 68–70.

23. 24. Spencer Weart, "Government: The View from Washington, DC," *Discovery of Global Warming,* May 2023, online.

8. Changing Directions in the Reagan Era

1. Cannon, *President Reagan,* 93.

2. Keeling, "Rewards and Penalties," 56–59.

3. Robert Reinhold, "Science Panel Urges Reagan to Bolster Arms and Industrial Technology," *New York Times,* November 12, 1980.

4. Keeling, "Rewards and Penalties," 59–60.

5. Baldwin memo, February 28, 1981, James Baker Files, box 1, Ronald Reagan Presidential Library, Digitized Textual Material.

6. "Reagan to Keep OSTP and CEQ," *Science,* May 27, 1981.

7. Robert Reinhold, "Los Alamos Physicist May Get Post as Science Advisor in White House," *New York Times,* May 8, 1981; Reinhold, "Physicist Is Named as Science Advisor," *New York Times,* May 20, 1981.

8. Letter from White House Office of Science and Technology to National Academy of Sciences, June 26, 1981, Revelle Papers, box 149.

9. Jason Goodell, "The Man Who Started Earth Day," *Rolling Stone,* April 22, 2020.

10. Hearing of the Subcommittee on Natural Resources, Agriculture Research and Environment and the Subcommittee on Investigations and Oversight on "Carbon Dioxide and Climate: The Greenhouse Effect," House of Representatives, 97th Congress, 1st Session, July 31, 1981.

11. Flohn, *Climate and Weather,* 156–61.

12. The Associated Press story on the hearing led with Lave's skepticism about synthetic fuels. See "Synthetic Fuels Called a Peril to the Atmosphere," *New York Times,* August 1, 1981.

13. Nathaniel Sheppard Jr., "Concern Grows Over Policy on National Labs," *New York Times,* December 8, 1981.

14. Stockman, *Triumph of Politics,* 114–15.

15. Keeling, "Rewards and Penalties," 66.

16. Comments delivered on the 80th anniversary of the Scripps Institution, October 13, 1983, Nierenberg Papers, box 169, file 17.

17. James Hansen et al., "Climate Impacts of Increasing Atmospheric Carbon Dioxide," *Science,* August 28, 1981, 957–66; Walter Sullivan, "Study Finds Warming Trend That Could Raise Sea Levels," *New York Times,* August 22, 1981.

18. Hearing of the Subcommittee on Natural Resources, Agriculture Research and Environment and the Subcommittee on Investigations and Oversight of the Committee on Science and Technology on "Carbon Dioxide and Climate: The Greenhouse Effect," House of Representatives, 97th Congress, 2nd Session, March 15, 1982.

19. Hansen's estimate of historic sea level rise would have been somewhat higher if he had not corrected for the sinking of shorelines, due to factors other than climate change.

20. Howe, *Behind the Curve,* 126–27.

21. William Nordhaus, "How Fast Should We Graze the Global Commons?," *American Economic Review* 72, no. 2 (May 1982): 242–46.

22. Seidel and Keyes, *Can We Delay a Greenhouse Warming?*

23. Ibid., iii–iv.

24. Ibid., iv–vii. The study focused on coal and shale oil because it assumed that supplies of conventional oil and gas would become depleted and play less of a role, even without policy responses to climate change.

25. Ibid., vii; section 7, 7.

9. Discounting the Future

1. Event documents, Revelle Papers, Files, box 154, file 10.

2. TV transcripts, Rockville, Maryland; Philip Shabecoff, "Haste of Global Warming Trend Opposed," *New York Times,* October 21, 1983.

3. Memorandum for Meese from Keyworth, October 21, 1983, RAC box 9, Reagan Library, Digitized Textual Material.

4. John Perry notes, Revelle Papers, box 154, files 4–5.

5. Dueling historical perspectives on the report can be found in Naomi Oreskes, Erik Conway, and Matthew Shindell, "From Chicken Little to Dr. Pangloss: William Nierenberg, Global Warming and the Social Deconstruction of Scientific Knowledge," *Historical Studies in the Natural Sciences* 38 (February 2008): 109–52; and Nierenberg, Tschinkel, and Tschinkel, "Early Climate Change Consensus," 318–49.

6. National Research Council, *Changing Climate,* 266–77, 292.

7. Ibid., 320. This chapter agreed that Hansen had demonstrated a close tie between carbon accumulations and temperature change but said his results had not been well replicated.

8. Ibid., 413.

9. Ibid., 36, 39.

10. Ibid., 32, 280.

11. Nierenberg, Tschinkel, and Tschinkel, "Early Climate Change Consensus," 338.

12. National Research Council, *Changing Climate,* 252–61.

13. Ibid., 260.

14. Ibid., 22, 95.

15. Ibid., 23, 96

16. Ibid., 22, 122, 146.

17. Ibid., 449.

18. Ibid., 469. Schelling appeared unaware of the international agreement to ban weather modification strategies because the control of such technologies would enhance the ability to inflict harm on other nations.

19. Ibid., 473–74.

20. Ibid., 477.

21. Weinberg to Nierenberg, July 1, 1983, Nierenberg Papers, box 86, file 7.

22. National Research Council, *Changing Climate,* 4.

23. Sharp, *Roger Randall Dougan Revelle: Preparations,* 78.

24. Michael Oppenheimer, "To Delay Global Warming," *New York Times,* November 9, 1983.

25. President's Science Advisory Committee, Environmental Pollution Panel, *Restoring the Quality of Our Environment,* 16.

26. American Presidency Project, February 23, 1966.

27. Roger Revelle, "Human Ecology and Ethics Are Inseparable," *New York Times,* January 12, 1970.

28. American Presidency Project, January 22, 1970.

29. American Presidency Project, March 27, 1979.

30. Council on Environmental Quality, *Global Energy Futures,* vii.

31. Lind et al., *Discounting for Time and Risk.*

32. Ibid., ix, xii.

33. Ibid., 151, 154.

34. John Perry to Members, Carbon Dioxide Assessment Committee, December 30, 1981, Revelle Papers, box 154, file 7.

35. Schelling made a brief reference to the issues raised by discounting in *Changing Climate*, 452–53, but didn't disclose how the assumptions in economic models influence their policy evaluations.

36. Philip Boffey, "Scientists Urged by Pope to Say No to War Research," *New York Times*, November 13, 1983.

37. Philip Boffey, "Once Hostile Vatican Forges Close Links with Scientists," *New York Times*, November 14, 1983.

38. Revelle Papers, box 121, file 8.

39. Marini-Bettolo, *Chemical Events in the Atmosphere*.

40. Hearing of the Subcommittee on Natural Resources, Agriculture Research and Environment and the Subcommittee on Investigations and Oversight on "Carbon Dioxide and the Greenhouse Effect," House of Representatives, Committee on Science and Technology, 98th Congress, 2nd Session, February 28, 1984.

10. Bipartisan Rebound

1. World Climate Programme, *Report of the International Conference*.

2. Ibid., 1, 3.

3. Philip Shabecoff, "Scientists Warn of Earlier Rise in Sea Levels," *New York Times*, November 3, 1985.

4. World Climate Programme, *Report of the International Conference*, 69–71.

5. Ibid., 31–33.

6. Ibid., 33.

7. Senate Environment and Public Works Subcommittee on Hazardous Wastes and Toxic Substances, Hearing of December 10, 1985, C-SPAN television archives.

8. "Action Is Urged to Avert Global Climate Shift," *New York Times*, December 11, 1985.

9. Gore got language requiring the president to report on any activities to organize such an international year inserted in the appropriations bill for the National Science Foundation. See Public Law 99-383, Sect. 9, August 21, 1986. In a message to Congress, however, Reagan reported there were no such plans. See American Presidency Project, January 26, 1988.

10. Hearings of the Subcommittee on Environmental Pollution on "Ozone Depletion, the Greenhouse Effect, and Climate Change," Senate, Committee on Environment and Public Works, 99th Congress, 2nd Session, June 10–11, 1986.

11. Rotberg and Rabb, *Climate and History*.

12. Rich, *Losing Earth*, 110–12.

13. *Congressional Record*, 99th Congress, 2nd Session, September 22, 1986.

14. American Presidency Project, November 1, 1986.

15. Congressional Quarterly, *Congress and the Nation, 1985–1988*, 453.

16. Philip Shabecoff, "Congress: Now, the Environment According to N. Dakota," *New York Times*, January 9, 1987.

17. Wallace Broecker, "Greenhouse Surprises," in Abrahamson, *Challenge of Global Warming,* 197.

18. Gordon MacDonald, "The Scientific Basis for the Greenhouse Effect," in Abrahamson, *Challenge of Global Warming,* 142–43. For a more recent analysis of the role "externalities" should play in the analysis of climate policy, see Brown and Sovacool, *Climate Change and Global Energy Security,* 147–59.

19. *Congressional Record,* 100th Congress, 1st Session, February 5, 1987.

20. American Presidency Project, February 26, 1987.

21. *Congressional Record,* March 3, 1987.

22. Congressional Quarterly, *Congress and the Nation, 1985–1988,* 493.

23. James Fleming, interview with Gordon MacDonald, March 21, 1994, Niels Bohr Library and Archives. Benedick also wrote an authoritative summary of the road to the Montreal Protocol and its implementation. See Benedick, *Ozone Diplomacy.*

24. Minutes of Domestic Policy Council Meeting, May 20, 1987, *Foreign Relations of the United States, 1981–1988,* "Global Issues II," document 363.

25. Robert Taylor, "Advice on Ozone May Be: 'Wear Hats and Stand in the Shade,'" *Wall Street Journal,* May 29, 1987.

26. Benedick, *Ozone Diplomacy,* 58–60.

27. George Shultz to Edwin Meese, June 1, 1987, *Foreign Relations of the United States* 41, document 364, tab A.

28. *Congressional Record,* June 5, 1987.

29. Minutes of Domestic Policy Council Meeting, June 11, 1987, *Foreign Relations of the United States* 41, document 369.

30. Minutes of Domestic Policy Council Meeting, June 18, 1987, *Foreign Relations of the United States* 41, document 370.

31. The EPA, like several other federal agencies, based its risk assessments on standards recommended by National Research Council, *Risk Assessment in the Federal Government.*

32. Benedick, *Ozone Diplomacy,* 3–4.

33. Ibid., 7.

34. Bolin, *History of the Science and Politics of Climate Change,* 46.

35. Michael Lemonick, "The Heat Is On," *Time,* October 19, 1987.

36. Benedick, *Ozone Diplomacy,* 23.

37. Hearings of the Committee on Energy and Natural Resources on "The Greenhouse Effect and Global Climate Change," Senate, 100th Congress, 1st Session, November 9–10, 1987.

38. *Congressional Record,* January 29, 1987.

39. H.R. 1777, Foreign Relations Authorization Act, Fiscal Years 1988 and 1989.

11. The Hottest Summer

1. Richard Smith (Acting Secretary) to Richard Hallgren, "Intergovernmental Panel on Climate Change," January 27, 1989, Tyrus Cobb Files, RAC box 2, Global Climate Change (1), Reagan Library, Digitized Textual Material.

2. The political balance of the Senate changed slightly on March 11, 1987, when the governor of Nebraska appointed Republican David Karnes to fill the seat of the late Democrat Edward Zorinsky.

3. Germond and Witcover, *Whose Broad Stripes and Bright Stars?*, 228.

4. "Conventional-Wisdom Watch," *Newsweek,* November 21, 1988. The airline ran out of cash and defaulted on its debt in 1990.

5. *Congressional Record,* 100th Congress, 2nd Session, May 19,1988.

6. "Toronto Economic Summit Economic Declaration," June 21, 1988.

7. Hearing of the Committee on Energy and Natural Resources on "The Greenhouse Effect and Global Climate Change—Part 2," Senate, 100th Congress, 1st Session, June 23, 1988.

8. Hansen's testimony was based on his recent refereed article, coauthored with Goddard atmospheric scientist Sergej Lebedeff, "Global Surface Air Temperatures: Update Through 1987," *Geophysical Research Letters* 15, no. 4 (April 1988): 323–26.

9. Philip Shabecoff, "Sharp Cut in Burning of Fossil Fuels Urged to Battle Shift in Climate," *New York Times,* June 24, 1988.

10. Bowen, *Censoring Science,* 223–24.

11. Hansen to "Academic and Other Colleagues," July 22, 1988, Revelle Files, box 126, folder 7.

12. John Noble Wilford, "His Bold Statement Transforms the Debate on Greenhouse Effect," *New York Times,* August 23, 1988.

13. Senate Committee on Energy and Natural Resources, "The Month Ahead (August)," August 7, 2009; Richard Besel, "Accommodating Climate Change Science: James Hansen and the Rhetorical/Political Emergence of Global Warming," *Science in Context* 26, no. 1 (2013): 138.

14. Glenn Kessler, "Setting the Record Straight: The Real Story of a Pivotal Climate Change Hearing," *Washington Post,* March 30, 2015.

15. Yergin, *The Quest,* 456–57.

16. United Nations Environment Programme and World Meteorological Organization, *Changing Atmosphere,* 298, 300. See also Abrahamson, *Challenge of Global Warming,* 44–62. Bert Bolin, a prominent participant in the conference, later argued that the goal of a 20 percent cut in emissions "was an unrealistic ad hoc recommendation" that was agreed to with little consideration of the importance of energy in society, the inherent inertia of government institutions and private corporations, and the needs of developing countries. See Bolin, *A History,* 48–49.

17. United Nations Environment Programme and World Meteorological Organization, *Changing Atmosphere,* 293.

18. Ibid., 295.

19. Written text found in American Association for the Advancement of Science, Climate Program Records, box 23.

20. American Presidency Project, July 18, 1988.

21. American Presidency Project, August 16, 1988.

22. American Presidency Project, August 20, 1988.

12. Maggie Thatcher's Science

1. "Speech to the Royal Society (climate change)," September 27, 1988, Margaret Thatcher Foundation, online.

2. Thatcher, *Downing Street Years,* 640.

3. Formerly withheld document 11713, October 31, 1988, George H. W. Bush Presidential Library.

4. Roger Revelle and David Burns, "Energy Options and Climatic Effects," August 9, 1988, Revelle Papers, box 209, folder 8.

5. Kay Mills, "As the World Warms: A Talk With UC's Revelle," *Los Angeles Times,* October 16, 1988.

6. Bolin, *History of the Science and Politics of Climate Change,* 49–52; Memorandum from Juanita Duggan to Boyden Gray, November 21, 1988, Ralph C. Bledsoe Files, 1985–1988, DPC Global Climate Change, folder 8, Reagan Library.

7. The actions of the Council on Environmental Quality during the Reagan administration are described in documents Hill circulated during the early months of the Bush Administration. See Speth, "Reference Page for *They Knew.*"

8. Brinkley, *Reagan Diaries,* 2:1006.

13. Bush in the Middle

1. Rich, *Losing Earth,* 138.

2. Meacham, *Destiny and Power,* 144–45.

3. Department of State, "Remarks of the Honorable James A. Baker III," annotated, January 30, 1989, formerly withheld document [FWD] 12354, George H. W. Bush Presidential Library.

4. Whipple, *Gatekeepers,* 160–62; American Institute of Physics, "Oral Interview with Gordon MacDonald," March 21, 1994, online.

5. Reilly to the President, April 28, 1989, FWD 2236, Bush Library.

6. G. H. W. Bush to John Sununu, May 1, 1989, FWD 2235, Bush Library.

7. Dan Corbett and Jim Mietus to Ken Schwartz et al., April 24, 1989, FWD 40879, Bush Library.

8. Stephen Farrar to Porter, May 1, 1989, FWD 2064, Bush Library.

9. Claudine Schneider, "Changing Climates, Changing Habits: Energy Efficiency and the Global Warming Prevention Act," in Minger, *Greenhouse Glasnost,* 67–69.

10. Ibid., 71–72.

11. Global Warming Prevention Act of 1989, H.R. 1078, 101st Congress (1989–1990).

12. International Climate Change Prevention Act of 1989, S. 1611, 101st Congress (1989–1990).

13. Hearing of the Subcommittee on Science, Technology, and Space on "Climate Surprises," Senate, 101st Congress, 1st Session, May 8, 1989.

14. Philip Shabecoff, "Scientist Says Budget Office Altered His Testimony," *New York Times,* May 8, 1989; "The White House and the Greenhouse," *New York Times,* May 9, 1989.

15. Hansen, *Storms of My Grandchildren,* xv–xvi.

16. American Presidency Project, May 12, 1989.

17. Bobbie Kilberg to David Demarest, "Meeting with Environmentalists," no date, FWD 5084, Bush Library.

18. Reilly to Sununu, June 23, 1989, FWD 15188 and 2141, Bush Library.

19. Attachment to Kenneth Yale et al. to Donald Carr, June 26, 1989, "Reference Page for *They Knew*"; Larry Stammer, "'Greenhouse Effect' Plan May Be Revived," *Los Angeles Times,* February 17, 1989.

20. Boyden Gray to General Scowcroft (copy to John Sununu), June 22, 1989, FWD 8352, Bush Library.

21. Bromley to the President, July 7, 1989, FWD 2055, Bush Library.

22. David Bates to the President and Secretary to the Cabinet, July 14, 1989, "Reference Page for *They Knew.*"

23. American Presidency Project, July 25, 1989.

24. *Noordwijk Declaration on Climate Change,* Atmospheric Pollution and Climatic Change Ministerial Conference held at Noordwijk, the Netherlands, November 6–7, 1989, National Technical Reports Library, online.

25. Eric Roston, "Lessons from the Early Days of Climate Diplomacy," *Bloomberg Green,* September 27, 2021.

26. "Speech to the United Nations General Assembly," November 8, 1989, Margaret Thatcher Foundation, online.

27. Bromley to Sununu, October 13, 1989, FWD 2066, Bush Library.

28. Bromley to the President, November 15, 1989, FWD 12366, Bush Library.

29. Richard Schmalensee to Task Force on Economic Costs of the Domestic Policy Council Working Group on Global Climate Change, November 28, 1989, FWD 2071 and 2072, Bush Library.

30. The President to Bromley (from Camp David), September 9, 1989, FWD 11329, Bush Library.

31. Bromley to Sununu, October 13, 1989, FWD 2066, Bush Library; C. M. R. Platt, "The Role of Cloud Microphysics in High-Cloud Feedback Effects on Climate Change," *Nature* 341 (October 5, 1989): 428–29.

32. William Stevens, "Skeptics Are Challenging Dire 'Greenhouse' Views," *New York Times,* December 13, 1989.

33. Bromley to Sununu, December 13, 1989, FWD 15534, with attachments, Bush Library.

34. Carl Weinberg and Robert Williams, "Energy from the Sun," *Scientific American,* September 1990, 149.

35. Sununu Oral History, June 8, 2000, Miller Center.

36. Smith and Tirpak, *Potential Effects of Global Climate Change.*

37. Remnick, *Fragile Earth,* ix–xi, 2–62; McKibben, *End of Nature.*

38. Thomas Lambrix, "Global Climate Change: A Retrospective View from Business," in Minger, *Greenhouse Glasnost,* 49–63.

39. Rich, *Losing Earth,* 81–82.

40. Pratt, *Exxon,* 252–55, 461–62. For later developments at Exxon, see Coll, *Private Empire.*

14. The World Speaks

1. Watkins and Reilly to Bromley, January 18, 1990, formerly withheld document 11475, Bush Library.

2. Daniel Esty to Robert Grady, January 18, 1990, Speech Backup Chron File, 1989–93, box 50, Bush Library.

3. Mark Plant to Task Force on Economic Cost of the DPC Working Group on Global Change, February 1, 1990, FWD 2078, Bush Library.

4. Bromley to Chriss Winston, February 1, 1990, Speech Backup Chron File, 1989–93, box 50, Bush Library.

5. "Administration Split on Warming Issue, Aides to Bush Say," *New York Times,* February 3, 1990.

6. American Presidency Project, February 5, 1990.

7. Philip Shabecoff, "Bush Asks Cautious Response to Threat of Global Warming," *New York Times,* February 6, 1990.

8. Nixon to Sununu, February 7, 1990, FWD 15584, Bush Library.

9. Beran to Houghton, March 23, 1990, Revelle Papers, box 163, folder 7–8.

10. Leggett, *Carbon War,* 2–3.

11. U.S. review prepared by Robert Watson, April 8, 1990, FWD 11719, Bush Library.

12. American Presidency Project, April 17–18, 1990.

13. Craig Whitney, "Scientists Urge Rapid Action on Global Warming," *New York Times,* May 26, 1990.

14. Jastrow, Nierenberg, and Seitz, *Scientific Perspectives,* 240–41.

15. Deland to Sununu, Scowcroft, Darman, and Gray, June 26, 1990, and Watkins to Sununu, July 3, 1990, FWD 2111 and 12276, Bush Library.

16. American Presidency Project, July 11, 1990.

17. Houghton, Jenkins, and Ephraums, *Climate Change.*

18. Bolin, *History of the Science and Politics,* 65.

19. Intergovernmental Panel on Climate Change, *Climate Change.* Available in IPCC's web archives.

15. Filibusters and PR Firms

1. U.S. Energy Information Administration, *Monthly Energy Review,* table 3.3a.

2. Hakes, *Energy Crises,* 333–35.

3. Holiday to Sununu, August 23, 1990, formerly withheld document 12051, Bush Library.

4. Crandall, *Regulating the Automobile,* 158.

5. *Congressional Record,* 101st Congress, 2nd Session, September 13–14, 1990.

6. Fred McClure to Sununu, September 19, 1990, FWD 12147, Bush Library.

7. *Congressional Record,* 101st Congress, 2nd Session, September 25, 1990.

8. Washington and Washington, *Odyssey,* 102–3.

9. Thatcher, "Speech at the 2nd World Climate Conference," November 6, 1990, Margaret Thatcher Foundation, online.

10. American Presidency Project, November 13, 1990.

11. Don Shannon, "Revelle Cited in White House Ceremony," *Los Angeles Times,* November 14, 1990.

12. American Presidency Project, November 14, 1990.

13. American Presidency Project, November 15, 1990.

14. Congressional Quarterly, *Congress and the Nation, 1989–1992,* 483.

15. Speth, "Reference Page for *They Knew.*"

16. "1990 Patrick Michaels Curriculum Vitae," climatefiles.com; *Washington Post,* June 15, 1986, and January 8, 1989.

17. Washington and Washington, *Odyssey,* 120.

18. Bromley to Sununu (with attachment), April 8, 1991, "Reference Page for *They Knew.*"

19. "Panel Says Nation Has Ability to Adapt to Global Warming," *New York Times,* September 7, 1991.

20. William Stevens, "Quick Steps Urged on Warming Threat," *New York Times,* April 11, 1991.

21. William Nordhaus, "How Much Should We Invest in Preserving Our Current Climate?" in Giersch, *Economic Progress and Environmental Concerns,* 255–99.

22. Johannes Heister and Frederick Schneider, "Ecological Concerns in a Market Economy: On Ethics, Accounting, and Sustainability," in Giersch, *Economic Progress and Environmental Concerns,* 25–48.

23. Bolin, *History of the Science and Politics of Climate Change,* 72.

24. Bromley to Sununu (with attachment), April 18, 1991, "Reference Page for *They Knew.*"

25. Nancy Maynard to Bromley, June 3, 1991, and Maynard to Sununu, June 7, 1991, "Reference Page for *They Knew.*"

26. David Anderson, Matt Kasper, and David Pomerantz, *Utilities Knew: Documenting Electric Utilities' Early Knowledge and Ongoing Deception on Climate Change from 1968–2017* (Energy and Policy Institute, 2017), 50–52, online. See also Rosenwald, *Talk Radio's America,* 1–2.

27. Matthew Wald, "Pro-Coal Campaign Disputes Warming Idea," *New York Times,* July 8, 1991.

28. Moore to Sununu, June 26, 1991, FWD 11563, Bush Library.

29. Gelbspan, *Heat Is On,* 33–46.

16. Road to Rio

1. David Smollar and Amy Wallace, "Revelle's Colleagues Recall 'Statesman of Science,'" *Los Angeles Times,* July 16, 1991; Walter Sullivan, "Roger Revelle, 82, Early Theorist in Global Warming and Geologist," *New York Times,* July 17, 1991.

2. William Stevens, "At Meeting on Global Warming, U.S. Stands Alone," *New York Times,* September 10, 1991.

3. David Rosenbaum, "A Man Who Was Raised to Be the President," *New York Times,* July 10, 1992.

4. Gore, *Earth in the Balance.*

5. National Research Council, *Automotive Fuel Economy,* 5–6. The study was released in April but its findings were available much earlier in draft form.

6. Doron Levin, "Detroit's Assault on Mileage Bill," *New York Times,* May 11, 1991.

7. *Congressional Record,* 102nd Congress, 2nd Session, February 6, 1992.

8. Washington and Washington, *Odyssey,* 93–101.

9. Ibid., 105–77; "Workshop on Global Climate Change (Sloan School 1992)—John Deutch, John Sununu, guests," Infinite MIT, infinite.mit.edu.

10. William Stevens, "Washington May Change Its Position on Climate," *New York Times,* February 18, 1992.

11. Skinner to Manuel Lujan et al., February 12, 1992, FWD 14112, Bush Library.

12. "Clayton Yeutter Oral History," *Presidential Oral Histories,* January 19, 2001, Miller Center, online.

13. Yeutter to the President (with annotations), March 12, 1992, FWD 1467, Bush Library.

14. American Presidency Project, March 24, 1992.

15. Yeutter to Skinner, April 20, 1992, and Yeutter to the President, May 1, 1992, FWD 2620 and 2621, Bush Library.

16. Leggett, *Carbon War,* 82; Oreskes, *Merchants of Doubt;* S. Fred Singer, "Chilling Out," letter to the editor, *Washington Post,* October 1, 1991.

17. GB to Yeutter, May 3, 1992, FWD 2623, Bush Library.

18. Yeutter to the President, May 8, 1992, FWD 1644, Bush Library.

19. Michael Weisskopf, "Bush Was Aloof in Warming Debate," *Washington Post,* October 31, 1992.

20. Leggett, *Climate War,* 94–95.

21. Roston, "Lessons from the Early Days of Climate Diplomacy."

22. Gelbspan, *Heat Is On,* 54; Sally Jacobsen, "OPEC Fears Attack on Oil Industry at Earth Summit," Associated Press, April 23, 1992; Richard Lindzen, "Global Warming: The Origin and Nature of the Alleged Scientific Consensus," *Regulation: The Cato Review of Business & Government,* Spring 1992: 87–98.

23. Bromley to Yeutter, April 23, 1992, FWD 14102, Bush Library.

24. Yeutter to Skinner, May 20, 1992, FWD 6051, Bush Library.

25. Yeutter to the President, June 9, 1992, FWD 5873, Bush Library; Leggett, *Climate War,* 95–99.

26. Darwell, *Age of Global Warming,* 142, 150; Vaclav Havel, "Rio and the New Millennium," *New York Times,* June 3, 1992.

27. Leggett, *Climate War,* 96–97.

28. *United Nations Framework Convention on Climate Change,* 1992, online.

29. American Presidency Project, June 13, 1992.

30. Boyden Gray to the President, June 11, 1992, FWD 1693, Bush Library.

31. Eugene Robinson and Michael Weisskopf, "Bush Pushes Tough Stand on Warming," *Washington Post,* June 9, 1992.

32. William Nitze, "A Failure of Presidential Leadership," in Mintzer and Leonard, *Negotiating Climate Change,* 187–200.

17. In the Gutter

1. Rosenbaum, "A Man Who Was Raised to Be the President."

2. Adam Clymer, "In Congress, Gore Has a Reputation for Taking the Long View," *New York Times,* July 12, 1992.

3. American Presidency Project, July 13, 1992.

4. "Senator Al Gore 1992 Acceptance Speech," July 16, 1992, *American History TV,* C-SPAN, cspan.org.

5. American Presidency Project, August 17, 1992.

6. Kevin Sack, "Quayle Draws Distinctions on Environment," *New York Times,* August 29, 1992.

7. "Al Gore and Mitch McConnell, Testimony Before the Senate Committee on Foreign Relations, September 18, 1992," in Howe, *Making Climate Change History.*

8. Clifford Krauss, "House Approves Energy Legislation," *New York Times,* October 6, 1992; *Congressional Record,* 102nd Congress, 2nd Session, October 5, 1992.

9. Congressional Quarterly, *Congress and the Nation, 1989–1992,* 297.

10. *Congressional Record,* 102nd Congress, 2nd Session, October 8, 1992.

11. Congressional Quarterly, *Congress and the Nation, 1989–1992,* 505–12.

12. Stokes, *Short Circuiting Policy,* 88–93.

13. "Gore, Quayle, Stockdale: The 1992 Vice Presidential Debate," *PBS NewsHour,* October 13, 1992. Available on YouTube.

14. March 2, 1990, Revelle Papers, box 191, file 5.

15. Matt Schudel, "S. Fred Singer, Scientist and Climate-Change Skeptic, Dies at 95," *Washington Post,* April 11, 2020; John Schwartz, "S. Fred Singer, a Leading Climate Change Contrarian, Dies at 95," *New York Times,* April 11, 2020.

16. Singer, *Changing Global Environment,* 5, 25.

17. Singer, *Global Climate Change,* 3.

18. Michael Isikoff, "Church Spends Millions on Its Image," *Washington Post,* September 17, 1984; "Washington Institute for Values in Public Policy," Center for Media and Democracy, online; Science and Environmental Policy Project, online.

19. Singer sworn court testimony, *In the Matter of: S. Fred Singer v. Justin Lancaster,* September 24, 1993, 1:224–27. These and other legal documents available at *DeSmog,* online.

20. Chauncey Starr, "Atmospheric CO_2 Residence Time and the Carbon Cycle," *Energy* 18 (1993): 1297–1310.

21. Revelle Papers, box 191, file 5.

22. S. Fred Singer, "What to Do About Greenhouse Warming," *Environmental Science and Technology* 24, no. 8 (August 1, 1990): 1138–39.

23. Chris Elfing to Revelle, December 13, 1992, Revelle Papers, box 209, folder 24. The NRC sent Revelle a transcript of his taped oral remarks, on which he made some minor edits for clarity. I have treated the amended transcript as the historic record of the event.

24. Affidavit of Christina Beran, August 2, 1993, *DeSmog,* online.

25. Affidavit of Justin Lancaster, May 20, 1993, *DeSmog,* online.

26. Gregg Easterbrook, "Everything You Know About the Environment Is Wrong," *New Republic,* April 30, 1990, 14–27.

27. Gregg Easterbrook, "Green Cassandras," *New Republic,* July 6, 1992, 23–25. The magazine hit newsstands more than a week earlier than the July 6 cover date.

28. Affidavit of Justin Lancaster, May 20, 1993.

29. George Will, "Al Gore's Green Guilt," *Washington Post,* September 3, 1992.

30. Carolyn Revelle Hufbauer, "Global Warming: What My Father Really Said," *Washington Post,* September 13, 1992.

31. *CBS Evening News with Dan Rather,* October 21, 1992.

32. "Gore's Warnings Bode No Good Will for Coal," *Bluefield Daily Telegraph,* September 6, 1992.

33. U.S. Energy Information Administration, *Annual Energy Outlook 1993,* table A15, and *Monthly Energy Review,* table 11.1, online.

34. See Cline, *Economics of Global Warming.*

35. The friendly jousting between the two academics can be found in William Nordhaus, "An Optimal Transition Path for Controlling Greenhouse Gases," *Science* 258 (November 20, 1992): 1315–19; and Stephen Schneider et al., "Pondering Greenhouse Policy," *Science* 259, no. 5100 (1993): 1380–81. However, some of Schneider's argument can be found only in his unedited submission to *Science,* "Comment on Nordhaus' DICE Model," December 22, 1992. This longer version can be found in the Kathleen McGinty papers, file Carbon Dioxide Tax, William J. Clinton Presidential Library, online.

36. Mildenberger, *Carbon Captured,* chapter 4.

18. After

1. Keeling, "Rewards and Penalties of Monitoring the Earth," 74–76.

2. The record of Keeling's correspondence and activities in 1997 can be found in various documents in Keeling Papers, box 7, folder 17, and box 82, file 4.

3. Kenneth Chang, "Charles D. Keeling, 77, Who Raised Global Warming Issue, Dies," *New York Times,* June 23, 2005.

4. NOAA, "American Chemical Society Honors Measurement Set at NOAA Observatory," April 23, 2015.

5. Broecker, *Fixing Climate,* especially xiv–xvi, 231–33.

6. Ari Natter and Brian Kahn, "Biden Bets Billions on Tech That Sucks Carbon Out of the Air," *Bloomberg Green,* August 11, 2023.

7. Hansen, *Storms of My Grandchildren.*

8. Speth, *They Knew,* x–xi, 197.

9. Zeke Hausfather et al., "Evaluating the Performance of Past Climate Model Projections," *Geophysical Research Letters* 47, no. 1 (January 16, 2020).

10. *In the Matter of: S. Fred Singer v. Justin Lancaster.*

11. Lancaster's perspective on the legal battle is available in Eli Rabett, "A Note About Roger Revelle, Justin Lancaster and Fred Singer," *Rabett Run,* September 13, 2014, online. Singer's version, "The Revelle-Gore Story: Attempted Political Suppression of Science," appeared in Gough, *Politicizing Science.*

12. Singer, *Hot Talk,* ix–x.

13. S. Fred Singer, "The Sea Is Rising, but Not Because of Climate Change," *Wall Street Journal,* May 15, 2018.

14. Michael Mann and Andrea Dutton, "Water's Rising Because It's Getting Warmer," *Wall Street Journal,* May 22, 2018. The paper did not include a line graph submitted with the letter, showing the clear rise in global mean sea level since the industrial revolution. The graph is available at Michael Mann, "Our Response to the Latest Climate Change–Denying *Wall Street Journal* Op-Ed," Michaelmann.net, May 22, 2018.

15. Patrick Michaels, "The Happy Warrior Saves His Best Climate Writing for Last," *Real Clear Politics,* April 6, 2021.

16. Schudel, "S. Fred Singer, Scientist and Climate-Change Skeptic, Dies at 95."

17. Schwartz, "S. Fred Singer, a Leading Climate Change Contrarian, Dies at 95"; Dessler, *Introduction to Modern Climate Change.*

18. William Nordhaus, "Discounting and Public Policies that Affect the Distant Future," in Portney and Weyant, *Discounting and Intergenerational Equity,* 145–62.

19. Stern, *Economics of Climate Change,* especially xiii–xviii, 394.

20. William Nordhaus, "A Review of the *Stern Review on the Economics of Climate Change,*" *Journal of Economic Literature* 55 (September 2007): 686–702.

21. William Nordhaus, "Why the Global Warming Skeptics Are Wrong," *New York Review of Books,* March 22, 2012; Claude Allegre et al., "No Need to Panic About Global Warming," *Wall Street Journal,* January 27, 2012.

22. Nordhaus's Nobel Prize lecture (with video and biography) can be found at nobelprize .org.

23. Nordhaus, *Spirit of Green,* 75, 130–31, 312–13.

24. Ethan Howland, "White House Bolsters Review Process for Power Sector, Other Rules with Expanded Cost-Benefit Analysis," *Utility Dive,* April 12, 2023, online.

25. Pew Charitable Trusts, "History of Fuel Economy: One Decade of Innovation, Two Decades of Inaction," fact sheet, May 1, 2007, online.

26. Matthew Wald, "Energy Dept. to Raise Efficiency of Air-Conditioners and Heat Pumps," *New York Times,* January 19, 2001; Dan Reicher, Andrew Delaski, and John Mandyck, "How Bill Richardson Helped Rescue the Planet," *The Hill,* September 14, 2023.

27. A critique of the trading system can be found in Victor, *Collapse of the Kyoto Protocol.*

28. Congressional Quarterly, *Congress and the Nation, 1997–2001,* 354–56.

29. Gregg Easterbrook, "Finally Feeling the Heat," *New York Times,* May 24, 2006.

30. United Nations Climate Change, *Paris Agreement,* online.

31. Obama, *Promised Land,* 506–16.

32. Eric Roston, "Planet's Breakneck Warming Likely to Pass 1.5°C," *Bloomberg Green,* April 4, 2022.

33. Siobhan Wagner, "Notes from the Ground," *Bloomberg Green,* December 5, 2023.

34. Bill McKibben, "What Can We Do with a Sentence?," McKibben subscriber newsletter, December 13, 2023; "The Long Goodbye?" *Economist,* December 16, 2023.

35. Congressional Quarterly, *Congress and the Nation, 2005–2008,* 484–90.

36. U.S. Energy Information Administration, *Monthly Energy Review,* table 11.5, online.

37. Ibid., table 11.2.

38. Alan Blinder, "Jimmy Carter Makes a Stand for Solar, Decades After Cardigan Sweater," *New York Times,* February 11, 2017. I attended the event and provided remarks on the history of solar energy.

39. "Explaining the Plummeting Cost of Solar Power," *MIT News,* November 20, 2018.

40. Akshat Rathi, "What 2023 Holds for the Fight Against Climate Change," *Bloomberg Green,* January 8, 2023.

41. David Gelles et al., "The Clean Energy Future Is Arriving Faster Than You Think," *New York Times,* August 16, 2023.

Works Cited

Books and Reports

Abrahamson, Dean, ed. *The Challenge of Global Warming*. Washington, DC: Island Press, 1989.

Alter, Jonathan. *His Very Best: Jimmy Carter, A Life*. New York: Simon & Schuster, 2020.

Ambrose, Stephen. *Eisenhower: The President*. New York: Simon & Schuster, 1984.

Benedick, Richard. *Ozone Diplomacy: New Directions in Saving the Planet*. Enlarged ed. Cambridge, MA: Harvard University Press, 1998.

Bird, Kai, and Martin Sherwin. *American Prometheus: The Triumph and Tragedy of Robert Oppenheimer*. New York: Vintage, 2006.

Bolin, Bert. *A History of the Science and Politics of Climate Change: The Role of the Intergovernmental Panel on Climate Change*. New York: Cambridge University Press, 2007.

Bowen, Mark. *Censoring Science: Inside the Political Attack on Dr. James Hansen and the Truth of Global Warming*. New York: Dutton, 2008.

Brinkley, Douglas, ed. *The Reagan Diaries*. Vol. 2. New York: Harper, 2009.

———. *Silent Spring Revolution: John F. Kennedy, Rachel Carson, Lyndon Johnson, Richard Nixon, and the Great Environmental Awakening*. New York: HarperCollins, 2022.

Broecker, Wallace, and Robert Kunzig. *Fixing Climate: What Past Climate Changes Reveal about the Current Threat—and How to Counter It*. New York: Hill and Wang, 2008.

Brown, Marilyn, and Benjamin Sovacool. *Climate Change and Global Energy Security: Technology and Policy Options*. Cambridge, MA: MIT Press, 2010.

Cannon, Lou. *President Reagan: The Role of a Lifetime*. New York: Public Affairs, 2000.

Carter, Jimmy. *White House Diary*. New York: Farrar, Straus and Giroux, 2010.

Christianson, Gale. *Greenhouse: The 200-Year Story of Global Warming.* New York: Penguin, 1999.

Cline, William. *The Economics of Global Warming.* Washington, DC: Institute for International Economics, 1992.

Coll, Steve. *Private Empire: ExxonMobil and American Power.* New York: Penguin, 2012.

Congressional Quarterly. *Congress and the Nation, 1945–1964.* Washington, DC: Congressional Quarterly, 1965.

———. *Congress and the Nation, 1965–1968.* Washington, DC: Congressional Quarterly, 1969.

———. *Congress and the Nation, 1969–1972.* Washington, DC.: Congressional Quarterly, 1973.

———. *Congress and the Nation, 1985–1988.* Washington, DC: Congressional Quarterly, 1990.

———. *Congress and the Nation, 1989–1992.* Washington, DC: Congressional Quarterly, 1993.

———. *Congress and the Nation, 1997–2001.* Washington, DC: Congressional Quarterly, 2002.

———. *Congress and the Nation, 2005–2008.* Washington, DC: Congressional Quarterly, 2010.

Conservation Foundation. *Implications of Rising Carbon Dioxide Content of the Atmosphere.* New York: Conservation Foundation, 1963.

Council on Environmental Quality. *Environmental Quality: The First Annual Report of the Council on Environmental Quality Together with the President's Message to Congress.* Washington, DC: U.S. Government Printing Office, 1970.

———. *Global Energy Futures and the Carbon Dioxide Problem.* Washington, DC: U.S. Government Printing Office, 1981.

———. *Solar Energy: Progress and Promise.* Washington, DC, 1978.

Council on Environmental Quality and Department of State. *The Global 2000 Report to the President: Entering the Twenty-First Century.* Washington, DC: U.S. Government Printing Office, 1980. Reprinted by Penguin, 1982.

Crandall, Robert, Howard Gruenspecht, Theodore Keeler, and Lester Lave. *Regulating the Automobile.* Washington, DC: Brookings Institution, 1986.

Dallek, Robert. *Flawed Giant: Lyndon Johnson and his Times, 1961–1973.* New York: Oxford University Press, 1998.

Darwell, Rupert. *The Age of Global Warming: A History.* New York: Quartet, 2013.

Day, Deborah. *Roger Randall Dougan Revelle Biography.* Scripps Institution of Oceanography Archives, Special Collections, University of California San Diego, no date. Online.

Dessler, Andrew. *Introduction to Modern Climate Change.* 3rd ed. New York: Cambridge University Press, 2022.

Divine, Robert, ed. *The Johnson Years.* Vol. 2: *Vietnam, the Environment, and Science.* Lawrence: University of Kansas Press, 1987.

Dorfman, Robert, and Peter Rogers, eds. *Science with a Human Face: In Honor of Roger Randall Revelle.* Cambridge, MA: Harvard University Press, 1997.

Duncan, Francis. *Rickover: The Struggle for Excellence.* Annapolis: Naval Institute Press, 2001.

Fleming, James Rodger. *The Callendar Effect: The Life and Work of Guy Stewart Callendar, 1898–1964.* Boston: American Meteorological Society, 2007.

Flippen, J. Brooks. *Nixon and the Environment.* Albuquerque: University of New Mexico Press, 2000.

Flohn, Hermann. *Climate and Weather.* New York: McGraw-Hill, 1969.

Gelbspan, Ross. *The Heat Is On: The Climate Crisis, the Cover-Up, the Prescription.* New York: Basic Books, 1998.

Germond, Jack, and Jules Witcover. *Whose Broad Stripes and Bright Stars? The Trivial Pursuit of the Presidency 1988.* New York: Warner Books, 1989.

Giersch, Herbert, ed. *Economic Progress and Environmental Concerns.* Berlin: Springer-Verlag, 1992.

Goldman, Eric. *The Tragedy of Lyndon Johnson.* New York: Alfred Knopf, 1968.

Gore, Al. *Earth in the Balance: Ecology and the Human Spirit.* New York: Houghton Mifflin, 1992.

———. *An Inconvenient Truth: The Planetary Emergency of Global Warming and What We Can Do About It.* New York: Rodale, 2006.

Gough, Michael, ed. *Politicizing Science: The Alchemy of Policymaking.* Stanford, CA: Hoover Institution Press, 2003.

Hakes, Jay. *Energy Crises: Nixon, Ford, Carter, and Hard Choices in the 1970s.* Norman: University of Oklahoma Press, 2021.

Halvorson, Charles. *Valuing Clean Air: The EPA and the Economics of Environmental Protection.* New York: Oxford University Press, 2021.

Hamblin, Jacob Darwin. *Arming Mother Nature: The Birth of Catastrophic Environmentalism.* New York: Oxford University Press, 2013.

Hansen, James. *Storms of My Grandchildren: The Truth about the Coming Climate Catastrophe and Our Last Chance to Save Humanity.* New York: Bloomsbury, 2009.

Helvarg, David. *Blue Frontier: Saving America's Living Seas.* New York: Henry Holt, 2001.

Hess, Stephen. *The Professor and the President: Daniel Patrick Moynihan in the Nixon White House.* Washington, DC: Brookings Institution Press, 2015.

Hewlett, Richard, and Jack Holl. *Atoms for Peace and War, 1953–1961.* Berkeley: University of California Press, 1989.

Houghton, John, G. J. Jenkins, and J. J. Ephraums, eds. *Climate Change: The IPCC Scientific Assessment.* New York: Cambridge University Press, 1990. Online.

Howe, Joshua. *Behind the Curve: Science and the Politics of Global Warming.* Seattle: University of Washington Press, 2014.

———, ed. *Making Climate Change History: Documents from Global Warming's Past.* Seattle: University of Washington Press, 2017.

Intergovernmental Panel on Climate Change. *Climate Change: The IPCC Response Strategies.* World Meteorological Organization/United Nations Environment Program, 1990. Online.

Isaacson, Walter. *Einstein: His Life and Universe.* New York: Simon & Schuster, 2007.

———. *Leonardo Da Vinci.* New York: Simon & Schuster, 2017.

JASON. *The Long Term Impact of Atmospheric Carbon Dioxide on Climate.* Technical Report JSR-78-07. Arlington, VA: SRI International for U.S. Department of Energy, 1979.

Jastrow, Robert, William Nierenberg, and Frederick Seitz, eds. *Scientific Perspectives on the Greenhouse Problem.* Ottawa, IL: Marshall Press, 1990.

Kistiakowsky, George. *A Scientist in the White House: The Private Diary of President Eisenhower's Special Assistant for Science and Technology.* Cambridge, MA: Harvard University Press, 1976.

Leggett, Jeremy. *The Carbon War: Global Warming and the End of the Oil Era.* New York Routledge, 2001.

Lifset, Robert, ed. *American Energy Policy in the 1970s.* Norman: University of Oklahoma Press, 2014.

Lind, Robert, Kenneth Arrow, Gordon Corey, Partha Dasgupta, Amartya Sen, Thomas Stauffer, Joseph Stiglitz, J. A. Stockfisch, and Robert Wilson. *Discounting for Time and Risk in Energy Policy.* Washington, DC: Resources for the Future, 1982.

Marini-Bettolo, G. B., ed. *Chemical Events in the Atmosphere and Their Impact on the Environment.* Rome: Pontificia Academia Scientiarum, 1985.

Maycock, Paul, and Edward Stirewalt. *Photovoltaics: Sunlight to Electricity in One Step.* Andover, MA: Brick House, 1981.

McCullough, David. *The Wright Brothers.* New York: Simon & Schuster, 2015.

McKibben, Bill. *The End of Nature.* New York: Random House, 1989.

Meacham, Jon. *Destiny and Power: The American Odyssey of George Herbert Walker Bush.* New York: Random House, 2015.

Mildenberger, Matto. *Carbon Captured: How Business and Labor Control Climate Policies.* Cambridge, MA: MIT Press, 2020.

Minger, Terrell, ed. *Greenhouse Glasnost: The Crisis of Global Warming.* New York: Ecco Press, 1990.

Mintzer, Irving, and J. A. Leonard, eds. *Negotiating Climate Change: The Inside Story of the Rio Convention.* New York: Cambridge University Press, 1994.

National Academy of Sciences. *Biographical Memoirs.* Vol. 84. Washington, DC: National Academies Press, 2004.

National Research Council. *Automotive Fuel Economy: How Far Should We Go?* Washington, DC: National Academies Press, 1992.

———. *Carbon Dioxide and Climate: A Scientific Assessment* [Charney report]. Washington, DC: National Academies Press, 1979.

———. *Changing Climate: Report of the Carbon Dioxide Assessment Committee* [Nierenberg report]. Washington, DC: National Academies Press, 1983.

———. *Energy and Climate: Studies in Geophysics.* Washington, DC: National Academies Press, 1977.

———. *Oceanography, 1960 to 1970.* Washington, DC: National Academies Press, 1959.

———. *Policy Implications of Greenhouse Warming: Mitigation, Adaptation, and the Science Base* [Evans report]. Washington, DC: National Academies Press, 1992.

———. *Risk Assessment in the Federal Government: Managing the Process.* Washington, DC: National Academies Press, 1983.

———. *Understanding Climatic Change: A Program for Action.* Washington, DC: National Academy of Sciences, 1975.

———. *Weather & Climate Modification: Problems and Progress.* Vols. 1–2. Washington, DC: National Academy of Sciences, 1966.

———. *Weather & Climate Modification: Problems and Progress.* Washington, DC: National Academies Press, 1973. (Update of 1966 report.)

Nemet, Gregory. *How Solar Energy Became Cheap: A Model for Low-Carbon Innovation.* New York: Routledge, 2019.

Nordhaus, William. *The Spirit of Green: The Economics of Collisions and Contagions in a Crowded World.* Princeton, NJ: Princeton University Press, 2021.

Obama, Barack. *A Promised Land.* New York: Crown, 2020.

Oreskes, Naomi. *Science on a Mission: How Military Funding Shaped What We Do and Don't Know about the Ocean.* Chicago: University of Chicago Press, 2021.

Oreskes, Naomi, and Erick Conway. *Merchants of Doubt: How a Handful of Scientists Obscured the Truth on Issues from Tobacco Smoke to Global Warming.* New York: Bloomsbury, 2010.

Pfeffer, Richard, ed. *Dynamics of Climate: The Proceedings of a Conference on Application of Numerical Integration Techniques to the Problem of General Circulation held October 26–28, 1955.* New York: Pergamon Press, 1960.

Portney, Paul, and John Weyant, eds. *Discounting and Intergenerational Equity.* Washington, DC: Resources for the Future, 1999.

Pratt, Joe, with William Hale. *Exxon: Transforming Energy, 1973–2005.* Austin: Briscoe Center for American History, University of Texas, 2013.

President's Science Advisory Committee, Environmental Pollution Panel. *Restoring the Quality of Our Environment.* Washington, DC: The White House, 1965. Later reprinted from the Collections of the University of California Libraries.

Remnick, David, and Henry Finder, eds. *The Fragile Earth: Writings from The New Yorker on Climate Change.* New York: HarperCollins, 2020.

Revelle, Roger, Ashok Khosla, and Maris Vinovskis, eds. *The Survival Equation: Man, Resources, and His Environment.* Boston: Houghton Mifflin, 1971.

Rhodes, Richard. *Energy: A Human History.* New York: Simon & Schuster, 2018.

Rich, Nathaniel. *Losing Earth: A Recent History.* New York: Farrar, Straus and Giroux, 2019.

Robinson, E. and R. C. Robbins. *Sources, Abundance, and Fate of Gaseous Atmospheric Pollutants.* Menlo Park, CA: Stanford Research Institute, 1969.

Rosenblith, Walter, ed. *Jerry Wiesner: Scientist, Statesman, Humanist.* Cambridge, MA: MIT Press, 2003.

Rosenwald, Brian. *Talk Radio's America: How an Industry Took Over a Political Party That Took Over the United States.* Cambridge, MA: Harvard University Press, 2019.

Roston, Eric. *The Carbon Age: How Life's Core Element Has Become Civilization's Greatest Threat.* New York: Walker, 2008.

Rotberg, Robert, and Theodore Rabb, eds. *Climate and History: Studies in Interdisciplinary History.* Princeton, NJ: Princeton University Press, 1981.

Schneider, Stephen. *Science as a Contact Sport: Inside the Battle to Save Earth's Climate.* Washington, DC: National Geographic Society, 2009.

Seaborg, Glenn. *The Atomic Energy Commission under Nixon.* New York: St. Martin's, 1993.

———. *Science, Man and Change: A Collection of Speeches.* Ann Arbor: University of Michigan Library, 1968.

Seidel, Stephen, and Dale Keyes. *Can We Delay a Greenhouse Warming?: The Effectiveness and Feasibility of Options to Slow a Build-Up of Carbon Dioxide in the Atmosphere.* Washington, DC: U.S. Environmental Protection Agency, Office of Policy Analysis, 1983.

Sharp, Sarah, interviewer. *Roger Randall Dougan Revelle: The International Scientist.* Berkeley: Regents of the University of California, 1998. Interview conducted 1984.

———. *Roger Randall Dougan Revelle: Preparation for a Scientific Career.* Berkeley: Regents of the University of California, 1988. Interview conducted 1984.

Singer, S. Fred, ed. *The Changing Global Environment*. Boston: Reidel, 1975.

———. *Global Climate Change: Human and Natural Influences*. New York: Paragon House, 1989.

———. *Hot Talk, Cold Science: Global Warming's Unfinished Debate*. Oakland, CA: Independent Institute, 1997.

Smith, Joel B., and Dennis Tirpak, eds. *The Potential Effects of Global Climate Change on the United State: Report to Congress*. EPA-230-05-89-050. Washington, DC: Environmental Protection Agency, Office of Policy, Planning, and Evaluation, 1989.

Speth, James Gustave. *Red Sky at Morning: America and the Crisis of the Global Environment*. New Haven, CT: Yale University Press, 2004.

———. *They Knew: The U.S. Federal Government's Fifty-Year Role in Causing the Climate Crisis*. Cambridge, MA: MIT Press, 2021.

Steele, Philip. *Galileo: The Genius Who Charted the Universe*. Washington, DC: National Geographic Society, 2008.

Stern, Nicholas. *The Economics of Climate Change: The Stern Review*. New York: Cambridge University Press, 2007.

Stockman, David. *The Triumph of Politics: Why the Reagan Revolution Failed*. New York: Harper & Row, 1986.

Stokes, Leah Cardamore. *Short Circuiting Policy: Interest Groups and the Battle Over Clean Energy and Climate Policy in the American States*. New York: Oxford University Press, 2020.

Thatcher, Margaret. *The Downing Street Years*. New York: HarperCollins, 1993.

Tolo, Kenneth, ed. *Beyond Today's Energy Crisis: Future of the American Environment*. Symposium Proceedings. Austin, TX: Lyndon B. Johnson School of Public Affairs, 1975.

United Nations Environment Programme and World Meteorological Organization. *The Changing Atmosphere: Implications for Global Security—Conference Proceedings*. WMO/OMM no. 710. Geneva: World Meteorological Organization, 1988.

U.S. Department of Energy, Carbon Dioxide Effects Research and Assessment Program. *Environmental and Societal Consequences of a Possible CO_2-Induced Climate Change: A Research Agenda*. DOE/EV/10019-01. Springfield, VA: National Technical Information Service, 1980.

———. *Workshop on Environmental and Societal Consequences of a Possible CO_2-Induced Climate Change*. Springfield, VA: National Technical Information Service, 1980.

U.S. Energy Information Administration. *Annual Energy Outlook 1993, with Projections to 2010*. DOE/EIA-0383(93). Washington, DC: Energy Information Administration, Office of Integrated Analysis and Forecasting, 1992.

U.S. Senate, Committee on Governmental Affairs. *Carbon Dioxide Accumulation in the Atmosphere, Synthetic Fuels and Energy Policy: A Symposium.* Washington, DC: U.S. Government Printing Office, 1979. Reprinted from the collections of the University of California Libraries.

U.S. Senate, Committee on Interior and Insular Affairs. *Congress and the Nation's Environment: Environmental Affairs of the 91st Congress.* Washington, DC: U.S. Government Printing Office, 1971.

Victor, David. *The Collapse of the Kyoto Protocol, and the Struggle to Slow Global Warming.* Princeton, NJ: Princeton University Press, 2001.

Walker, Samuel. *Three Mile Island: A Nuclear Crisis in Historical Perspective.* Berkeley: University of California Press, 2004.

Washington, Warren, and Mary Washington. *Odyssey in Climate Modeling, Global Warming, and Advising Five Presidents.* 3rd ed. Self-published, 2012.

Weart, Spencer. *The Discovery of Global Warming.* Cambridge, MA: Harvard University Press, 2003.

Whipple, Chris. *The Gatekeepers: How the White House Chiefs of Staff Define Every Presidency.* New York: Crown, 2017.

Woodwell, George, Gordon MacDonald, Roger Revelle, and Charles (Dave) Keeling. *The Carbon Dioxide Problem: Implications for Policy in the Management of Energy and Other Resources, A Report to the Council on Environmental Quality (July 1979).* Reprint. Woods Hole, MA: Woods Hole Research Center, 2008.

World Climate Programme. *Report of the International Conference Assessment of the Role of Carbon Dioxide and of Other Greenhouse Gases in Climate Variations and Associated Impacts* [Villach report]. WMO No. 661. Geneva: World Meteorological Organization, 1986.

Yergin, Daniel. *The Quest: Energy, Security, and the Remaking of the Modern World.* New York: Penguin, 2010.

Document Collections and Libraries

American Association for the Advancement of Science, Climate Program Records, Washington, DC. aaas.org/archives

American Presidency Project, University of California, Santa Barbara. presidency.ucsb.edu

Charles David Keeling Papers, Special Collections and Archives, University of California, San Diego, La Jolla.

Clarence E. Larson Science and Technology Oral History Collection, Special Collections Research Center, no. C0079, George Mason University Libraries, Fairfax, VA. ead.lib.virginia.edu

Foreign Relations of the United States series, "Historical Documents," U.S. Department of State, Office of the Historian. history.state.gov/historicaldocuments

George H. W. Bush Presidential Library, College Station, TX.

George Mason University Libraries, Fairfax, VA. library.gmu.edu

Industry Documents Library, University of California, San Francisco Library. industrydocuments.ucsf.edu

Intergovernmental Panel on Climate Change, "Documentation." ipcc.ch/documentation

Jimmy Carter Presidential Library, Atlanta.

John C. Whitaker files, White House Central Files: Staff Member and Office Files, Richard Nixon Presidential Library, Yorba Linda, CA.

John F. Kennedy Presidential Library, Boston. jfklibrary.org

Lyndon Baines Johnson Presidential Library, Austin, TX. lbjlibrary.org

Miller Center, University of Virginia, Charlottesville. millercenter.org

National Technical Reports Library, U.S. Department of Commerce. ntrl.ntis.gov/NTRL/

Niels Bohr Library and Archives, American Institute of Physics, College Park, MD. aip.org/history-programs/niels-bohr-library

Nierenberg (William) Papers, Special Collections and Archives, University of California, San Diego, La Jolla.

Ralph C. Bledsoe Files, 1985–1988, Ronald Reagan Presidential Library, Simi Valley, CA.

"Reference Page for *They Knew,*" comp. James Gustave Speth, Our Children's Trust, ourchildrenstrust.org/speth-they-knew-reference

Richard Nixon Presidential Library, Yorba Linda, CA.

Roger Revelle Papers, History of Oceanography and Marine Science collection, Library Digital Collections, University of California, San Diego, La Jolla. library.ucsd.edu/dc/collection/bb0581792g

Ronald Reagan Presidential Library, Simi Valley, CA.

Ronald Reagan Presidential Library, Digitized Textual Material. reaganlibrary.gov/archives/digitized-textual-material

William J. Clinton Presidential Library, Little Rock, AR. clintonlibrary.gov

Index

Gray, Boyden, 208, 234

Green political parties, 255, 305

Greenpeace, 227, 261–62

Griffin, Robert, 64

Hadley Centre for Climate Prediction and Research, 236

Hagel, Chuck, 304

Hair, Jay, 207

Hansen, James, 129–31, 158, 174, 178, 183–89, 192–93, 204–7, 218–21, 228, 240, 256, 291, 294–95, 330n19, 331n7, 334n8

Hart, Gary, 112

Harvard University, 39, 42–43, 55–59, 61, 67, 106, 113, 122, 135, 141, 199, 219, 250, 258, 283, 285, 291, 296, 305

Havel, Vaclav, 265

Hayes, Denis, 104, 120–21

Hayes, Philip, 74

Heffner, Hubert, 60–61

Heartland Institute, 281, 288

Heinz, John, 203, 211

Hidy, George, 177

Hill, Alan, 135, 177, 197–98, 207–9

Hodel, Donald, 169–71

Hoffman, John, 149, 151

Hofstadter, Richard, 39

Hogan, Bill, 258

Hornig, Donald, 38, 41, 54

Houghton, John, 226, 291

Howe, Joshua, 97

Hubbard, H. M. (Hub), 226

Hufbauer, Carol Revelle, 286

Humble Oil Company. *See* Exxon

Humphrey, Hubert, 297

ice (melting of), 11, 13, 24–25, 35, 46, 51, 63–64, 84, 93, 102, 124, 131, 138, 148, 206, 213, 297, 327n3

Ichimura, Shinichi, 99

Idso, Sherwood, 281

Ikard, Frank, 47–48

India, 34, 56, 148, 153

Inflation Reduction Act (2022), 309–10, 313

Informed Citizens for the Environment, 246–47

Institute for Advanced Study, 10, 97

Industrial College of the Armed Forces, 62

Intergovernmental Panel on Climate Change (IPCC), 179, 182, 186, 196–97, 199, 204, 207, 210, 220, 223–27, 229–31, 236, 243–45, 249, 251, 256, 259, 264–65, 279, 281–83, 285–86, 291, 297, 299, 304, 306–7

International Geophysical Year, 13–14, 17, 19, 21, 123, 157

Interior, Department of, 32–34, 39, 61, 63, 169, 171, 197

Iowa State University, 63

Iraq, 233

Isaacson, Walter, 2

Italy, 104, 148, 268

Japan, 9, 22, 99, 104, 153, 202, 211, 249, 264, 293, 304, 313

JASON report, 92–98, 105, 107–8, 113, 119, 156, 193, 285

Jastrow, Robert, 228

Johns Hopkins University, 5, 43

Johnson, Lyndon, 31, 39, 49, 52–54, 57; and environmental protection, 43, 47, 49–52, 54; and the "Great Society," 41–43; and climate change, 43, 47

Johnston, J. Bennett, 162–63, 165, 167, 182–83, 252–53

Kaplan, Lewis, 12

Keeling, Charles (Dave), 17, 34–36, 44, 84, 92, 104, 123, 177, 186, 291–93, 302; implications of research, 3–4, 21, 38, 45, 50, 57, 129, 295; measurements of atmospheric carbon, 2–3, 15–17, 20, 31, 46, 55, 61, 81, 118–19, 122, 127, 129, 182–83, 239, 249, 279,

Nemet, Gregory, 313

Netherlands, 141, 210, 228, 249

Nevada, 273

New Jersey, 10, 36, 125, 174

Newton, Isaac, 191

Nickles, Don, 235–36

Nierenberg, William, 112–13, 115, 118–20, 128–29, 132–37, 140, 142–43, 147, 149, 158, 193, 228, 230, 240, 245, 248, 255, 257, 279, 297

Nitze, Paul, 268

Nitze, William, 268

Nixon, Richard, 58–72, 76, 82, 92–93, 102, 119, 145, 161, 186, 199–200, 225

Nobel Prize, 62, 121, 297, 301–2, 306

Nongovernmental Panel on Climate Change, 296

Noordwijk declaration, 210–13

Nordhaus, William, 80, 83, 89, 97, 213–14, 226, 243–45, 255, 258–59, 279, 298–302; articles on global warming, 81–83, 132–33, 289–90; and Nobel Prize, 301–2; and National Academy reports, 113, 115, 136, 140, 147, 240

North Atlantic Treaty Organization (NATO), 113

North Carolina, 2, 208

North Dakota, 183

Northwestern University, 15, 283

nuclear power, 27, 66, 68, 71–72, 79, 83, 87, 90–91, 94, 97–98, 100–101, 103–4, 109, 111–12, 115–16, 119, 121, 123, 130, 143, 164, 176–77, 188–90, 192, 194–95, 215–16, 222–23, 228–29, 232, 257, 259, 271, 274, 277–80; in twenty-first century, 294, 310–11, 313; breeder reactor, 66, 112; early development, 22–25, 34, 50–51; Nuclear Regulatory Commission, 111, 327n15

Oak Ridge National Laboratory, 25, 83, 98–100, 142, 279

oceans: influence on climate, 3, 8, 10, 12, 14, 18–21, 25, 29–32, 37, 43, 45, 61, 63, 71, 84, 98, 100, 102, 107, 123, 149, 158, 205, 217–18, 231, 237, 254, 256, 293, 295; sea level rise due to climate change, 14, 24–25, 35, 46, 51, 60, 65, 70, 81, 90, 93, 102, 109, 124, 131, 138, 141, 148, 163, 195, 197, 203, 206, 218, 231, 297, 307

oceanography, 9, 12–13, 18–21, 28–33, 37, 61, 70, 73, 248, 256

Office of Management and Budget, 82, 105, 128, 146, 174, 201, 204, 219, 223, 227, 284, 302–3

Office of Technological Assessment, 160, 253

oil, 2–3, 5–6, 15, 21, 40, 44, 47, 51, 55, 59–60, 62, 64–65, 67, 72, 78–80, 83, 88–89, 93–94, 102, 104–5, 112, 130, 133, 146, 167, 173, 183, 190, 195–96, 199, 201, 208, 220–21, 227, 233–35, 252, 257, 263–64, 271, 278, 285–86, 303, 308, 314, 330n24

Olson, Jerry, 99

Oppenheimer, Michael, 143, 185–86, 207, 243, 255–57, 259, 308

Oppenheimer, Robert, 11, 97

Oregon State University, 89–90, 242

Oreskes, Naomi, 277

Oxford University, 285

Organization of Petroleum Exporting Countries (OPEC), 104, 196, 264

Our Children's Trust, 295

ozone layer, 90, 115, 154, 158, 160–61, 164, 168–72, 177, 179–80, 188–90, 197, 200, 202, 214, 250–53, 257, 277, 289, 314

Pakistan, 33–34, 55–56

Paris agreement (2015), 306–7

"peak oil" theory, 59

Pelosi, Nancy, 103

Pennsylvania Electric Company, 44

Perot, Ross, 275, 287

Perry, Harry, 84